GENERAL
AND
INORGANIC CHEMISTRY

GENERAL AND INORGANIC CHEMISTRY

[For Undergraduate & Postgraduate students of all leading South Asian universities]

Bidhan Chandra Ray

MSc (Gold-medalist), PhD; Professor,
Department of Chemistry, Jadavpur University, Kolkata

Satyanarayan Das

MSc, PhD; Reader, Dapartment of Chemistry (Retd), Tamralipta Mahavidyalaya,
Tamluk, Purba Medinipur; Ex-Reader in Chemistry, Surendranath Evening College, Kolkata

Jaya Mukherjee

MSc, PhD; Visiting Teacher in Chemistry, Postgraduate Evening Course,
Jadavpur University, Kolkata; Srisikshyatan College, Kolkata;
Ex-Lecturer in Chemistry, Gurudas College, Kolkata

New Central Book Agency (P) Ltd

LONDON

HYDERABAD ERNAKULAM BHUBANESWAR

DELHI KOLKATA PUNE GUWAHATI

NCBA

REGD OFFICE
8/1 Chintamoni Das Lane, Kolkata 700 009, India
email: ncbapvtltd@eth.net

OVERSEAS
NCBA (UK) Ltd, 149 Park Avenue North, Northampton, NN3 2HY, UK
email: ncbauk@yahoo.com

EXPORT
NCBA Exports Pvt. Ltd, 212 Shahpur Jat, New Delhi 110 049
email: ncbaexp@ncbapvtltd.com

BRANCHES

212 Shahpur Jat
New Delhi 110 049
email: ncbadel@ncbapvtltd.com

Shop Nos. 3 & 4, Vinayak Towers
681/B Budhwar Peth, Appa Balwant Chowk
Pune 411 002
email: ncbapune@ncbapvtltd.com

Shop No. 15, Bharati Towers, Ground Floor
Forest Park, Bhubaneswar 751 001
email: ncbabub@ncbapvtltd.com

House No. 3-1-315, 1st Floor
Nimboliadda, Kachiguda, Hyderabad 500 027
email: ncbahydb@ncbapvtltd.com

GSS Shopping Complex, 1st Floor
Opposite College Post Office, Convent Road
Ernakulam 682 035
email: ncbaernk@ncbapvtltd.com

Radhamayee Complex (Second Floor)
Opp. SBI, Panbazar, Jaswanta Road
Guwahati 781 001
email: ncbaguwahati@ncbapvtltd.com

GENERAL AND INORGANIC CHEMISTRY • Bidhan Chandra Ray, Satyanarayan Das, Jaya Mukherjee

First Published: April 2008
Revised Reprint: May 2013

PUBLISHER AND TYPESETTER
New Central Book Agency (P) Ltd
8/1 Chintamoni Das Lane, Kolkata 700 009

PRINTER
New Central Book Agency (P) Ltd
Web-Offset Division, Dhulagarh, Sankrail, Howrah

COVER DESIGNER
Soumen Paul

TECHNICAL EDITOR
Pradip Kr. Chowdhury

PROJECT TEAM
Suman Dutta, Mithu Karmakar

ILLUSTRATOR
Suman Dutta

ISBN 978 81 7381 580 5

Price: ₹ 295.00

CONTENTS

Preface

Without the concept of mathematics, the study of science is incomplete. Clear knowledge of chemistry cannot be obtained if the ideas are not based on mathematical concept. Hence, attempts have been made to explain the scientific results based on mathematics.

This book has been written mainly to serve **B.Sc. (Hons.) undergraduate students** and some chapters, for **M.Sc. students**. The correlation between the properties of elements and compounds and their structures with their electronic arrangements has been shown as far as possible using orbital hybridization, molecular orbital and crystal field concepts.

We are thankful to Dr Suchandra Biswas of Bankim Sardar College, Dr Ranjan Datta, Dr Arup Gayen, Dr Kausiki Sankar Pramanik and Dr Jnan Prakash Naskar of Jadavpur University for their timely help and cooperation to complete the book. They have gone through the manuscript and given us their valuable suggestions and sometimes, notes, to include in this book. So we are indebted to them. We are thankful to Ms Mamita Chowdhury and Mr S. K. Saha for their help in computation.

We are fully aware of our limitations as no human being can have perfect knowledge in all sections. So some errors might have crept in this book. Hence suggestions and criticisms from our learned colleagues and also from inquisitive students for the betterment of this book will be thankfully received for appropriate upgradation of the book in future.

Prof Bidhan Chandra Ray
Dr Satyanarayan Das
Dr Jaya Mukherjee

GENERAL AND INORGANIC CHEMISTRY

Ray • Das • Mukherjee

ERRATA

Page No.	Line No.	Printed	Corrected
[IV]	15	Pradip Kr. Chowdhury	Dipan Roy (Technical Editor)
4	17	5.27×10^{7} esu/gm	5.27×10^{17} esu/gm
4	23	Comparison of value of H^+	Comparison of e/m value of H^+
5	Last line	$\frac{mg(V_1+V_2)}{V_1E}$	$\frac{mg(V_1+V_2)}{V_1E}=e_x$
6	Fig. 1.3	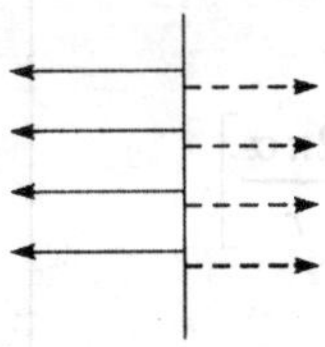	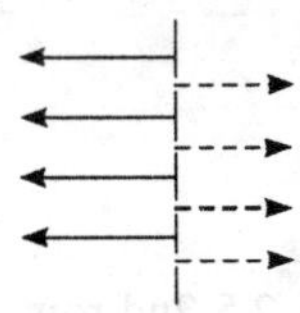
7	23	Neutrino and	Neutrino ν and
8	Table: Column 1 row-11	if not captured by ${}^{1}_{1}p$	not captured by ${}^{1}_{1}p$
9	1st heading	NEUTRINO AND	NEUTRINO (ν) AND
29	10	$\frac{2\pi e^4\mu_{He}}{h^3c}\left[\frac{4}{n_1^2}\ \frac{4}{n_2^2}\right]$	$\frac{2\pi^2e^4\mu_{He^+}}{h^3c}\left[\frac{4}{n_1^2}-\frac{4}{n_2^2}\right]$
29	12	$\bar{\nu}=\frac{2\pi^2e^4\mu_{He}}{h^3c}\left[\frac{1}{n_1^2}-\frac{1}{n_2^2}\right]$	$\bar{\nu}=\frac{2\pi^2e^4\mu_{H}}{h^3c}\left[\frac{1}{n_1^2}-\frac{1}{n_2^2}\right]$
39	4	$\left[\text{as } e^2=\frac{P_\theta{}^2}{ma\left(1-\varepsilon^2\right)^2}\right]$	$\left[\text{as } e^2=\frac{P_\theta^2}{ma\left(1-\varepsilon^2\right)}\right]$
40	31 (last equation)	$E=-\frac{2\pi^2Z^2c^4m}{\dot{n}^2h^2}$	$E=-\frac{2\pi^2Z^2e^4m}{\dot{n}^2h^2}$
42	29	where m	where m_l
43	Fig. 1.25a	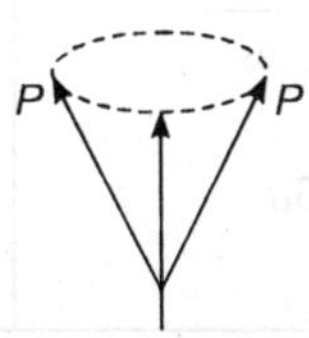	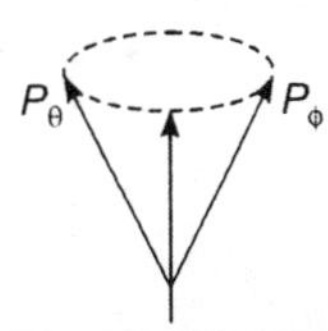

Page No.	Line No.	Printed	Corrected
44	Fig. 1.26	d_{x^2-y}	$d_{x^2-y^2}$
44	Fig. 1.26 (Caption)	$3s$	$3d$
46	9	n_1	n
53	14	$E_{ex} \propto \frac{2(2-1)}{2} = 01$	$E_{ex} \propto \frac{2(2-1)}{2} = 1$
55	Table 1.3	$\cdot\, d^2s^2$	d^ns^x
55	Line 14	$n\lambda = 2d$ $\sin\theta$	$n\lambda = 2d \sin\theta$ in one line (to be written)
56	Fig. 1.33	L no printed	L in the place of lense
61	34	i.e., $\nu = 1$	i.e., $P = 1$
68	17	$\left[-\frac{l(l+1)}{r^2} - \alpha^2 + \frac{2\pi\alpha}{r}\right]$	$\left[-\frac{l(l+1)}{r^2} - \alpha^2 + \frac{2\lambda\alpha}{r}\right]$
73	Fig. 2.5 2nd row	$2p_r$	$2p_x$
73	Fig. 2.5 last row	$3d_{r^2-y^2}$ $3d_{x^2}$	$3d_{x^2-y^2}$ $3d_{z^2}$
76	Fig. 3.1 Caption	a particle	α-particle
76	Fig.3.2a	Recovery of Thorium	Recovery of Uranium
91	11	Avogadro's number can	Avogadro's number N can
92	10	where $\lambda_p N_p$ is the	where $\lambda_d N_d$ is the
98	10	emission	${}^4_2$He-emission
115	19	$= K \times \frac{H^2}{x} . \frac{e}{m}$	$= K \times \frac{H^2}{X} . \frac{e}{m}$
123	31	of the portion taken	of the portion A′ taken
125	3	Equal signs (=) misaligned	all equal signs (=) to be printed in the same level
126	5	deleted	detected
127	1	$\frac{N_{Th} - D}{N_{Th}}$ or, $\frac{N_{Ac} - D}{N_{U^{235}}}$	$\frac{N_{Th-D}}{N_{Th}}$ or, $\frac{N_{Ac-D}}{N_{U^{235}}}$
137	6	$d = \sqrt{3}a^2 = \sqrt{3}a$	$d = \sqrt{3a^2} = \sqrt{3}a$

Page No.	Line No.	Printed	Corrected
138	5	$a = \frac{\sqrt{2}r^-}{\sqrt{2}}$	$a = \frac{2r^-}{\sqrt{2}}$
138	7	$r^+ \frac{\sqrt{6}r^-}{2} - r^-$	$r^+ = \frac{\sqrt{6}r^-}{2} - r^-$
155	9	impart, extra	they impart extra
155	10	and is called	and this is called
155	33	$U_T = -\frac{e^2}{r_0^2} + \frac{e^2 r_0^{n-1}}{n r_0^n} = -\frac{e^2}{r_0^2}$	$U_T = -\frac{e^2}{r_0} + \frac{e^2 r_0^{n-1}}{n r_0^n} = -\frac{e^2}{r_0}$
161	20	latter	later
162	First one of the two figures of Fig. 7.12(a)	Ψ_a^2	Ψ^{*2}
166	Fig. 7.15a	$2\pi_a$	$2\pi^*$
167	Fig. 7.16, row–4	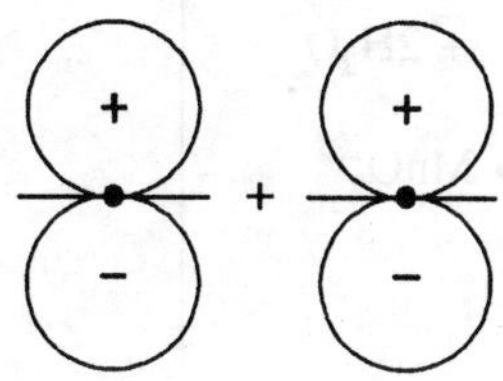	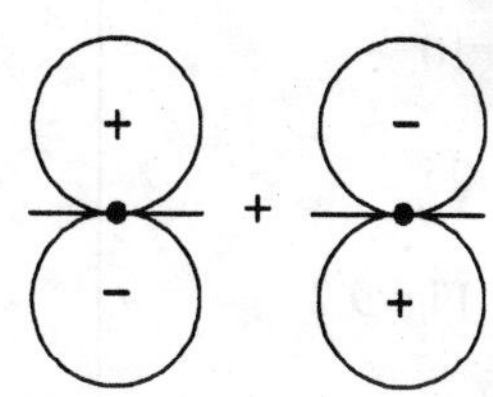
169	First four equations, last term	not distinct	Ψ_{p_z}
171	After table 7.5 2nd line	O undergo (wrong font)	O undergo (normal font to be used)
172	Fig. 7.23a	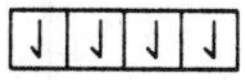	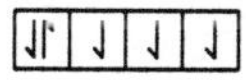
179	15	The *lp-bp* for and	The *lp-bp* in NH_3 and
186	4 (middle)	Na–Na | | Na–Na → Na^+ Na^+	Na Na | | Na Na → Na^+ Na
186	Fig. 8.19 (b & c)	bond	band
190	24	MnO_4^{2-}	MnO_4^-
190	26	MnO^{4-}	MnO_4^-
191	10	$E = E_0 + \frac{0.0591}{n}\log C_M^{n+}$	$E = E_0 + \frac{0.0591}{n}\log C_{M^{n+}}$

Page No.	Line No.	Printed	Corrected
201	3	$= E_{0\ (Ag+ \mid Ag\ )} - E_{0\ (Cu^{2}+ \mid Cu)}$ $= 0.34 - 0.80 = -0.46$ volt.	$= E_{0\ (Ag+ \mid Ag\ )} - E_{0\ (Cu^{2+} \mid Cu)}$ $= 0.80 - 0.34 = +0.46$ volt.
203	1	$= \frac{0.0591}{2} \log \frac{[Sn^{4+}]\,[Fe^{3+}]^2}{[Sn^{2+}]\,[Fe^{2+}]^2}$	$= \frac{0.0591}{2} \log \frac{[Sn^{4+}]}{[Sn^{2+}]} \Big/ \frac{[Fe^{3+}]^2}{[Fe^{2+}]^2}$
203	5	$\log K = n(E_0' - E_0'')\,/0.059$	$\log K = \frac{n(E_0' - E_0'')}{0.059}$
203	12	$K = \frac{[Fe^{2+}]^2 [Sn^{4+}]}{[Fe^{2+}]\,[Sn^{2+}]}$	$K = \frac{[Fe^{2+}]^2 [Sn^{4+}]}{[Fe^{3+}]\,[Sn^{2+}]}$
207	1	$K = \frac{[I^-][CuI]}{[Cu^{2+}][I^-]}$	$K = \frac{[I_2][CuI]}{[Cu^{2+}][I^-]}$
209	4 from bottom	$2Cl^-$	H_2O
210	10	$2OCl^- + 2\overset{+}{C}l^- + 2H_2O$	$2\overset{+}{O}Cl^- + 2Cl^- + 2H_2O$
210	25	$Mn^{3+} - e \rightarrow MnO_2^-$	$Mn^{3+} - e \rightarrow MnO_2$
211	Fig. 9.2	$KMnO_4$	$K_2Cr_2O_7$
211	Fig. 9.2 (Caption)	permanganate	dichromate
214	16	concept was by	concept was given by
217	Table 10.1 row–14 column–Acid$_2$	$H_2N.CO.NH_2^+$	$H_2N.CO.NH_3^+$
223	10 from bottom	HgI_2	HgF_2
225	3	$K = (C_{H_3O^+.A^-}) / C_{HA}.C_{H_2O}$	$K = (C_{H_3O^+}.C_{A^-}) / C_{HA}.C_{H_2O}$
225	4	$K' = KC_{H_2O} = (C_{H_3O^+.A^-}) / C_{HA}$	$K' = KC_{H_2O} = (C_{H_3O^+} C_{A^-}) / C_{HA}$
243	Last line	$En = \frac{h^2 n^2}{8mL^2}$	$E_x \approx E_n = \frac{h^2 n^2}{8mL^2}$

CHAPTER

1

EARLY HISTORY OF ATOM

1.01 INTRODUCTION

More than 2500 years before, ancient Hindu philosophers were of the opinion that matters we see are not continuous but consist of small particles. About the same time the Greek scholars also thought in this way about matter. During 5th century B.C., Leucippus and his student Democritus (eminent Greek School of Atomists) thought that materials are made up of ultimately small, indivisible units, called *atoms*. The word 'atom' derived from two Greek words, *a* (not) + *temnien* (to cut) means something that is not divisible. Because of the objection of Greek philosopher Aristotle (384–322 B.C.), this idea about atom was suppressed and did not come to light for several hundreds of years, till the Renaissance in Europe. In 16th and 17th century A.D., scholars and philosophers of different Western countries favoured the atomic concept. Although atomic concept was in the mind of ancient scholars and philosophers, the credit for modern atomic theory was given to John Dalton because he quantitatively and scientifically developed the idea. According to Dalton,

1. atoms are smallest particle of matter and the properties of these particles are alike with matter.
2. these smallest particles are indivisible, they cannot be created or destroyed or transformed.
3. atoms of the same matter are alike in properties, weights, etc.
4. atoms of different matters are different in physical and chemical properties as well as in weights.
5. when chemical combination between matters takes place, it takes place between atoms in definite numerical ratios.

The modern scientific developments questioned the validity of this concept, but they formed the basis of modern thinking. Atoms are now considered to be made up of fundamental particles like electrons, protons, neutrons, and can be transformed from one atom to another; they can be destroyed and converted to energy. Atoms of same elements have different weights (isotopes) and of different elements have same weights (isobars).

Dalton's concept can best explain the laws of chemical combinations but failed to explain the Gay–Lussac's law of gaseous volume that deals with the chemical combination between gases to form gaseous compounds. Ultimately, this was explained by introducing the idea of molecules by Avogadro. Molecules are the ultimate state of matter and have free existence, their properties are identical with that of matter and are formed by the union of atoms of the constituting elements. Hydrogen is the simplest and lightest element. Next to Dalton, the important idea regarding the composition of matter was proposed by Prout (1815). According to him, all other elements are composed of the hydrogen atoms and their weights will be the simple multiple of the weight of hydrogen atom. Thus, all elements had complex atomic structure. Although Prout's idea was discarded at that time as other elements have atomic weights not integral multiple of the weight of hydrogen atom but the idea gets momentum at present due to the development of the conception of nuclear reactions and matter-energy relationship. The development of modern atomic theory is presented step by step as follows:

1.02 FUNDAMENTAL PARTICLES

Dalton (1803) considered that matters are composed of very tiny, indivisible particles having the properties like that of matter, called *atoms*. Towards the end of nineteenth century and beginning of twentieth century, many discoveries and experiments revolutionized our idea about the atoms. Atoms are now not indivisible but can be shown to be composed of some still indivisible smaller particles like electron, proton, etc. These smallest indivisible particles that constitute atom are called *fundamental particles*. Depending on their stability, they can be classified as stable particles (longevity range from few seconds to few minutes) and the unstable particles (stability ranges from 10^{-16} sec to $^{-6}$sec). Their discoveries and characteristic properties, etc., are discussed as follows:

STABLE PARTICLES

Stable particles are three in numbers, viz., electron (e^-), proton $\left(p^+ \text{ or } H^+\right)$ and neutron $\left({}_0^1n\right)$.

1.03 DISCOVERY OF ELECTRON (DISCHARGE TUBE PHENOMENA)

The discharge tube is a strong, 30-cm long, 4-cm in diameter, provided with two metal electrodes sealed at the two ends. The side tube is connected to a pump to create very low pressure, and electrodes are connected to the secondary induction coil capable of producing potential drop of 50 000 volts or more. At ordinary pressure, the gas molecules attached to the electrodes get ionized but lose their charge on collision with other gas molecules and thus, become non-conductor. With gradual lowering of pressure, the following observations are obtained:

1. At 10 cm of Hg pressure, irregular streak of light with cracking noise.
2. At 1 cm of Hg pressure, cracking streaks broaden to luminous column with buzzing sound called *positive column*.
3. At 3-4 mm of Hg pressure, a blue glow near the cathode, called *cathode glow* or *negative glow*, the dark space called *Faraday dark space* and then striated positive column near the anode appear.
4. At 1mm of Hg pressure, the positive column shifts towards the anode and gets shortened, Faraday dark space extends to a greater length.
5. At 0.1mm of Hg pressure, cathode glow shifts and Crookes dark space appears, the positive column breaks to alternate bright and dark bands, called *striations*.
6. At 0.01 mm of Hg pressure, Crookes dark space fills the tube, blue or green fluorescence (depending on the gas in the tube) appears on the wall of the tube due to X-ray caused by

the impinge of some agency in Crookes dark space and glass wall. The agencies in the Crookes dark space are produced from the surface of the cathode, called *cathode rays.*

Careful study of the cathode rays reveals that

(a) cathode rays emerge normally from the cathode surface and travel in straight line since they cast shadow of an opaque obstacle placed in its path;

(b) it produces mechanical motion to the light mica wheel placed in its path, indicating particle nature of the rays;

(c) it produces heat when it strikes matter and it can be reflected. All these prove that the rays are of particle nature with inertia;

(d) the rays affect photographic plate like ordinary light;

(e) they ionize the gases through which they pass, like electric field;

(f) they can be deflected by electric and magnetic fields and from the nature of deflection it can be argued that the cathode-ray particles are negatively charged;

(g) the cathode ray can charge the plate of an electrometer negatively, indicating the particles are negatively charged.

Using any metal as cathode, the same cathode rays are obtained. Thomson named the cathode rays as "stream of negative corpuscles", while Stoney called them electron. Thus, electron becomes the common constituent of matter. Thus, it is said that electron is the fundamental constituent of matter.

DETERMINATION OF SPECIFIC CHARGE OF AN ELECTRON

Prof. J. J. Thomson (1897) determined specific charge, i.e., charge per unit mass of electron in the following experiment:

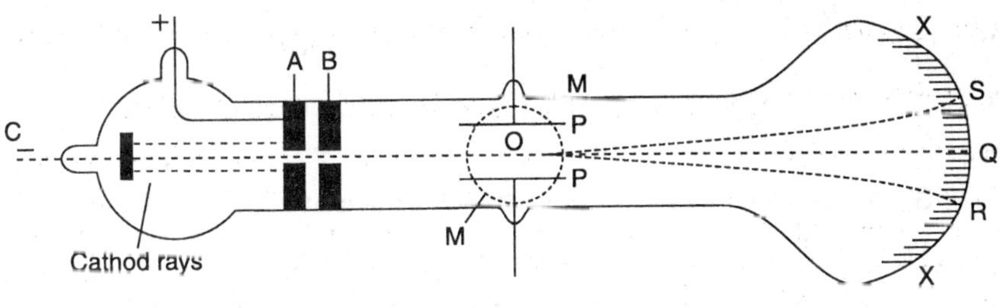

Fig. 1.1 Thomson's apparatus for determination of *e/m* value of cathode-ray particles.

Thomson's apparatus consists of a highly evacuated discharge tube with cathode at *C* and an anode *A* and very narrow metal slit *B* such that a very narrow pencil beam of cathode rays finds its way to reach the fluorescent screen *XX* at the point *Q* in absence of any field. *P-P* are the plates to apply electric field perpendicular to the direction of motion of the cathode rays and are on the plane of the diagram. The magnetic field *MM* acts perpendicular to the plane of the diagram and also to the direction of motion of cathode rays. The two fields are adjusted in such a way that the cathode rays will be under the influence of the fields over the same length of path.

Let strength of the electric field = X; so, electrical force = Xe, e being electronic charge and acceleration, $f = \frac{Xe}{m}$ where m is the mass of electron and the electron will be deflected along the path of a parabola.

Let t be the time of flight, i.e., time taken by electron to travel the distance, l (from centre of the fields to the screen) with velocity v, i.e., $t = l/v$.

Thus, the deflection, $QS = \frac{1}{2}ft^2 = \frac{1}{2}\frac{Xe}{m}\frac{l^2}{v^2} = \frac{1}{2}\frac{Xel^2}{mv^2}$ (1.1)

Now, the magnetic field, H is applied in such a way that the deflection is just balanced, i.e., the deflection due to magnetic field is equal and opposite to that of the first. So, when both the fields act at a time, the electron will suffer no deflection.

Under the influence the magnetic force = Hev, the path will be circular and so, $Hev = \frac{mv^2}{r}$

or, $\frac{e}{m} = \frac{v}{Hr}$. (1.2)

Again, the two fields being equal, $Hev = Xe$, hence $v = X/H$.

Therefore, $\frac{e}{m} = \frac{X}{H^2 r}$ (1.3)

Now, the radius of curvature, r of the circular path will be

$$QS\,(2r - QS) = OQ^2 \quad (1.4)$$

QS being small compared to OQ, QS^2 can be neglected.

So, $2rQS = OQ^2$.

Here, QS and OQ are measured from the apparatus.

So, r can be determined.

The value of $\frac{e}{m}$ = 1.76×10^7 emu /gm = 5.27×10^7 esu /gm.

[Or, it can be shown as follows:

$$QS = \tfrac{1}{2}\frac{Xe}{m}.\frac{l^2}{v^2} = \tfrac{1}{2}\frac{Xe}{m}.\frac{l^2H^2}{X^2} = \tfrac{1}{2}\frac{l^2H^2}{X}.\frac{e}{m}$$

or, $\frac{e}{m} = 2QS\,\frac{X}{l^2H^2} = \frac{2QS}{OQ^2}\frac{X}{H^2} = \frac{1}{r}\frac{X}{H^2}$]

Thus, $\frac{e}{m}$ can be measured by measuring field strengths and deflections.

This value is independent of the nature of gas in the tube, the nature of the metal used as cathode. Comparison of value of H^+ and that for an electron is 1840 times less. From Faraday's laws, 96 500 coulombs or 2.89×10^{14} esu electricity deposits or eliminates one gm-equivalent of any substance at the electrode and hence Faraday is the charge carried by one gm-equivalent or Avogadro number of H^+.

So, $F = Ne$ or $2.89 \times 10^{14} = N \times 4{\cdot}8 \times 10^{-10}$

or $N = 6.023 \times 10^{23}$

Thus, Avogadro number denoted by N, can be calculated.

DETERMINATION OF CHARGE OF AN ELECTRON BY MILLIKAN'S OIL-DROP EXPERIMENT

Several scientists measured the charge of an electron by different methods, but Millikan's method is more satisfactory because he used

1. non-volatile, non-polar oil instead of polar solvent, water to avoid evaporation and
2. higher field to keep the drop suspended for longer time in between the plates to make its rise or fall at will.

The apparatus is like this:

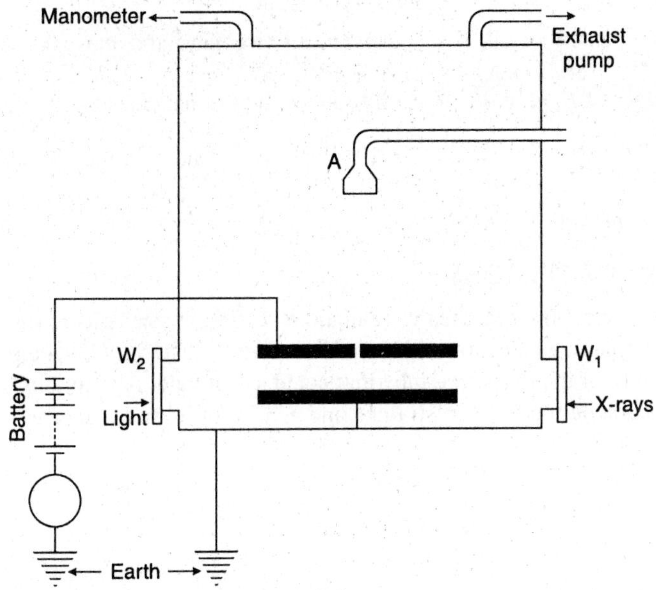

Fig. 1.2 Millikan's apparatus for determination of electronic charge (oil-drop experiment).

It is a cylindrical vessel at constant temperature, air inside the vessel at very low pressure, measured by manometer. It is fitted with an atomiser *A* to spray oil in very fine droplet that generally becomes positively charged by friction in the atomising process and could be charged by shining X-rays or gamma-rays into the neighbourhood of the drops. The droplets are illuminated by passing light through the window W_2 and are viewed by another window fitted perpendicularly to this with a telescope (not shown in the diagram). At the bottom there are two circular copper plates each of 22-cm diameter, upper one is perforated and connected to the positive pole of a battery and the lower one is earthed. The two plates are 1.6-cm apart. About 5 000–10 000 volts is applied to the plate. The time of fall and rise to a distance of 0.5222 cm of the droplets is viewed and recorded automatically by a high precision chronograph. When a particular droplet finds its way through the hole of the upper plate, it is viewed, and time to fall through the distance 0.5222 cm is noted. When it is just to touch the lower plate, the upper plate is charged and due to electrostatic attractions, the droplet will be pulled up and the time required for upward journey is noted. Again, switch is put off and X-rays are passed to charge the drop differently, droplet comes down. Several times the time for upward journey and downward journey is noted. Each case time for downward journey will be the same as this occurs due to gravitational pull. But time for upward journey will be different indicating the attainment of negative charge on the drop by passing X-ray at different times; as for example, the times recorded are 12.5, 21.8, 34.8, 84.5, 34.6 sec, etc. Let downward velocity and upward velocity are V_1 and V_2 respectively, and m is the mass of oil drop.

$V_1 \propto mg$ and $V_2 \propto Ee_x - mg$ where E is the electrical force field and e_x is the charge on the oil drop.

$$\frac{V_1}{V_2} = \frac{mg}{Ee_x - mg} \text{ or, } \frac{V_2}{V_1} = \frac{Ee_x}{mg} - \frac{mg}{mg} \text{ or, } \frac{V_1 + V_2}{V_1} = \frac{Ee_x}{mg} \text{ or, } \frac{mg(V_1 + V_2)}{V_1 E} \quad (1.5)$$

Now, the mass of the oil drop m, can be obtained from Stokes' equation when rate of fall v of spherical oil drop under gravity in viscometer apparatus and of radius r of the drop are given by

$$v = \frac{2gr^2d}{9\eta} \text{ [} d = \text{density of oil, } \eta = \text{coefficient of viscosity] and } m = \frac{4}{3}\pi r^3 d.$$

Putting the values of E, m, g, V_1, V_2 each time, the additional charge of the oil drop was found to be multiple of 4.77×10^{-10} esu or 1.602×10^{-19} coulomb. Knowing the e/m as $5.27 \times 10^{17} \frac{\text{esu}}{\text{gm}}$, the mass of electron, m_e is 9.1×10^{-28} gm.

1.04 DISCOVERY OF PROTONS

After the discovery of electron, searches were made by different scientists to discover the counterpart of electron in the atom as the atom as a whole is neutral. In 1886, Goldstein used a perforated metal disc at the centre of the discharge tube that acted as cathode. A luminous beam emerged out of the holes of the cathode travelled in straight line and away from the anode.

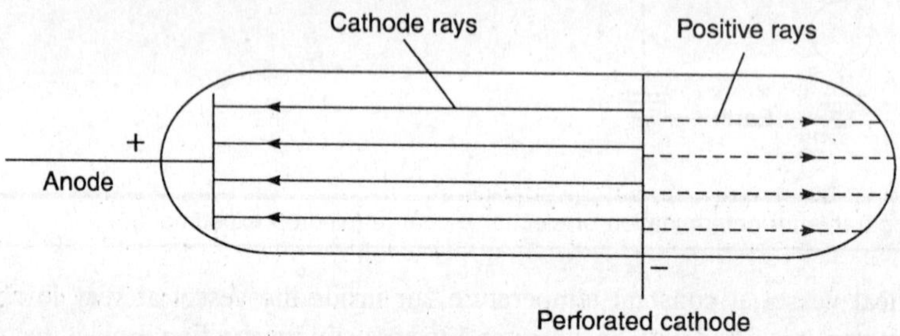

Fig. 1.3 Positive rays.

First, these were called canal rays as they passed through the holes or channels and this nature indicated that they were positively charged and proved by observing their deflections in electric and magnetic fields. Hence, Thomson named them positive rays. The determination of specific charge *e/m* was found to be very small in comparison to that of electron indicating their masses are very large and they are actually ions of the residual gas in the tube. If the tube initially filled with hydrogen gas, the *e/m* ratio becomes maximum and 1840 times than that of electron. Thus, the positive rays differ from electron or cathode rays in two respects: (a) they are ions of atoms or molecules and (b) they are dependent on the nature of the residual gas present in the tube. In spite of several researches, no such particle parallel to electron but with (+)ve charge was found, but with hydrogen gas in the tube the (+)ve particle has the mass like that of hydrogen atom from which electron has been removed.

This positively charged hydrogen atom called *proton* is found to be important unit in the structure of other atoms. It is represented by ${}^1_1\text{H}$ or ${}^1_1p$ with mass 1836 m_e and 1.007277 amu.

1.05 DISCOVERY OF NEUTRON, ${}^1_0n$

Rutherford (1920) postulated the existence of neutral particle having a mass equal to that of proton within the nucleus. He also suggested the name 'neutron' for such particle as it is formed by the neutralization of charges of electron and proton. In 1930, Bothe and Backer allowed α-particles from polonium to fall on beryllium, when a very high penetrating radiation, assumed to be γ-rays, was emitted. In 1932, F. Joliot and I. Curie observed that when paraffin was interposed in the path of this radiation, proton was emitted with high velocity. Emission of proton as a result of impact with γ-rays and paraffin was difficult to explain but it becomes easier to explain if we consider the radiation of particle nature with mass equal to that of a proton. From the analysis of the velocity and masses of the particles concerned J. Chadwick (1932) concluded that the so-called radiation consisted of particles having the mass of proton but no charge, i.e., ${}_0n^1$, the neutron which Rutherford postulated.

$$ {}_4^9\text{Be} + {}_2^4\text{He} \rightarrow {}_6^{12}\text{C} + {}_0^1n $$

The detection of neutron is very difficult as it has no charge and so mass cannot be determined easily because it will not ionize the cloud in cloud-chamber experiment. However, its mass can be determined by collision with other nucleus by the equation, $v_2 = \dfrac{2m_1 v}{m_1 + m_2}$, where m_1 and v is the mass and velocity of neutron before collision and m_2 is the mass of the target nucleus at rest and v_2 its velocity after collision. Again, if the same neutron collides with other particle of mass m_3 at rest and after collision the velocity is v_3, then

$$ v_3 = \frac{2m_1 v}{m_1 + m_3}. $$

So,

$$ \frac{v_3}{v_2} = \frac{m_1 + m_2}{m_1 + m_3}. $$

This relation gives the value of m_1 = 1.00867 amu. (1.6)

Neutron has magnetic property and behaves as an elementary magnet. It possesses some magnetic moment like spinning electron. Thus, it is assumed that ${}_0^1n$ contains some positive charge at the center surrounded by thin layer of negative charge. Proof of this supposition is that for higher elements greater number of ${}_0^1n$ required for balancing the *p-p* repulsion to have stable nucleus.

1.06 UNSTABLE PARTICLES

(A) Positron (e^+ or β^+), (B) Meson $\left(\pi^{\pm},\ \pi^0, \mu^{\pm}, K^{\pm},\ 2K^0\right)$, and (C) Neutrino and Antineutrino, $\bar{\nu}$.

(A) POSITRON

P. A. M. Dirac in 1928, based on relativistic wave mechanics, presented theoretical arguments for the existence of a new particle of mass, which is equal to that of electron with positive charge. But C. D. Anderson gave the proof of its existence in 1932 during his study of the effect of cosmic ray on matter. Millikan's cloud chamber is a vessel placed in high-intensity magnetic field in which artificially cloud is made. The charged particles when pass through this cloud, become charged and their tracks can be photographed. The intensity of the track provides information of the mass of the particle and from the deflection of the track their charge can be determined. Analysis of the cloud-chamber photographs shows a curvature of a new track. This indicates (i) the particle is positively charged, (ii) length of the track due to proton is much less, about $\frac{1}{10}$th of the new track, and (iii) intensity of this track is very much less than that produced by proton but equal to that produced by electron. Anderson named this positively charged particle with mass equal to that of electron as positron and negative particle as negatron, although the name electron is widely used. This pair is called positron-electron pair. The specific charge *e*/*m* for positron (β^+ or e^+) was found to be the same as that of electron.

The average life of β^+ is only 10^{-9} sec. They are produced by the action of high-energy rays on matter or as a result of artificial radioactivity. They immediately disappear as a result of combination with electron producing γ-rays called *positron-electron annihilation radiation*. Due to its very short life, its discovery was so much delayed.

(B) MESON $\left(\pi^{\pm}, \pi^{0}, \mu^{\pm}, K^{\pm}, 2K^{0}\right)$

The Greek word 'meso' means intermediate, the new unstable particle has mass in between that of electron and proton that transmits the strong interaction binding nucleus together. The new particle was called at the time of discovery as Mesotron, but Dr. H. J. Bhaba (1939) named them as meson as this is adopted.

In 1937, Anderson and Neddermeyer and simultaneously Street and Stevenson examined the cloud chamber photograph of cosmic ray and they proved the existence of some strange particles whose mass is greater than that of electron but less than that of proton. These particles are either positively or negatively charged and some are neutral. It is evident from the curvature of the track. According to the mass, charge and stability, they are classified as follows:

$\pi^{\pm}, \pi^{0}$ Pion	$\mu^{\pm}$ Muon	$K^{\pm}$, $2K^{0}$ Kaon
Mass: $m_{\pi^0} = 263m_e$ $m_{\pi^\pm} = 274m_e$	Mass: $m_{\mu^\pm} = 260m_e = \frac{1}{9} m_{{}^1_1p}$	Mass: $m_{K^\pm, 2K^0} = 966.6m_e$
Average life: $\pi^\pm = 2.6 \times 10^{-8}$ sec $\pi^0 = 1 \times 10^{-16}$ sec	Average life: 2×10^{-6} sec So more stable than $\pi^\pm$	Average life: 1.24×10^{-8} sec
Change: π^- captured by ${}^1_1p$ $\pi^- + {}^1_1p \to {}^1_0n + \gamma$ or ${}^1_0n + \pi^0$ if not captured by ${}^1_1p$ $\pi^\pm \to \mu^\pm + \frac{\nu}{\overline{\nu}}$ energy $\pi^0 \to \gamma$-rays	Change: $\mu^- + {}^1_1p \to {}^1_0n +$ energy if not taken by ${}^1_1p$ μ^+ (or μ^-) $\to e^+ \left(e^-\right) + \nu$ or $\overline{\nu}$ + energy No μ^0 is known Found in cosmic ray	Change: They decay in six ways but the common ways are K^+ or $K^- \to \pi^+$ or π^- $+\mu^+$ or $\mu^- + e^+$ or e^- $+ \nu$ or $\overline{\nu}$
Production: ${}^4_2He^{+2} \xrightarrow{380\text{ MeV}} Be/C/Cu \to \pi^+$	Production: It does not react readily, α-bombardment does not produce muon.	Production: ${}^1_1p \to Cu \to K^\pm, 2K^0$ + energy very high energy.
Diff.: It reacts readily with matter. α-bombardment with matter produces π-meson.		

(C) NEUTRINO AND ANTINEUTRINO ($\bar{\nu}$)

Fermi (1934) postulated the existence of these particles of very negligible mass (~ 0) without charge. They are hypothetical particles. Their existence was supposed in this way: The β-emission of radioactive element is always associated with liberation of energy. Careful observation shows that about $\frac{2}{3}$ of this energy disappears violating the law of conservation, which is followed everywhere. Again, the law of conservation of momentum of the emitted β-particle and the recoiled matter nucleus is not followed. To explain the mass-balance and energy-balance during β-emission Fermi supposed the existence of hypothetical particle neutrino ν and another particle antineutrino $\bar{\nu}$. These can be represented as follows:

$${}_0^1n \rightarrow {}_1^1p + e^- + \nu \text{ change of } {}_0^1n \rightarrow {}_1^1p$$

$${}_0^1p \rightarrow {}_1^1n + e^+ + \nu \text{ (positron emission)}$$

$${}_1^1p + e^- \rightarrow {}_1^1n + \nu \text{ (}K\text{-capture)}$$

Recent developed techniques successfully detected ν in 1953 by Reines and Cowan.

(D) ANTIPROTON

The cosmic-ray study indicates that every fundamental particle must have an antiparticle like electron-positron pair and, therefore, there must be a proton-antiproton pair. The antiproton must have mass like that of proton with unit negative charge. The energy required to materialize a proton-antiproton pair is very high that can be obtained only in cosmic ray. Again, analysis of the photograph of cosmic ray in cloud chamber experiment is highly complex in nature. O. Chamberlain, E. Segre et al. first successfully prepared in 1955 by impinging copper-block with high-energy proton obtained from Bevatron instrument and the particles emitted are then passed through magnetic field when negative particles, including antiproton, are separated from the rest, and using precise instrument, the existence of antiproton in the particles were noted. The average life is 5.2×10^{-8} sec and its velocity is 80% velocity of light, denoted by $\bar{p}$ The proton-antiproton annihilation occurs with the liberation of tremendous energy. Again, these antiproton will bind up with positron to form a new type of hydrogen atom, unlike the ordinary hydrogen atom produced by binding up proton with electron.

(E) ANTINEUTRON

As stated before, cosmic-ray study on matter indicates the existence of positron and antiproton; similarly, this study also indicates the presence of antineutron. All these particles are not the common constituent of matter. When a proton and an antiproton come very close to each other but not very close to cause annihilation, then charge of the proton will be transferred to antiproton to create neutron and antineutron, and both particles will be neutral. This can be written as proton + antiproton → neutron + antineutron. Allowing antiproton from Bevatron apparatus to enter into a vessel containing liquid hydrogen when about 3 in 1000 ${}_1^1p$ in hydrogen is converted to antineutron. The antineutron is then annihilated with neutron, with liberation of energy. Now neutron, although electrically neutral, has magnetic moment associated with spinning of negative charge. Hence, a neutron can be considered as a central positive core surrounded by thin negative charge so anti- neutron will have central negative core surrounded by positive charge. Hence, if the neutron and antineutron rotate in the same direction, they will induce opposite magnetic field.

1.07 ATOMIC MODEL

THOMSON MODEL (*WATERMELON MODEL*): J. J. Thomson from his X-ray scattering experiment pictured the atomic model like this. The positive charge is uniformly distributed in the sphere representing the atom. As the atom is stable, electrons are held by this positive charge at different positions such that electron-electron repulsion is balanced by the force of attraction towards the centre of the sphere just like an ornament box where valuable gems (electrons) are placed in cotton pad (positive charge). So, for a single-electron system (hydrogen), the electron is at the centre of the positive sphere. For 2-electron system (helium), the 2 electrons are symmetrically situated on opposite side of the centre at a distance half the radius of the positive sphere. For 3-electron system, the three electrons are at the corners of an equilateral triangle of each side equal to the radius the positive sphere whose centre will be the orthocentre of the triangle. In this way, Thomson gave the electronic arrangement for so many atoms. This theory describes the stability of the atom but fails to explain many other observations like atomic spectra, etc. Again, the idea of positive sphere is also unacceptable. Some radioactive elements emit α-particles of high energy. Calculations show the existence of positive charge confined in small region of 10^{-8} cm. On the basis of this picture α-particles scattered greater than 90° will be zero. But 1 in 20 000 is scattered greater than 90°. Thus, Thomson model was rejected.

1.08 RUTHERFORD NUCLEAR MODEL (SOLAR MODEL)

α-RAY SCATTERING EXPERIMENT: Modern view accepts nuclear model of an atom based on the famous α-ray scattering experiment carried out by Lord Rutherford in 1911. In this experiment α-ray from a radio element (Radium-C) was allowed to strike on a 0.0004 mm thin gold foil with a ZnS screen behind when about 1 in 20 000 particles were deflected greater than 90°. This scattering can be due to strong repulsion experienced by double positive charge and heavy mass α-particles by the more heavy mass and higher charge at the point of the target metal at its closest approach.

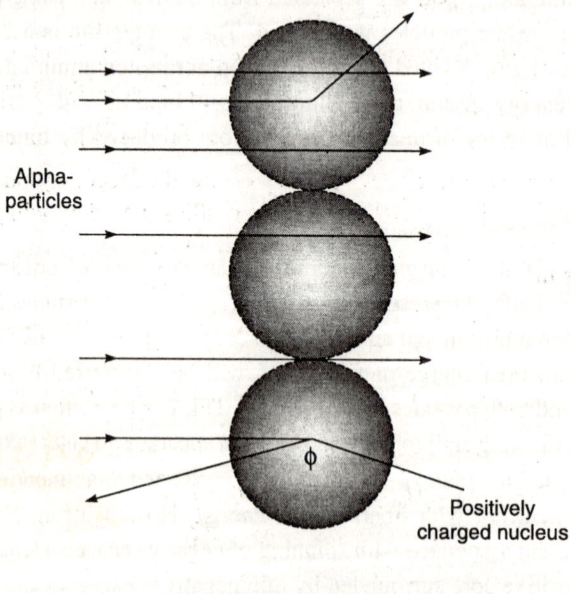

Fig. 1.4 Scattering of α-particles as interpreted by Rutherford.

This central mass and positive charge of the atom are concentrated in a small volume, called *nucleus* of the atom. Thus, Rutherford pictured atom as:

(a) The central nucleus contains the total mass and total positive charge in a volume of 10^{-36} cc.

(b) As the dimension of the atom is of 10^{-24} cc, so most part of the atom is vacant.

(c) The total electrons are outside the nucleus and the total number is equal to the total positive charge to make the atom electrically neutral.

(d) If the electrons were static, they would be pulled by electrostatic force of the nucleus. So, Rutherford considered extremely rapid motion of the electrons around the nucleus just like the solar system such that the resulting centrifugal force of the electrons will balance the inward electrostatic force to make the atom stable. These electrons being negatively charged and of low masses will not interact with α-particle. So most of the α-particles will pass un-deviated. Now the path of the deflected particles can be ascertained by considering Coulomb's inverse square law that shows it to be a hyperbola with the nucleus at the focus. The fraction of α-particles, $f(\phi)$ scattered by angle ϕ, is given by

$$f(\phi) = nt\left(\frac{Ze^2}{mv^2 r}\right)\text{cosec}^4\frac{\phi}{2}. \tag{1.7}$$

n = number of atoms per unit volume of scattering material

t = thickness of the scattering material (0.0004 mm)

Z = number of unit charges on the nucleus

e = electronic charge (4.8×10^{-10} esu)

m and v = mass and velocity of α-particle

ϕ = scattering angle

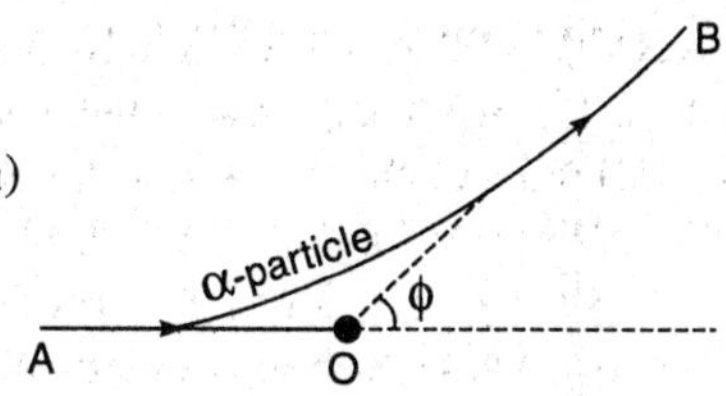

Fig. 1.5 Path of deflected α-particle.

1.09 DEMERITS OF RUTHERFORD'S MODEL

In Rutherford's model, like planetary system, the electrons are revolving in orbits when its direction and velocity both changes and the electron will be accelerated towards the centre and acceleration will be $\frac{v^2}{r}$, where v is the linear velocity of the electron and r is the radius of the electronic orbit. According to classical electromagnetic theory, such a system will emit continuous radiation. So, the electronic path will be circular with gradually decreasing radius, i.e., they will form a spiral path round the nucleus and ultimately drop into the nucleus, which will lead to the destruction of the atom or matter. This leads the instability of the atom, which is not true.

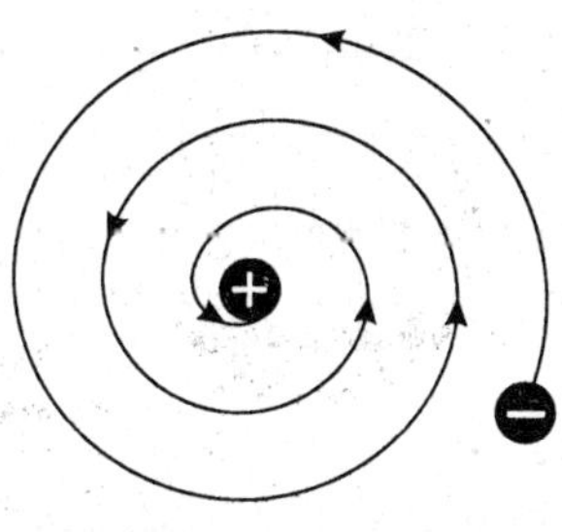

Fig. 1.6 The fate of an electron according to the electromagnetic theory.

The other demerit of the concept is that according to electromagnetic theory, the frequency of the energy radiated or absorbed by moving electron will be equal to the frequency of revolution. So, frequency of radiation will change continuously. We know, $v = \omega r$ and $\frac{mv^2}{r} = \frac{e^2}{r^2}$

$$\text{or, } m\omega^2 r^2 = \frac{e^2}{r} \quad \text{or, } \omega^2 = \frac{e^2}{mr^3}.$$

i.e., as r decreases, ω increases or radiation emitted will have continuous increasing frequency and spectrum will be continuous one.

From this point, atomic property is divided into two sections—(i) the property of the nucleus that deals with radioactivity, isotopes, etc., and (ii) the extra-nuclear section that deals with chemical properties of the elements.

1.10 MOSELEY'S WORK ON X-RAY SPECTRA AND DETERMINATION OF ATOMIC NUMBER

In an X-ray tube, when the cathode-ray electrons are allowed to strike on anti-cathode (an element), a high-penetrating short wavelength radiation is emitted, called *X-ray*. When this is allowed to fall on various materials, then secondary X-rays are produced. It depends upon the characteristic of the target material called *characteristic X-rays*. Moseley in 1913 resolved the characteristic X-ray by passing through $K_4[Fe(CN)_6]$ which acts as a crystal grating (slit) and then photographed. The spectrum of each element has two lines, one is sharp K_α-line that changes regularly from element to element; the frequency of K-series of elements (first transition series and potassium and calcium) shifts to the left with increasing group position in the Periodic Table as shown in the figure:

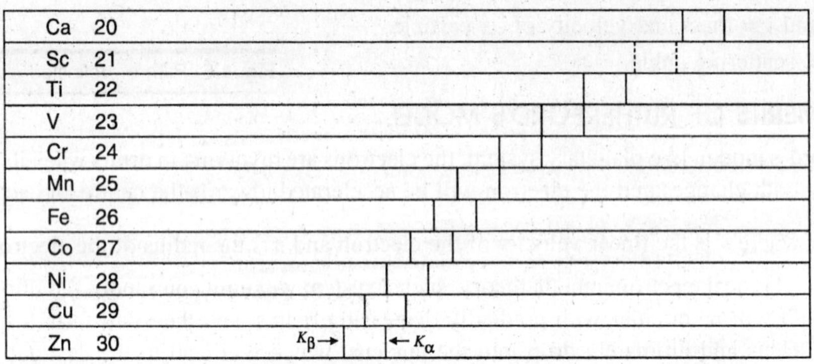

Fig. 1.7 X-ray spectra (*K*-series).

With still heavier elements another set of lines, called *L*-series are obtained and so on for still higher elements. Moseley compared the frequency ν of the most prominent line of the spectrum of a series with the serial number or ordinal number (Z) of the element and gave the relation between them as $\nu = a(Z-b)^2$ or, $\sqrt{\nu} = Z\sqrt{a} - b\sqrt{a}$, where a is the proportionality constant and its value is 2.47×10^{15} and b depends on the characteristic of the series (for K-series, $b = 1$; for L-series, $b = 7.4$ and so on). The plot of $\sqrt{\nu}$ against ordinal number Z is linear, but with atomic weight such relation is not observed.

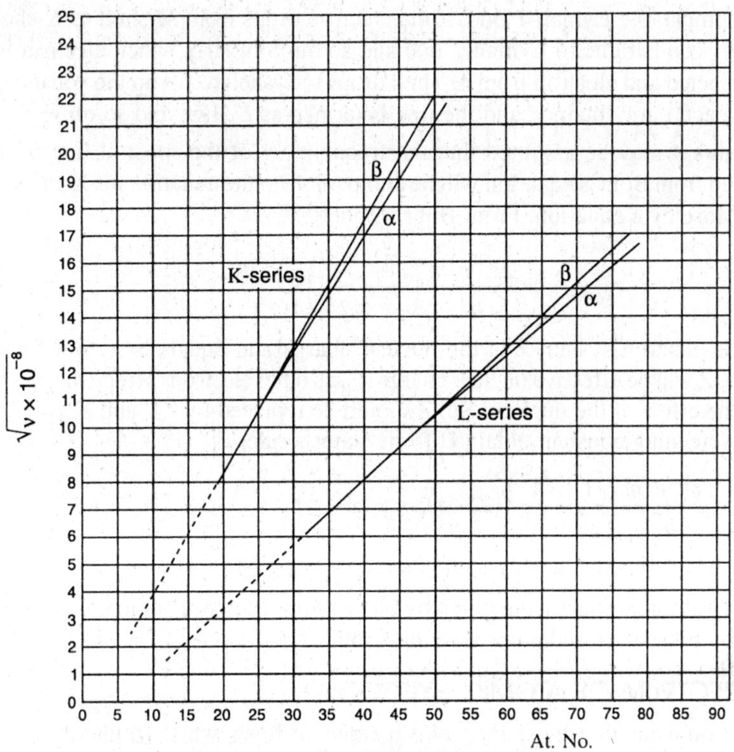

Fig. 1.8 Relation of frequency of X-ray spectral line with atomic number.

Thus, atomic number Z of an element is more significant and it is more fundamental property of an element. Chadwick (1920) calculated the atomic number of different elements from X-ray scattering of elements and the results are in good agreement with this value within ±2% tolerance. The physical and chemical properties of elements depend upon the atomic number and the periodic law modified in this respect.

The importance of the study of X-ray spectra can be mentioned as below:

(a) It justifies the placement of some elements in the periodic table, e.g., Ar and K, Co and Ni etc., in accordance with their atomic number.

(b) The gaps in the x-ray spectra graphs indicate some new elements yet to be discovered. In this way the missing elements Tc(43), Pm(61), Hf(72) and Re(75) were discovered.

The X-ray spectra and atomic spectra differ in these ways:

(i) X-ray spectra are very simple than atomic spectra.

(ii) The former is similar for different elements, differs only in frequency but the latter differs from one group of elements to other.

(iii) The X-ray spectra concern the excitation energy that exceeds a minimum value characteristic to the series but atomic spectra concerns the gradual increasing exciting energy required for promotion of electron from one energy level to the other. It will be clear if we consider the origin of X-ray spectra. When an atom absorbs high energy, one electron from K-shell is promoted to higher shell, which is filled up. Hence, if electron absorbs very high energy a limiting value, it will escape from the atomic system leaving a vacancy in K-shell. Then, if electron from L-shell drops to K-shell to fill the vacancy, it will emit X-ray of K-series and the line will be termed as K_α-line comparable Lyman-1 line, if the electron drops from *M*-shell to *K*-shell, it will give K_β comparable to Lyman-2 line and so on. Similarly, when electron from *L*-shell is ejected and electron from *M*-shell drops to *L*-shell to fill up the vacancy, then *L*-series spectra are obtained and the line is termed as L_α-line and so on.

- From this it may be assumed that the frequency ν of the spectral line of atomic spectra obtained from Bohr's equation will be comparable to the equation for X-ray spectra obtained from Moseley's equation. From Bohr's equation,

$$\nu_{\text{Lyman-1}} = \frac{2\pi^2 e^4 m}{h^3}\left(\frac{1}{1^2} - \frac{1}{2^2}\right) Z^2 . \tag{1.8}$$

For one-electron system Z is the nuclear charge and equals to 1. But for multi-electron system, *Z* will be effective nuclear charge Z_{eff} as other electrons screen the electron concerned from the effect of the nucleus. So, *Z* should be replaced by Z_{eff} and $Z_{eff} = (Z - \sigma)$, where σ is the screening constant obtained from Slater's rule.

So, $\nu = \frac{2\pi^2 e^4 m}{h^3}\left(\frac{1}{1^2} - \frac{1}{2^2}\right)(Z-\sigma)^2 = a\,(Z-\sigma)^2$

which is identical to Mosley's equation, $\sqrt{\nu} = \sqrt{a}\,(Z-\sigma)$.

(iv) The atomic spectra are given by outer orbital electron but the X-ray spectra are given by both inner and outer level electron.

1.11 THE SPECTRUM OF ATOMIC HYDROGEN

Hydrogen atom on excitation by spark yields a series of lines, which gradually converge towards higher wavelength side. It consists of six such series—each one named after their discoverers—Lyman, Balmer, Paschen, Brackett, Pfund and Humphreys of which the first one is in the ultraviolet range, the second one is in the visible range and the rests are in the infrared region. Balmer carefully studied the second series and observed many lines and named them as H_α, H_β, H_γ and H_δ and so on. Balmer arbitrarily gave an equation to the wave number $\bar{\nu}$ of the lines of the second series as

$\bar{\nu} = \frac{1}{\lambda} = R\left(\frac{1}{2^2} - \frac{1}{n^2}\right)$, where $n > 2$, i.e., 3, 4, 5, etc., where *R* is a constant, called *Rydberg constant*;

the experimentally determined value of it is 109677.75 cm^{-1} for hydrogen atom and termed as R_H. Similarly, Lyman gave the equation for 1st series and so on for others as:

Lyman series $\frac{1}{\lambda} = \bar{\nu} = R_H\left(\frac{1}{1^2} - \frac{1}{n^2}\right)$, when n = 2, 3, 4, ...

Paschen series $\frac{1}{\lambda} = \bar{\nu} = R_H\left(\frac{1}{3^2} - \frac{1}{n^2}\right)$, when n = 4, 5, ...

Brackett series $\frac{1}{\lambda} = \bar{\nu} = R_H\left(\frac{1}{4^2} - \frac{1}{n^2}\right)$, when n = 5, 6, 7, ...

Pfund series $\frac{1}{\lambda} = \bar{\nu} = R_H\left(\frac{1}{5^2} - \frac{1}{n^2}\right)$, when n = 6, 7, 8, ...

Humphreys series $\frac{1}{\lambda} = \bar{\nu} = R_H\left(\frac{1}{6^2} - \frac{1}{n^2}\right)$, when n = 7, 8, 9, ...

From general observation of these equations, it appears that the wave number $\bar{\nu}$ for any spectral series can be expressed as combination of two terms. One of them is constant for a series, called *series term* and the other term is variable called *current term* and combining them Ritz gave a general equation,

$$\frac{1}{\lambda} = \bar{\nu} = R_H\left(\frac{1}{n_1^2} - \frac{1}{n_2^2}\right), \text{ where } n_2 > n_1, \ \frac{R_H}{n_1^2} \text{ is the series term and } \frac{R_H}{n_2^2} \text{ is the current term,}$$

$n_1 = 1$ for Lyman, 2 for Balmer and so on. This is known as *Ritz combination principle*, and

$$\bar{\nu}_{H_\alpha} = R_H\left(\frac{1}{2^2} - \frac{1}{3^2}\right), \ \bar{\nu}_{H\beta} = R_H\left(\frac{1}{2^2} - \frac{1}{4^2}\right), \ \bar{\nu}_{H_\gamma} = R_H\left(\frac{1}{2^2} - \frac{1}{5^2}\right) \text{ and so on.}$$

Subtracting $\bar{\nu}_{H_\beta} - \bar{\nu}_{H_\alpha} = R_H\left(\frac{1}{3^2} - \frac{1}{4^2}\right)$, this equation gives the $\bar{\nu}$ for first Paschen series and so on. Thus, Ritz Combination Principle gained importance for all types of spectra.

1.12 PLANCK'S QUANTUM THEORY (BLACK-BODY RADIATION)

We know, when a body is heated, it first becomes red, then yellow and finally white hot. During this process it emits thermal radiation in the form of electromagnetic wave which depends on temperature and spread over a wide range of frequency or wavelength. Under the same condition a dull black surface emits more energy per second than a polished surface. Maximum emission occurs in case of a perfect black body, which may have the following characteristics:

1. It will be a hollow enclosure having a very small hole in one side. Size of the hole will be negligible compared to the surface of the body so that all radiation entering into it will be completely absorbed, no part will be reflected from it.
2. It will absorb radiation of all frequencies.
3. It will emit the same type of radiation it absorbs and so a good absorber will be a good radiator also. Inside wall of the enclosure should be painted black. Any electromagnetic disturbance inside it can be considered as superposition of various standing waves, which may behave like simple harmonic oscillator. The thermal spectrum of such a black body at different temperatures against λ is shown in figure. The plots show that a small amount of energy is radiated at very short and long wavelengths, the curves pass through maxima and intense radiation is obtained at higher temperature.

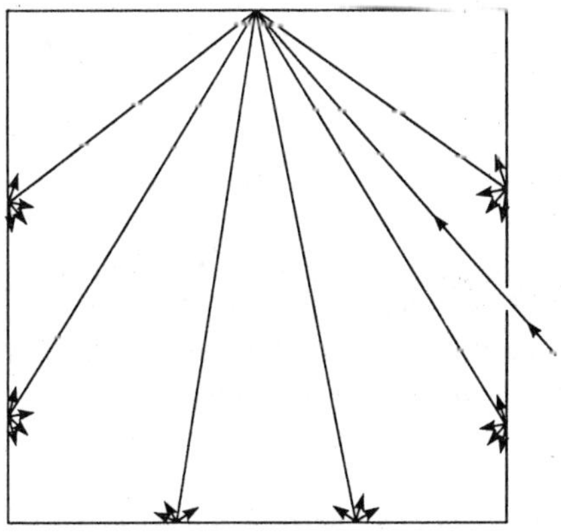

Fig. 1.9 A black body enclosure with a small opening in one wall.

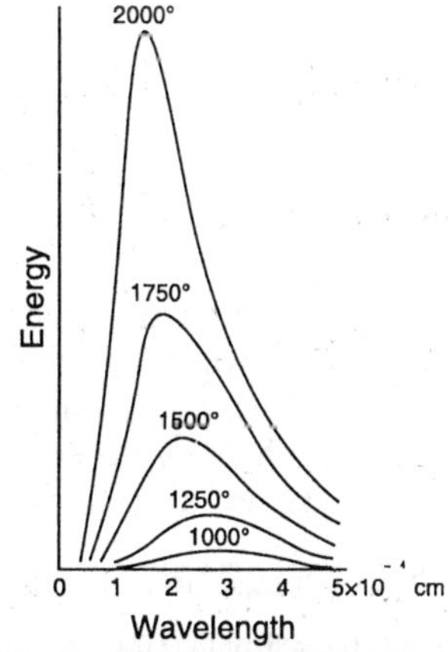

Fig. 1.10 Energy distribution for black-body radiation at different absolute temperatures.

From the study of the radiation spectra Stefan and Boltzmann stated that radiation density is proportional to the fourth power of temperature and $E = e\,\sigma\,T^4$, (1.9)
where E is the rate of emission of radiant energy per unit area, T is absolute temperature, σ is the Stefan-Boltzmann's constant [5.699 $\times 10^{-5}$ erg sec^{-1} cm^2 (K-deg)^4] and e is the emissivity E/E_0 of the surface and for black body $e = 1$. This equation is applicable to lower wavelength but quite unsatisfactory in the longer wavelength region.

Wien stated that maximum radiation intensity at a wavelength λ, given out by a black body is inversely proportional to absolute temperature,

$$\text{i.e., } \lambda_{max} \propto \frac{1}{T} \qquad (1.10)$$

or, $\lambda_{max} T$ = constant = 2.897 $\times 10^{-3}$ m(k).

This is our common experience that more we heat a body more it radiates energy. The plot of radiation from Wien's equation is in good agreement with that of Stefan-Boltzmann's equation.

Rayleigh and Jeans tackled the problem on the basis of classical principles of equipartition of energy and derived the equation like this:

$$E_\lambda = \frac{2\pi kT}{C\lambda^4} \qquad (1.11)$$

This equation holds good for longer wavelength but inapplicable to shorter wavelength, i.e., high- energy region. The general expression for the radiation energy of the black body over entire range of wavelength from $\lambda = 0$ to ∞ is

$$E = CT^{5-\mu}\lambda^{-\mu}e^{-b/(\lambda T)^{\nu}} \qquad (1.12)$$

From this equation, Wien's equation can be obtained by putting $\mu = 5$ and $\nu = 1$ and Rayleigh-Jeans' equation can be obtained by putting $\mu = 4$ and $b = 0$. All these are not quite satisfactory to explain the energy distribution of a black-body radiation.

In 1901, Max Planck proposed a new hypothesis called *quantum theory* to explain the radiation spectra. He considered each standing wave in the cavity radiator as a simple harmonic oscillator. According to classical theory these oscillators can absorb or radiate energy continuously. He discarded this concept and proposed that the energy changes could take place discontinuously and in discrete amounts, always an integral multiple of a small unit of energy, called *quantum*. With this new quantum idea, explanation of various experimental results was obtained. Thus emitted energy will be ε_0, $2\varepsilon_0$, $3\varepsilon_0$, $5\varepsilon_0$, $8\varepsilon_0$ and so on and $\varepsilon_0 = h\nu$, where ν is the frequency of radiation and h is called *Planck's constant* whose value is 6.62 $\times 10^{-27}$ erg-sec or 6.62 $\times 10^{-34}$ J-sec. With this idea Planck derived the equation for energy,

$$E = \frac{2\pi hc^2}{\lambda^5} \cdot \frac{1}{e^{\frac{ch}{\pi kT}} - 1} \qquad (1.13)$$

where k is Boltzmann's constant, c the velocity of light. This equation fits the experimental curves very well. For short wavelength, i.e., high energy $e^{ch/\lambda kT}$ becomes greater than 1 and the equation takes the form of Wien's equation. For longer wavelengths, this equation takes the form of Rayleigh-Jeans' equation. When this packet of energy or quantum is emitted in the form of light, it is called *photon*. This illustrates the particle nature of radiation.

1.13 PHOTOELECTRIC EFFECT

Albert Einstein established the quantum or corpuscular nature of light for explaining the photoelectric effect. In 1887 during his experiment on electromagnetism, Einstein observed that when light is allowed to fall on some metals like Zn, Mg, Li, Na, K, etc., as cathode in an evacuated tube, negatively charged particles are emitted from cathode and produce current. The determination of *e/m* for these negatively charged particles definitely proves them to be electrons. Thus emission of electrons from metal surface by the action of light is called photoelectric effect, the electrons are called *photoelectrons* and the current produced is called *photoelectric current.* In this connection the following things are important:

(a) The photoelectric current is directly proportional to the intensity of light used. If I be the current and e be the charge of an electron, then number of photoelectrons emitted per second is I/e, e being constant, I is proportional to intensity of illumination or number of photoelectrons emitted per second is proportional to the intensity of illumination, i.e., light.

(b) Velocity and kinetic energy of photoelectron vary from zero to a certain maximum value and this value depends on the wavelength or frequency of the incident light, not on the intensity of light. When a monochromatic light of different intensity is used and potential V is applied to the cathode is changed from V^+ through V_0 to V^-, the following graphs are obtained:

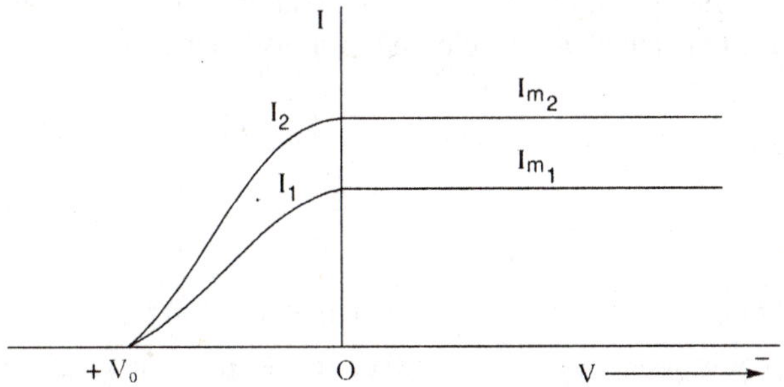

Fig. 1.11 The effect of potential on the photoelectric current produced by a monochromatic light beam.

From the graph we come to know that, when V^+ potential at the cathode is used, no photo-current is obtained, this is known as *stopping potential*. As V^+ increases, photocurrent increases bit by bit, becomes maximum at $V = 0$ and no increase in current when V^- is applied. We can see that $I_2 > I_1$, where illumination intensity is $I_{m_2} > I_{m_1}$.

Again, if amount of current produced using monochromatic light of different λ is plotted against V potential at the cathode, the graph is as follows:

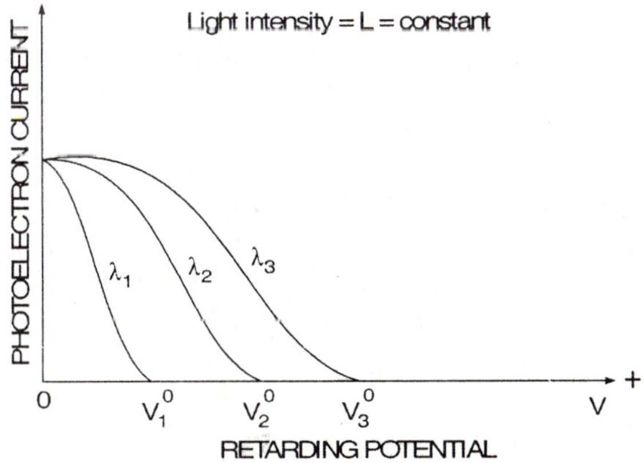

Fig. 1.12 The extinction voltage V^0 depends upon the frequency ν of the light. When the retarding potential is $V = 0$, the photoelectron current is the same for light of a given intensity L regardless of its frequency.

When $\lambda_1 > \lambda_2 > \lambda_3$, then potential applied to the emitter is $V_3^+ > V_2^+ > V_1^+$. This shows kinetic energy of the photoelectron increases with frequency of the incident light.

(c) *Nature of the metal used as emitter*: Nature of the metal and their surface condition plays an important role in photoelectric effect. Alkali metals emit electrons when visible light falls on them, i.e., threshold energy is low in the infrared region; the alkaline earth metals are photosensitive in ultraviolet region and so on. Again, absorbed gases in the upper layer or interior of the metal increase the emission of electron and also shift the threshold value to longer wavelength region. Again, when the incident radiation is capable of penetrating into the inner body of the metal (in case of insulators) or a body of high resistance (in case of semiconductors), some electrons are liberated at the interior of the body and will increase their conductivity. This is known as *photoelectric conductivity*.

(d) The photoelectrons are emitted instantaneously and time interval between irradiation and emission of electron is only 10^{-9} sec.

(e) For electron emission a minimum energy is required, called *threshold energy*, below which no electron from the metal will be emitted whatever be its intensity. Emission will occur in vacuum also.

If we try to explain this in the light of classical electromagnetic theory, minimum energy required for emission of electron will be

$$E = V + \frac{1}{2}mv^2 \tag{1.14}$$

where E is the energy of the light radiation falling on the metal surface, V is the potential energy of the electron with which it is bounded to metal surface and $\frac{1}{2}mv^2$ is the kinetic energy with which the electron is liberated. Now E is obtained from the incident light and is dependent on intensity of the light and E is continuously absorbed. When sufficient energy is absorbed, electron will be emitted after some time. If intensity is increased, time interval will decrease. This conclusion does not explain all experimental observations.

Einstein in 1905 tried to explain this effect in the light of Planck's quantum theory. According to his suggestion, light is composed of some discrete unit, called *quantum*, having energy $h\nu$, also called *corpuscles*, i.e., small particle or photon. Radiation or emission and absorption by electron take place in this discrete unit. It can pass through space with a velocity comparable to the velocity of light. When this light beam strikes the metal surface, an electron will absorb whole energy. If this energy is greater than the binding energy, W of electron, to the nucleus, then the electron will be emitted with kinetic energy equal to the difference of energy absorbed and the potential energy.

$$\text{So, } h\nu = W + \frac{1}{2}mv^2 \quad \text{and} \quad W = h\nu_0 . \tag{1.15}$$

This W is called the *photoelectric work function* of the metal. Greater intensity means the number of quanta per cross-section or volume of the incident beam is greater and so number of electrons emitted and hence, current will be higher. As long as $h\nu$ is constant, K.E. for electron for a metal will be constant. It is evident that the work function of different metals will be different depending on its electronic configuration or more correctly effective nuclear charge that regulates or controls the atomic radius. Hence, W will increase along a period and decrease down a group. Again, in the same group and same period, the transition metals will have higher value of photoelectric work function than the representative elements.

From curve V^+ to V^- against I in Fig. 1.11 at different intensity of light, V^+ potential applied to the cathode to stop emission of electron called *stopping potential*, Planck's constant h can be calculated. At stopping potential, electric energy will be equal to the K.E. of electron.

So, $Ve = \frac{1}{2}mv^2$ and Einstein's equation

$$h\nu = h\nu_0 + \frac{1}{2}mv^2 = h\nu_0 + Ve$$

or, $$\nu = \nu_0 + \left(\frac{e}{h}\right)V \tag{1.16}$$

Hence, plot of ν against stopping potential V will be a straight line and intercept on the ordinate is characteristic of the metal. The slope of this line is $\left(\frac{e}{h}\right)$, which is universal constant, i.e., all

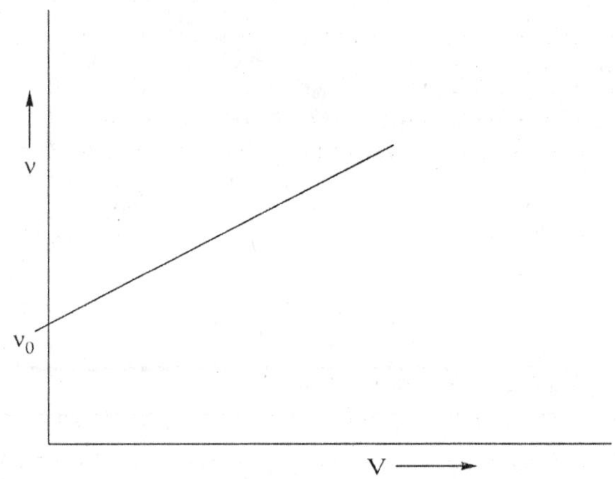

Fig. 1.13 Plot of ν against stopping potential *V*.

photoeffective metals will have same slope but ν_0 will be different, i.e., the electronic charge is constant and from the slope or gradient of this plot, *h*, Planck's constant can be calculated, which is in good agreement with the values calculated from other methods.

1.14 COMPTON EFFECT

Compton in 1925 established the particle nature of light. When X-ray is allowed to strike an electron, scattering of X-ray occurs. On the basis of classical theory

(a) the scattered X-ray will have same wavelength as the incident X-ray;

(b) the scattering coefficient will be independent of the wavelength of incident beam ($\sigma = 0.2$);

(c) distribution intensity of scattered beam will be symmetrical.

But the observed fact is that

(i) the scattered X-ray has higher wavelength than the incident X-ray;

(ii) the scattering coefficient, σ is dependent on the wavelength of incident beam, σ decreases as the wavelength of the incident beam decreases;

(iii) the scattering is not symmetrical but takes place in the forward direction.

This increase in wavelength of scattered X-ray beam from the incident beam due to collision with the electron is called *Compton Effect* and the difference between the two beams is called *Compton Shift*.

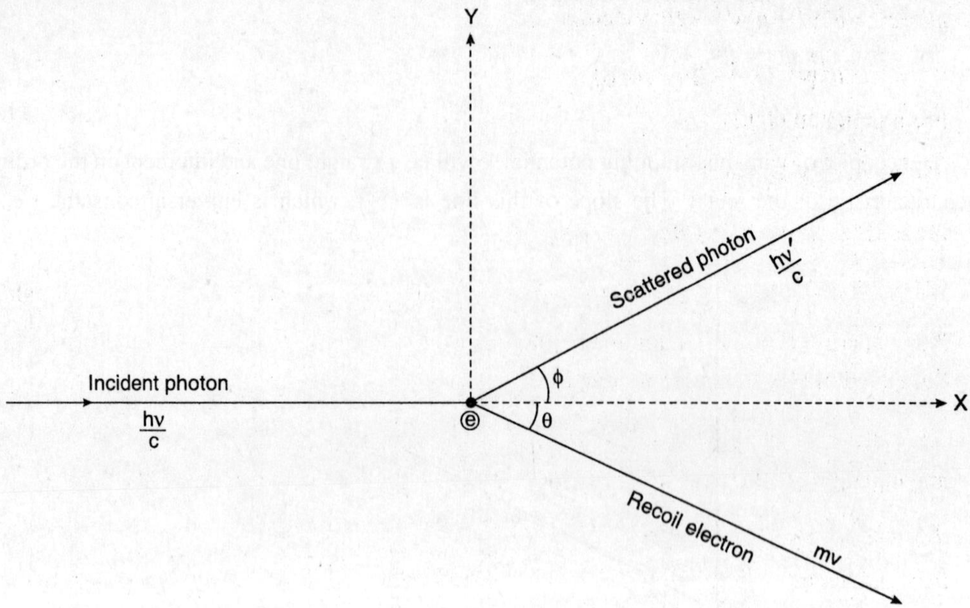

Fig. 1.14 Schematic representation of the Compton effect.

Before collision

Energy of the incident beam = $h\nu$

Energy of the electron = m_0c^2 (where m_0 is the mass of electron at rest)

Momentum of the beam = $h\nu/c$

Momentum of the electron = 0

After collision

Energy of scattered beam = $h\nu'$

Energy of the recoiled electron = mc^2 (where m is the mass of the electron in motion)

Momentum of the scattered photon = $h\nu'/c$

Momentum of the recoiled electron = mv (where v is the velocity)

Following the law of conservation of energy,

$$h\nu + m_0c^2 = h\nu' + mc^2 \tag{1.17}$$

From the law of conservation of momentum in the forward direction

$$\frac{h\nu}{c} + 0 = \frac{h\nu'}{c}\cos\phi + mv\cos\theta \tag{1.18}$$

and that in the direction perpendicular to the above component

$$\frac{h\nu'}{c} \times \sin\phi = mv\sin\theta \quad \text{[being equal and opposite]} \tag{1.19}$$

Squaring and adding equations (1.18) and (1.19), we get

$$m^2v^2c^2 = h^2\nu^2 + h^2\nu'^2 - 2h^2\nu\nu'\cos\phi$$

$$= h^2\left(\nu^2 + \nu'^2 - 2\nu\nu'\cos\phi\right) \tag{1.20}$$

From equation (1.17)

$$mc^2 = h(\nu - \nu') + m_0c^2$$

$$\text{or, } m^2c^4 = h^2(\nu-\nu')^2 + m_0^2c^4 + 2hm_0c^2(\nu-\nu')$$

$$= h^2\left(\nu^2 - \nu'^2 - 2\nu\nu'\right) + 2hm_0c^2(\nu-\nu') + m_0^2c^4 \tag{1.21}$$

Now, energy, $E = mc^2$ and momentum $p = mv$.

So, equation (1.20) can be written as

$$p^2c^2 = h^2\left(\nu^2 + \nu'^2 - 2\nu\nu'\cos\phi\right) \tag{1.22}$$

and equation (1.21) takes the form

$$E^2 = h^2\left(\nu^2 + \nu'^2 - 2\nu\nu'\right) + 2hm_0c^2(\nu-\nu') + m_0^2c^4 \tag{1.23}$$

Now subtracting equation (1.23) from equation (1.22)

$$E^2 - p^2c^2 = 2h^2\nu\nu'(\cos\phi - 1) + 2h(\nu-\nu')m_0c^2 + m_0^2c^4 \tag{1.24}$$

The relation between momentum and energy

$$E^2 - p^2c^2 = m_0^2c^4 \tag{1.25}$$

From (1.24) and (1.25) we can write

$$2h^2\nu\nu'(\cos\phi - 1) = -2h(\nu-\nu')m_0c^2$$

$$\text{or, } \frac{\nu-\nu'}{\nu\nu'} = \frac{h}{m_0c^2}(1-\cos\phi)$$

$$\text{or, } \frac{1}{\nu'} - \frac{1}{\nu} = \frac{h}{m_0c^2}(1-\cos\phi)$$

$$\text{or, } \lambda' - \lambda = \frac{h}{m_0c^2}(1-\cos\phi) = \Delta\lambda$$

The term $\frac{h}{m_0c^2}$ is called the *Compton wavelength of the electron.*

From this equation $\Delta\lambda$ is independent of λ or ν of incident beam and nature of the scattering material, but depends only on the scattering angle ϕ. Now, we can say, when

(a) $\phi = 0°$, then $\lambda = \lambda'$; so no scattering

(b) $\phi = 90°$, then $\lambda' > \lambda$

(c) $\phi = 180°$, then $\lambda = \lambda'$.

Thus, ϕ being 0° to 180°, the scattered frequency changes from maximum to minimum value. This scattering occurs only with the free electron. If the electron is bound to the atom, then X-ray will collide with the atom as a whole and m_0, the rest mass will be the mass of the atom, which is many times larger than that of electron. So, the Compton shift $\Delta\lambda$ will be negligible.

If low energy, e.g., visible light falls on metal surface, all electrons are bound to the atom. So, scattering will not occur but photoelectric effect may be possible with increasing energy. When energy is sufficient to make the electron free, scattering may occur. Thus, photoelectric effect occurs with low energy photon or X-ray up to 0.1 MeV energy, but Compton scattering occurs with high energy photon or X-ray of the order of 1MeV energy. Again, photoelectric effect occurs with bound electrons, but scattering occurs with free electrons.

X-ray diffraction pattern also clearly demonstrates the particle nature of radiation.

1.15 BOHR MODEL OF HYDROGEN ATOM

To explain the stability of an atom in Rutherford model concept and also to explain atomic spectra, Bohr in 1913 proposed some revolutionary suggestions. He kept aside the classical electromagnetic theory of radiation. His assumptions are as follows:

1. The electron in the atom always moves in a closed circular orbit and as long as it does so, it will not radiate nor absorb energy. These orbits are stable and hence called *stationary states* and the electron in these stationary states has fixed energy. More than one such stationary state are possible, and which states are possible and what will be their size have been given in his second assumption. Thus, first postulate gives only pictorial nature of the atom. The term stationary refers to the energy only, not the static electron.
2. In the second assumption he stated that of all theoretically possible energy states for an electron, only those are stable for which the orbital angular momentum of the electron is quantized and this value is an integral multiple of $h/2\pi$.

If the electron of mass m moves with normal velocity v in a circular orbit of radius r, then angular velocity, $\omega = vr$ and so angular momentum will be mvr.

Accordingly, $mvr = \frac{nh}{2\pi}$ (h is Planck's constant), where n is called *principal quantum number* and may have any positive integer values except zero. Thus, it designates the order of the orbit from the nucleus with greater and greater radius or in other words, n describes the distance of the electron from the nucleus. This is merely an assumption although it is the natural outcome of wave mechanics.

[Electron has wave-particle duality and the relation is according to de Broglie equation

$$\lambda = \frac{h}{mc} \quad (E = mc^2 = h\nu \text{ and } \nu = \frac{c}{\lambda})$$

The electron wave is forced to move round the nucleus in complete cycle, only when the circumference of the circular orbit contains an integral number of waves. If not in integral number, successive waves will not superimpose exactly on one another and the waves will destroy each other by interference.

So, $2\pi r = n\lambda$

or, $2\pi r = \frac{nh}{mv}$ [as $v \approx c$]

or, $mvr = \frac{nh}{2\pi}$]

As the electron waves become smaller, the orbit becomes circular as a limiting case. The electrons are observed in the fixed energy zone, not in the fixed energy orbit.

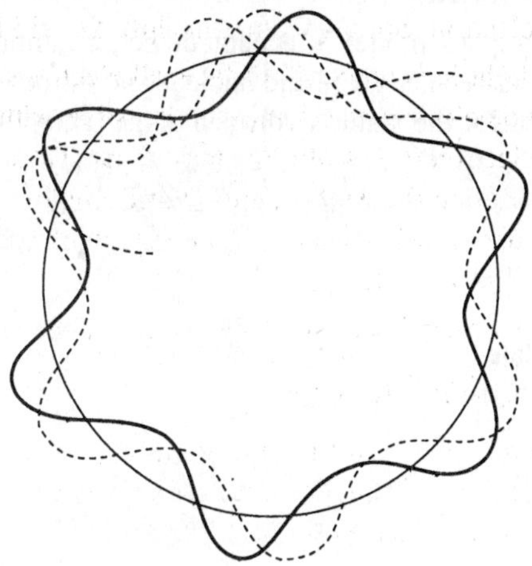

Fig. 1.15 A Bohr orbit in terms of electron waves. As the waves become smaller, the circular Bohr orbit is obtained as a limiting case. For an orbit of radius r, $2\pi r = n\lambda$, where n is an integer and λ is the wavelength. Substituting $\lambda = h/mv$, one gets $mvr = nh/2\pi$.

Thus, Bohr's second postulate introduces the quantum idea in the atomic picture.

3. In the third postulate Bohr explained the origin of spectral lines. Although the electrons while staying in the stationary state it will neither radiate nor absorb any energy. But the electron transition from higher energy (E_2) outer orbit to lower energy (E_1) inner orbit may take place. In that case the electron would radiate energy depending on the energy difference $\Delta E = h\nu$ between the orbits and will give rise to emission spectra (here h is Planck's constant and ν is the frequency of radiation). Likewise, transition from lower energy (E_1) of inner orbit to higher energy (E_2) outer orbit will absorb energy and will give rise to absorption spectra.

   However, if this transition takes place, then

   $\Delta E = E_2 - E_1 = h\nu = hc\bar{\nu}$

   or $\nu = \Delta E/h$ and $\bar{\nu} = \Delta E/hc$

The success of Bohr's assumptions depends on how these can be applied to hydrogen-like atom to calculate

(a) radius of the electronic orbit
(b) angular velocity of electron in the orbit
(c) energy of the electron in stationary orbit
(d) the frequency of spectral lines
(e) introduction of the concept of rotation of nucleus, i.e., reduced mass
(f) evaluation of Rydberg's constant
(g) evaluation of the ratio of the mass of an electron to that of a proton
(h) calculation of ionisation potential.

In the following mathematical deductions the notations with usual meaning are used:

m = mass of an electron, Z = atomic number, e = charge of an electron, v = velocity of electron, r = radius of the electron orbit, c = velocity of light, ε_0 = permitivity of vacuum.

(a) RADIUS OF ELECTRONIC ORBIT

In this calculations basic Rutherford model is accepted, i.e., condition of equilibrium between electrostatic force of attraction between electron and nucleus and the centrifugal force due to circular motion of electron is considered for the stable hydrogen atom. Therefore,

$$\underset{\text{(Centrifugal force)}}{\frac{mv^2}{r}} = \underset{\text{(Coulombic force)}}{\frac{Ze^2}{r^2}} \quad \text{or,} \quad mv^2 = \frac{Ze^2}{r} \tag{1.26}$$

Again, from Bohr's postulate

$$mvr = \frac{nh}{2\pi} \quad \text{or,} \quad v = \frac{nh}{2\pi mr} \tag{1.27}$$

$$\text{or, } v^2 = \frac{n^2h^2}{4\pi^2m^2r^2} \tag{1.28}$$

Putting the value of (1.28) in (1.26), we get

$$m\frac{n^2h^2}{4\pi^2m^2r^2} = \frac{Ze^2}{r}$$

$$\text{or, } \frac{n^2h^2}{4\pi^2mr} = Ze^2$$

$$\text{or, } r = \frac{n^2h^2}{4\pi^2mZe^2} \tag{1.29}$$

Thus, since other terms being constant $r \propto n^2$ and putting n as 1, 2, 3, etc., radius for the first, second and third Bohr orbit can be obtained. With the same value for n, r decreases with increasing Z, i.e., along a period the radius, i.e., size of an atom will decrease. It is evident that the distance of the successive radius gradually increases although the energy difference between them will gradually decrease as will be evident from the calculation of their energy. Putting the values of h in ergs, m in gram and e in esu, the value of r becomes 0.053Å (for the first orbit) and this is the unit of atomic length, called *Bohr* and sometimes denoted by a_0.

Using new electrostatic unit we get the equation as $\frac{mv^2}{r} = \frac{Ze^2}{4\pi\varepsilon_0 r^2}$.

Hence, $r = \frac{\varepsilon_0 n^2h^2}{\pi mZe^2}$.

(b) VELOCITY OF ELECTRON IN AN ORBIT

From equation (1.27) and (1.29)

$$v = \frac{nh}{2\pi mr} = \frac{nh}{2\pi m}\frac{4\pi^2mZe^2}{n^2h^2} = \frac{2\pi Ze^2}{nh} \tag{1.30}$$

Here, since other terms being constant, $v \propto \frac{1}{n}$, i.e., as the value of n increases or the distance between the electron and the nucleus increases, the electronic velocity decreases. It is evident, as with increasing distance electrostatic force of attraction decreases and the balancing centrifugal force will decrease.

Next, Bohr calculated the energy of the electron in different orbits and tried to show that the frequencies of the spectral lines are related to the energy difference between the electronic orbits.

(c) ENERGY OF THE ELECTRON

The electron in an atom possesses kinetic energy due to its revolution around the nucleus and potential energy due to electrostatic attraction between the electron and the nucleus.

From equation (1.26)

$$\text{Kinetic Energy, K.E.} = \frac{1}{2}mv^2 = \frac{Ze^2}{2r} \tag{1.31}$$

$$[\frac{1}{2}mv^2 = \frac{Ze^2}{8\pi^2\varepsilon_0 r} \text{ using new electrostatic unit.}]$$

The potential energy of the electron in an atom is the work done in bringing the electron from infinity to place it in the orbit of radius r. As at infinity there is no attraction and hence, P.E. is zero at infinite distance.

$$\text{Therefore, P.E.} = \int_{r=\infty}^{r=r} \frac{Ze^2}{r^2}dr = -\frac{Ze^2}{r} \left[\text{P.E.} = -\frac{Ze^2}{4\pi\varepsilon_0 r}\right]. \tag{1.32}$$

Total energy, E = P.E.+ K.E.

From equations (1.31) and (1.32)

$$E = -\frac{Ze^2}{r} + \frac{Ze^2}{2r}$$

$$= -\frac{Ze^2}{2r} = -\frac{2\pi^2 Z^2 e^4 m}{n^2 h^2}. \tag{1.33}$$

[Using new electrostatic unit, $E = \dfrac{Ze^2}{8\pi\varepsilon_0 r} - \dfrac{Ze^2}{4\pi\varepsilon_0 r} = -\dfrac{Ze^2}{8\pi\varepsilon_0 r} = -\dfrac{Z^2e^2m}{8\varepsilon_0^2 n^2 h^2}$.]

From equation (1.33) the following generalizations can be drawn:

(a) The negative sign in the energy term indicates that the electron is bound to the nucleus with this amount of energy and the same amount of energy is to be supplied to the electron to make it free.

(b) Other terms in the equation (1.33) being constant, $E \propto -\frac{1}{n^2}$, i.e., with increasing value of n, the energy of the orbit increases or in other words, the electron closer to the nucleus will have less energy.

(c) n remaining same, when Z increases, E being more negative will be more stable, that is why along a period, when orbit is more filled up, atom becomes more stable.

(d) With increasing value of n, the energy of the orbits increases in the order $-1 : -\frac{1}{4} : -\frac{1}{9}$ and so on, and the difference in energy between successive orbits gradually decreases with higher value of n and ultimately will diffuse or crowd together for higher values of n.

For $n = 1$, $E_1 = -21.79 \times 10^{-19}$ J
$= -13.6$ eV.

$$n = 2,\ E_2 = \frac{E_1}{2^2} = -5.45 \times 10^{-19}\,\text{J}$$
$$= -3.4\ \text{eV}.$$

$$\therefore\ E_2 - E_1 = 10.2\ \text{eV}$$

$$E_3 = \frac{E_1}{9} = -2.42 \times 10^{-19}\,\text{J} = -151\ \text{eV}$$

$$E_3 - E_2 = -1.51 + 3.4 = 1.89\ \text{eV}$$
$$\text{and } E_3 - E_1 = 12.1\ \text{eV}.$$

	Energy (eV)	*Excitation Energy* (eV)
$n = \infty$		
$n = 5$		
$n = 4$	$-13{\cdot}6/4^2 = -0{\cdot}85$	12·75
$n = 3$	$-13{\cdot}6/3^2 = -1{\cdot}51$	12·1
$n = 2$	$-13{\cdot}6/2^2 = -3{\cdot}4$	10·2
$n = 1$	$-13{\cdot}6$	

Fig. 1.16 Energy-levels corresponding to the Bohr orbits (not to scale). The ground state corresponds to $n = 1$. The state corresponding to $n = 2$ is called the *first excited state*. The energy required to promote an electron from the ground state to an excited state is called the *excitation energy*. The energy difference between the levels $n = 1$ and $n = \infty$ is the *ionization energy*.

(d) INTERPRETATION OF SPECTRUM

In his third assumption, Bohr stated that the transition of electron from one state to the other gives rise to atomic spectra. Let in hydrogen-like atom, the electron transition takes place from energy state E_2 of quantum number n_2 to the energy state E_1 of quantum number n_1, where $E_2 > E_1$, then from third postulate,

$$h\nu = E_2 - E_1 = \Delta E = -\frac{2\pi^2 Z^2 e^4 m}{n_2^2 h^2} + \frac{2\pi^2 Z^2 e^4 m}{n_1^2 h^2} = \frac{2\pi^2 Z^2 e^4 m}{h^2}\left(\frac{1}{n_1^2} - \frac{1}{n_2^2}\right)$$

$$\therefore\ \nu = \frac{2\pi^2 Z^2 e^4 m}{h^3}\left(\frac{1}{n_1^2} - \frac{1}{n_2^2}\right) \text{ and } \bar{\nu} = \frac{2\pi^2 Z^2 e^4 m}{h^3 c}\left(\frac{1}{n_1^2} - \frac{1}{n_2^2}\right) = RZ^2\left(\frac{1}{n_1^2} - \frac{1}{n_2^2}\right) \qquad (1.34)$$

where $R = \dfrac{2\pi^2 Z^2 e^4 m}{h^3 c}$.

[Using new electrostatic unit, $\bar{\nu} = \frac{\nu}{c} = \frac{me^4Z^2}{8\varepsilon_0 ch^3}\left(\frac{1}{n_1^2} - \frac{1}{n_2^2}\right)$.]

For hydrogen atom, $Z = 1$ and so equation (1.34) reduces to $\bar{\nu} = R_H\left(\frac{1}{n_1^2} - \frac{1}{n_2^2}\right)$.

And putting $n_1 = 2$, the equation becomes identical to that of Balmer. Here, R_H is called *Rydberg constant* for hydrogen. Substituting the values of *e*, *m*, etc., R_H becomes 109677.5 cm^{-1} and this value is in good agreement with the value assigned by Balmer that gets theoretical support from Bohr's theory.

Now transition of electron to $n_1 = 1$ from higher 2nd, 3rd or 4th quantum states gives rise to 1st, 2nd or 3rd, etc., lines of the Lyman series which occurs in the ultraviolet region. Transition from 3rd, 4th, 5th quantum states to the 2nd quantum state gives rise to the 1st, 2nd or 3rd Balmer line; this occurs in the visible region.

Likewise, Paschen, Brackett and Pfund series of spectral lines are obtained in the near- and far-infrared regions. During transition when electron enters a fixed energy zone, it will stop absorption or radiation of energy and hence, the spectra become discontinuous one. With increasing value of *n*, the energy states come closer and closer, so spectral lines will also come closer and the spectrum becomes continuous one for higher values of *n*. The nuclear field disappears in this region and electron will absorb or radiate energy continuously. Under this condition behavior of the electron will not be guided by quantum restrictions. Bohr's theory thus successfully accounts for the main features of hydrogen spectra; frequencies calculated theoretically are in good agreement with the experimental value. The identity of Bohr's equation with that of Ritz and Balmer establishes firmly the validity of Bohr's theory.

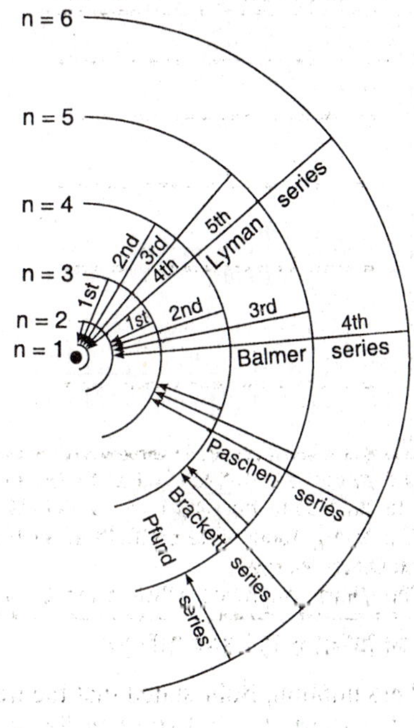

Fig. 1.17 Transitions leading to the five spectral series of the hydrogen atom.

(e) INTRODUCTION OF THE CONCEPT OF REDUCED MASS

Although experimental values of *E*, $\bar{\nu}$ and *R* are in the good agreement of the theoretical values, but the exact agreement can be found by considering the law of dynamics to the atomic system. It states that wherever a particle moves in a circle around another of finite mass, the central mass also move in a circle about the common centre of gravity.

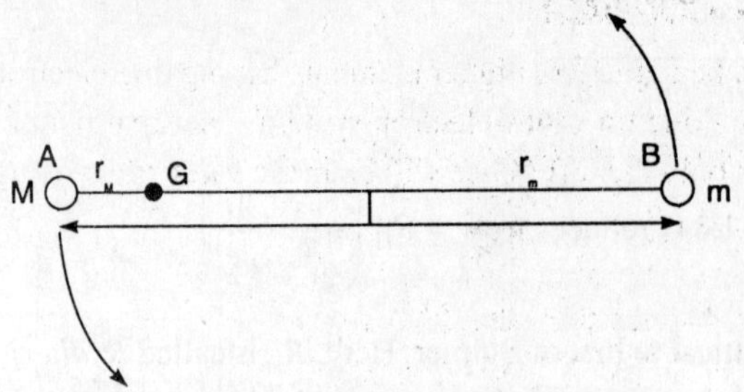

Fig. 1.18 Motion around the center of mass of a hydrogen atom.

Here m and M are the mass of electron and nucleus respectively and G is the centre of gravity around which electron and the nucleus revolve, r is the distance between the electron and the nucleus. In spite of these revolutions, the moment of M around G and moment of m around G balance each other.

So, $M \times AG = m \times BG = m\,(AB - AG)$

$$= m\,(r - AG)$$

$$AG\,(M + m) = mr$$

$$\text{or, } AG = \frac{mr}{M+m} \text{ and so } BG = \frac{Mr}{M+m}.$$

If V is the velocity of the nucleus and v the velocity of electron, then kinetic energy

$$\text{K.E.} = \frac{1}{2}MV^2 + \frac{1}{2}mv^2$$

$$= \frac{1}{2}M(\omega AG)^2 + \frac{1}{2}m(\omega BG)^2 \quad [\text{since } V = \omega r]$$

$$= \frac{1}{2}M\omega^2\frac{m^2r^2}{(M+m)^2} + \frac{1}{2}m\omega^2\frac{M^2r^2}{(M+m)^2}$$

$$= \frac{1}{2}Mm\omega^2r^2\left[\frac{m}{(M+m)^2} + \frac{M}{(M+m)^2}\right]$$

$$= \frac{1}{2}Mm\omega^2r^2 \times \frac{1}{M+m}$$

$$= \frac{1}{2}\omega^2r^2 \times \frac{Mm}{M+m}$$

$$= \frac{1}{2}\omega^2r^2 \times \mu \qquad (1.35)$$

Here, $\mu = \dfrac{mM}{M+m}$ is called the *reduced mass.*

So in all equations we should write μ instead of m such that

$$\frac{1}{\mu} = \frac{1}{M} + \frac{1}{m}.$$

Using the value of μ better agreement with the experimental value of $\bar{\nu}$, r and R can be obtained. Thus,

$$\bar{\nu} = \frac{1}{\lambda} = \frac{2\pi^2Z^2e^4}{h^3c}\frac{Mm}{M+m}\left(\frac{1}{n_1^2} - \frac{1}{n_2^2}\right) = \frac{Mm}{M+m}Z^2R\left(\frac{1}{n_1^2} - \frac{1}{n_2^2}\right)$$

and $R = 109677.8\ \text{cm}^{-1}$ has been found.

(f) FURTHER CONSIDERATION OF RYDBERG CONSTANT

All these calculations based on Bohr's theory can be applied to higher elements having one electron like He^+, Li^{2+}, Be^{3+} and so on, but this cannot be done for multi-electron system. The treatment of the higher ions will be like that of hydrogen atom; only the difference will be in the value of Rydberg constant, i.e., in other words, their reduced mass is μ.

$$\frac{1}{\mu_H}=\frac{1}{m_e}+\frac{1}{M_H}, \qquad \frac{1}{\mu_{He^+}}=\frac{1}{m_e}+\frac{1}{M_{He^+}}, \qquad \frac{1}{\mu_{Li^{+2}}}=\frac{1}{m_e}+\frac{1}{M_{Li^{+2}}}.$$

Now the $\bar{\nu}$ for He^+ and that for hydrogen atom are

$$\bar{\nu}_{He^+}=\frac{1}{\lambda}=\frac{2\pi^2 Z_{He^+}^2 e^4 \mu_{He}}{h^3 c}\left(\frac{1}{n_1^2}-\frac{1}{n_2^2}\right)$$

$$=\frac{2\pi^2 4e^4 \mu_{He^+}}{h^3 c}\left[\frac{1}{(n_1^2)}-\frac{1}{(n_2^2)}\right]$$

$$=\frac{2\pi e^4 \mu_{He}}{h^3 c}\left[\frac{4}{n_1^2}-\frac{4}{n_2^2}\right]$$

$$=\frac{2\pi^2 e^4 \mu_{He^+}}{h^3 c}\left[\frac{1}{\left(\frac{n_1}{2}\right)^2}-\frac{1}{\left(\frac{n_2}{2}\right)^2}\right] \qquad (1.36)$$

$$\bar{\nu}=\frac{2\pi^2 e^4 \mu_{He}}{h^3 c}\left[\frac{1}{n_1^2}-\frac{1}{n_2^2}\right]$$ [From equation (1.34) putting Z = 1 for hydrogen]

Therefore, $R_{He^+}=\dfrac{2\pi^2 e^4 \mu_{He^+}}{h^3 c}$ and $R_H=\dfrac{2\pi^2 e^4 \mu_H}{h^3 c}$.

So, $$\frac{R_{He^+}}{R_H}=\frac{\mu_{He^+}}{\mu_H}=\frac{\frac{1}{\mu_H}}{\frac{1}{\mu_{He^+}}}=\frac{\frac{1}{m_e}+\frac{1}{M_H}}{\frac{1}{m_e}+\frac{1}{M_{He^+}}}$$

$$=\frac{\frac{1}{m_e}+\frac{1}{M_H}}{\frac{1}{m_e}+\frac{1}{4M_H}}$$ [As mass of He = 4 × mass of H]

$$= \frac{\dfrac{M_H + m_e}{m_e M_H}}{\dfrac{4M_H + m_e}{4m_e M_H}} = \frac{M_H + m_e}{m_e M_H} \times \frac{4m_e M_H}{4M_H + m_e}$$

$$= \frac{4(M_H + m_e)}{4M_H + m_e} = \frac{4\left(1.67 \times 10^{-24} + 9 \times 10^{-28}\right)}{\left(4 \times 1.67 \times 10^{-24} + 9 \times 10^{-28}\right)}$$

$$= 1.0004088 \quad \text{or,} \quad R_{He^+} = 1.0004088 R_H. \qquad (1.37)$$

The experimental value is 1.0004080. The value is in excellent agreement. Likewise the $R_{Li^{+2}}$ and $R_{Be^{+3}}$ can be calculated.

From equation (1.36) another important conclusion regarding the frequency of spectral lines for H and He^+ can be drawn.

$$\bar{\nu}_{He^+} = \frac{2\pi^2 e^4 \mu_{He^+} \times 2^2}{h^3 c}\left(\frac{1}{n_1^2} - \frac{1}{n_2^2}\right)$$

$$= 4\frac{2\pi^2 e^4 \mu_{He^+}}{h^3 c}\left(\frac{1}{n_1^2} - \frac{1}{n_2^2}\right)$$

$$= 4R_H\left(\frac{1}{n_1^2} - \frac{1}{n_2^2}\right) \left(R_{He^+} \approx R_H\right)$$

$$= R_H\left[\frac{1}{\left(\dfrac{n_1}{2}\right)^2} - \frac{1}{\left(\dfrac{n_2}{2}\right)^2}\right]$$

$$\text{and } \bar{\nu}_H = R_H\left[\frac{1}{n_1^2} - \frac{1}{n_2^2}\right].$$

Therefore, putting $n_1 = 2$ and $n_2 = 3, 4, 5$, etc., for He^+ we shall get Lyman series for He^+ which has the similar frequency as that for hydrogen atom. Similarly, putting $n_1 = 4$ and n_2 greater than 4 we shall get Balmer series for He^+ having same frequencies as that of hydrogen atom. But here the difference will be like this; for odd set of values for n_2, we shall get a set of spectral lines and for even set of values for n_2, we shall get another set of spectral lines. This shows that for higher atoms the spectra will become complicated and number of lines for a particular series for He^+ will be double the number for that series for hydrogen atom.

From equation (1.37) we know that with increasing atomic numbers R will increase. But this is not the case.

$$\frac{1}{\mu_A}=\frac{1}{m_e}+\frac{1}{M_A}; \quad \text{when} \quad M_A >>> m_e, \quad \text{then} \quad \frac{1}{M_A} <<< \frac{1}{m_e}.$$

So we can neglect the value of $\frac{1}{M_A}$ and the equation becomes $\frac{1}{\mu_A} \approx \frac{1}{m_e}$ and as the mass of electron is constant, so μ_A is constant for higher elements.

(g) TO EVALUATE THE MASS RATIO OF ELECTRON TO PROTON

$$\frac{R_{He^+}}{R_H}=\frac{\mu_{He^+}}{\mu_H}=\frac{\frac{1}{m_e}+\frac{1}{M_H}}{\frac{1}{m_e}+\frac{1}{M_{He^+}}}=\frac{\frac{M_H+m_e}{M_H \times m_e}}{\frac{M_{He^+}+m_e}{m_e M_{He^+}}}$$

$$=\frac{(M_H+m_e)\,M_{He^+}}{\left(M_{He^+}+m_e\right)M_H}=\frac{\frac{M_{H^+}+m_e}{M_H}}{\frac{M_{He^+}+m_e}{M_{He^+}}}=\frac{1+\frac{m_e}{M_H}}{1+\frac{m_e}{4M_{He^+}}}$$

Putting the value for $\frac{R_{He^+}}{R_H}$, we get the value of $\frac{m_e}{M_H}=\frac{1}{1839}$ and this ratio is in good agreement with the experimental value.

(h) CALCULATION OF IONIZATION ENERGY

$$\Delta E = h\nu = \frac{2\pi^2 Z^2 e^4 m}{h^2}\left(\frac{1}{n_1^{\;2}}-\frac{1}{n_2^{\;2}}\right)$$

Ionization energy is the energy required to remove a most loosely bound electron of an isolated gaseous neutral atom to infinite distance so that the atom becomes an ion. Here, most loosely bound electron means the electron in the valance shell. This value can be calculated from Bohr's theory.

Here $n_2 = \infty$; so ionization energy

$$E_H = \frac{2\pi^2 e^4 \mu_H}{h^2 n^2} \quad (n = 1, Z = 1)$$

$= 13.6$ eV or 21.79×10^{-19} J for single e^-

For 1 mole of H_2 it will be

$E = 21.79 \times 10^{-19} \times 6.023 \times 10^{23}$ J/mole

$= 1312$ kJ/mole.

1.16 LIMITATION OF BOHR'S THEORY

In the above discussions, we see that the theory is successful in respect of accounting for many observations. Still it has some limitations:

1. Bohr proposed his theory based on some assumptions only to explain the discontinuous nature of the hydrogen spectrum; there was no theoretical basis. But modern wave mechanics provided its justification.
2. This theory is successfully applicable to hydrogen-like atom, i.e., single-electron systems, viz., He^+, H_2^+, Li^{2+}, Be^{3+}, but totally fails when applied to multi-electron system.
3. Bohr treated electron as a particle but modern concept accepts its wave-particle duality.
4. The ionization energy obtained by using this calculation does not tally with the experimental value.
5. When spectral lines are viewed by spectroscope, each single line appears to have more single lines closely spaced together, called *fine structure*, indicating variation of electronic energy in the same quantum states. Bohr's theory fails to explain this.
6. Bohr considered electron as particle revolving in an orbit and calculated its energy accurately. But modern development in physics showed that the position and momentum of an electron could not be calculated simultaneously with accuracy. In doing so, there must be some error (uncertainty principle).

In spite of these limitations, Bohr's theory is very helpful to us in understanding and explaining many scientific results.

1.17 SOMMERFELD EXTENSION OF BOHR'S THEORY

The spectroscopic view of fine structure of the hydrogen spectrum indicates the existence of more than one energy state within the same quantum states. To explain these fine structures Sommerfeld solved the problem by introducing two modifications:

(a) The first modification deals with the geometry of the orbit and

(b) The second modification regarding the relativistic correction of the orbital velocity.

(a) ORBITAL GEOMETRY

Sommerfeld was of the opinion that the electronic orbit cannot be circular but elliptical which may be developed as follows:

In performing spectroscopic observation, the substance taken always has an aggregation of a large number of electrons and proton for which the electronic motion is disturbed to a large extent. Due to the resultant of a large number of attractions and repulsion in the central force field, the electronic orbit will be elliptical, and not circular just as the solar system, where the planetary motions are in elliptical orbit due to different gravitational pull. The nucleus is at one of its focus. The major and minor axes will be different, but during revolution the major and minor axes will change and sometimes may both be equal when circular orbit results. Thus, circular orbit will be a special moment of the elliptical orbit when major and minor axes are equal. When the electron moves in circular orbit, only the angle of revolution or azimuthal angle θ varies and the second coordinate r, the distance of electron from the nucleus, is constant. So for one degree of freedom, Bohr introduced one quantum condition $mvr = \frac{nh}{2\pi}$. But in elliptical motion two degrees of freedom are there as the azimuthal angle θ and the radius vector r both vary. To specify the position of electron in an elliptical orbit, Sommerfeld introduced two quantum conditions. The quantum conditions can be derived as follows:

The velocity v of the electron in the elliptical path may be resolved into two vectors, one along the radius vector $\dot{r}$ and the other azimuthal vector $r\dot{\theta}$ at right angle to the radius vector such that

$$v^2 = \dot{r}^2 + r^2\dot{\theta}^2 \qquad \left[\text{where } \dot{r} = \frac{dr}{dt}\right]$$

Like velocity, momentum P also being a vector quantity can be resolved into two analogue components P_r (radial angular momentum vector) and P_θ (azimuthal angular momentum vector) such that

$$P = P_r + P_\theta$$

Now kinetic energy,

$$E_k = \frac{1}{2}v^2 m = \frac{1}{2}m\dot{r}^2 + \frac{1}{2}mr^2\dot{\theta}^2$$

$$P_r = \frac{dE_k}{d\dot{r}} = m\dot{r} \qquad (1.37a)$$

$$\text{and } P_\theta = \frac{dE_k}{d\dot{\theta}} = mr^2\dot{\theta}. \qquad (1.37b)$$

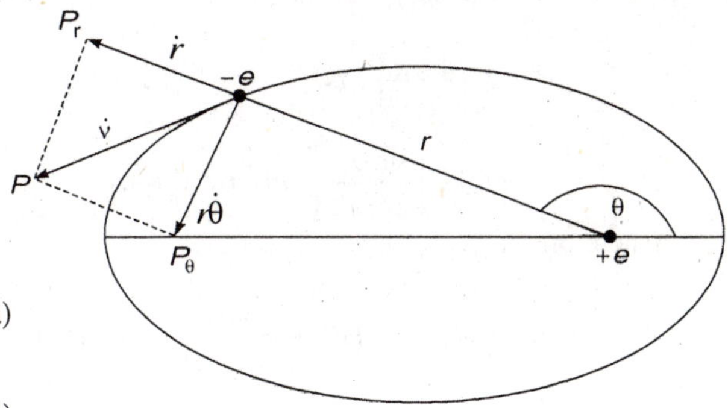

Fig. 1.19 Sommerfeld's elliptical orbit.

Sommerfeld extended Bohr's quantum condition here and postulated that

$$\int P_r dr = rh \text{ where } r = 1, 2, 3, \ldots, \text{etc} \qquad (1.38)$$

$$\text{and } \int P_\theta\, d\theta = kh \text{ when } k = 1, 2, 3, \ldots, \text{etc} \qquad (1.39)$$

r is called the radial quantum number and k is the azimuthal quantum number. The equation (1.39) can be solved easily with the help of Kepler's Law; here P_θ is independent of θ and for one complete revolution it changes from 0 to 2π and so we can write

$$\int_0^{2\pi} P_\theta\, d\theta = kh \qquad \text{or, } P_\theta \times 2\pi = kh \qquad \text{or, } P_\theta = \frac{kh}{2\pi}$$

The shape of the elliptical orbit and whether any elliptical orbit is possible or not, can be obtained from the solution of P_r. The equation for P_r is complicated but can be done by considering the equation of ellipse in polar coordinate.

$$r = a\left(\frac{1-\varepsilon^2}{1+\varepsilon\cos\theta}\right); \quad a = \text{semi-major axis}, \quad b = \text{semi-minor axis},$$

$$\varepsilon = \text{eccentricity of ellipse}.$$

$$\text{Now, } \frac{dr}{d\theta} = \frac{a\left(1-\varepsilon^2\right)\varepsilon\sin\theta}{(1+\varepsilon\cos\theta)^2} = r\frac{\varepsilon\sin\theta}{1+\varepsilon\cos\theta} \qquad \text{or, } \frac{1}{r}\frac{dr}{d\theta} = \frac{\varepsilon\sin\theta}{1+\varepsilon\cos\theta}. \qquad (1.40)$$

$$\text{Again, } P_r = m\dot{r} = m\times\frac{dr}{dt} = m\times\frac{dr}{d\theta}\times\frac{d\theta}{dt} = m\dot{\theta}\frac{dr}{d\theta} = m\dot{\theta}r^2\times\frac{1}{r^2}\times\frac{dr}{d\theta}$$

$$= P_\theta\frac{1}{r^2}\frac{dr}{d\theta} \quad [\text{by (1.37b)}] \qquad (1.41)$$

So, $P_r\,dr = P_r \dfrac{dr}{d\theta} d\theta$

$$= P_\theta \frac{1}{r^2} \times \frac{dr}{d\theta} \times \frac{dr}{d\theta} \times d\theta \text{ [by (1.41)]}$$

$$= \frac{P_\theta}{r^2}\left(\frac{dr}{d\theta}\right)^2 d\theta$$

$$= P_\theta\left(\frac{1}{r} \times \frac{dr}{d\theta}\right)^2 d\theta$$

$$= P_\theta\left[\frac{\varepsilon \sin\theta}{1+\varepsilon\cos\theta}\right]^2 d\theta \text{ [by (1.40)]} \tag{1.42}$$

Therefore,

$$\int_0^{2\pi} P_r dr = \int_0^{2\pi} P_\theta\left[\frac{\varepsilon\sin\theta}{1+\varepsilon\cos\theta}\right]^2 d\theta = P_\theta \int_0^{2\pi}\left[\frac{\varepsilon\sin\theta}{1+\varepsilon\cos\theta}\right]^2 d\theta$$

$$= P_\theta \int_0^{2\pi} \varepsilon\sin\theta \times \frac{\varepsilon\sin\theta}{(1+\varepsilon\cos\theta)^2} \times d\theta$$

or, $$\int_0^{2\pi} P_r dr = P_\theta\left[\varepsilon\sin\theta \int \frac{\varepsilon\sin\theta}{(1+\varepsilon\cos\theta)^2} d\theta - \int\left\{\frac{d\,(\varepsilon\sin\theta)}{d\theta}\int\frac{\varepsilon\sin\theta}{(1+\varepsilon\cos\theta)^2} d\theta\right\} d\theta\right]_0^{2\pi}$$

$$\left[\text{ As } \int (uv)\,dx = u\int v dx - \int\left\{\frac{du}{dx}\int v \times dx\right\}dx\right]$$

or, $$\int_0^{2\pi} P_r dr = P_\theta\left[\frac{\varepsilon\sin\theta}{1+\varepsilon\cos\theta} - \int\left\{\varepsilon\cos\theta \times \frac{1}{1+\varepsilon\cos\theta}\right\} d\theta\right]_0^{2\pi}$$

Let $\quad 1+\varepsilon\cos\theta = x, \quad \therefore dx = -\varepsilon\sin\theta\, d\theta.$

$$\therefore \frac{\varepsilon\sin\theta d\theta}{(1+\varepsilon\cos\theta)^2} = -\frac{1}{x^2}dx \text{ and } \int -\frac{dx}{x^2} = \frac{1}{x}.$$

Again, $$\int \frac{\varepsilon\cos\theta}{1+\varepsilon\cos\theta} d\theta = \int \frac{1+\varepsilon\cos\theta-1}{1+\varepsilon\cos\theta} d\theta = \int\left(1 - \frac{1}{1+\varepsilon\cos\theta}\right) d\theta$$

$$= \int d\theta - \int \frac{d\theta}{1+\varepsilon\cos\theta}$$

$$= P_\theta\left[\frac{\varepsilon\sin\theta}{1+\varepsilon\cos\theta} - \int d\theta + \int \frac{d\theta}{\cos^2\frac{\theta}{2} + \sin^2\frac{\theta}{2} + \varepsilon\cos^2\frac{\theta}{2} - \varepsilon\sin^2\frac{\theta}{2}}\right]_0^{2\pi}$$

$$= P_\theta \left[\frac{\varepsilon \sin\theta}{1+\varepsilon\cos\theta} - \theta + \int \frac{d\theta}{(1+\varepsilon)\cos^2\frac{\theta}{2} + (1-\varepsilon)\sin^2\frac{\theta}{2}} \right]_0^{2\pi}$$

$$= P_\theta \left[\frac{\varepsilon \sin\theta}{1+\varepsilon\cos\theta} - \theta + \int \frac{\sec^2\frac{\theta}{2}\,d\theta}{(1+\varepsilon) + (1-\varepsilon)\tan^2\frac{\theta}{2}} \right]_0^{2\pi}$$ [Dividing numerator and denominator by $\cos^2\frac{\theta}{2}$]

$$= P_\theta \left[\frac{\varepsilon \sin\theta}{1+\varepsilon\cos\theta} - \theta + \int \frac{2dx/\sqrt{1-\varepsilon}}{(1+\varepsilon)+x^2} \right]_0^{2\pi}$$

[Put $\tan\frac{\theta}{2}\sqrt{1-\varepsilon} = x, \quad \therefore\ dx = \sqrt{1-\varepsilon}\sec^2\frac{\theta}{2} \times \frac{1}{2} \times d\theta$

or, $\dfrac{2dx}{\sqrt{1-\varepsilon}} = \sec^2\frac{\theta}{2}\,d\theta$]

$$= P_\theta \left[\frac{\varepsilon \sin\theta}{1+\varepsilon\cos\theta} - \theta + \frac{2}{\sqrt{1-\varepsilon}} \int \frac{dx}{\left(\sqrt{1+\varepsilon}\right)^2 + x^2} \right]_0^{2\pi}$$

$$= P_\theta \left[\frac{\varepsilon \sin\theta}{1+\varepsilon\cos\theta} - \theta + \frac{2}{\sqrt{1-\varepsilon}} \times \frac{1}{\sqrt{1+\varepsilon}} \tan^{-1} \frac{\sqrt{1-\varepsilon}\tan\frac{\theta}{2}}{\sqrt{1+\varepsilon}} \right]_0^{2\pi}$$

[since, $\displaystyle\int \frac{dx}{a^2+x^2} = \frac{1}{a}\tan^{-1}\frac{x}{a}$]

$$= P_\theta \left[\frac{\varepsilon \sin\theta}{1+\varepsilon\cos\theta} - \theta + \frac{2}{\sqrt{1-\varepsilon^2}} \tan^{-1} \left\{ \sqrt{\frac{1-\varepsilon}{1+\varepsilon}} \tan\frac{\theta}{2} \right\} \right]_0^{2\pi}$$

$$= P_\theta \left[0 - 2\pi + \frac{2}{\sqrt{1-\varepsilon^2}} \tan^{-1} \left\{ \sqrt{\frac{1-\varepsilon}{1+\varepsilon}} \tan\pi \right\} \right]$$

$$= P_\theta \left[0 - 2\pi + \frac{2}{\sqrt{1-\varepsilon^2}} \tan^{-1} \left\{ \sqrt{\frac{1-\varepsilon}{1+\varepsilon}} \times 0 \right\} \right]$$

$$= P_\theta \left[0 - 2\pi + \frac{2}{\sqrt{1-\varepsilon^2}} \tan^{-1} 0 \right].$$

$$\therefore \quad \int P_r dr = P_\theta \left[-2\pi + \frac{2\pi}{\sqrt{1-\varepsilon^2}} \right] = P_\theta \times 2\pi \left[\frac{1}{\sqrt{1-\varepsilon^2}} - 1 \right] = rh$$

or, $$rh = \frac{kh}{2\pi} \times 2\pi \left[\frac{1}{\sqrt{1-\varepsilon^2}} - 1 \right]$$

So, $$r = k \left[\frac{1}{\sqrt{\frac{b^2}{a^2}}} - 1 \right] = k \left[\frac{a}{b} - 1 \right]$$

or, $$\frac{r}{k} + 1 = \frac{a}{b}$$

or, $$\frac{r+k}{k} = \frac{a}{b}.$$

Here, $a = \dfrac{n^2 h^2}{4\pi^2 mZe^2}$ and $b = \dfrac{nkh^2}{4\pi^2 mZe^2}$.

Now as r and k have integral values, so their sum $r + k$ will be an integer (say $= n$).

So, $$\frac{n}{k} = \frac{a}{b} = \frac{\text{semi-major axis}}{\text{semi-minor axis}}, \tag{1.43}$$

r and k are known as the radial quantum number and azimuthal quantum number and values of k, i.e., b indicates the shape of the elliptical orbit which is also quantized, i.e., all ellipses are not possible.

Now, let us see the different values of n, k, r and corresponding shape of the orbit. For a given value of n, k will have any integral value 0, 1, 2, 3. ..., n, etc.; corresponding r will be n, $(n-1)$, $(n-2)$, $(n-3)$, ..., 3, 2, 1, 0, etc. It is evident from the equation (1.43) $\frac{n}{k} = \frac{a}{b}$ and $n \geq k$ or $n \nless k$ or r. But $k \neq 0$, since $\frac{n}{k} = \frac{a}{b}$, when $k = 0$, $b = 0$, i.e., minor axis is absent. The electronic path will be a straight line passing through the nucleus, which is impossible, so $r \neq n$. When $n = k$, then $a = b$, i.e., the electronic orbit will be a circular. When k becomes smaller and smaller than n, elliptical orbit with greater and greater eccentricity will develop. So for different values of n, the number of orbits and their shapes are as follows:

$n = 1$	$n = 2$	$n = 2$
$k = 1$	$k = 1$	$k = 2$
$a = b$	$a = 2b$	$a = b$

$n = 3$	$n = 3$	$n = 3$	$n = 4$	$n = 4$	$n = 4$	$n = 4$
$k = 1$	$k = 2$	$k = 3$	$k = 1$	$k = 2$	$k = 3$	$k = 4$
$a = 3b$	$2a = 3b$	$a = b$	$a = 4b$	$a = 2b$	$3a = 4b$	$a = b$

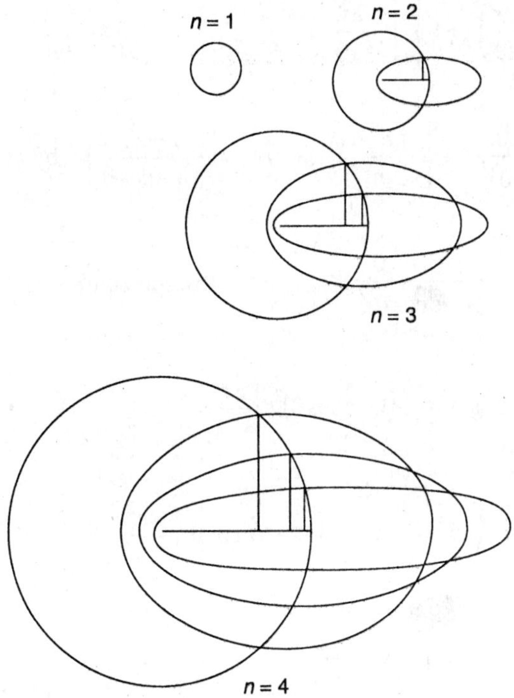

Fig. 1.20 Permitted electronic orbits of H-atom.

Thus Bohr's single orbit actually gets subdivided into number of sub-orbits. Now, let us calculate the energy of these sub-levels. Energy of the electron in elliptical orbit will be

$$E = \text{K.E.} + \text{P.E.} = \frac{1}{2}m\left(\dot{r}^2 + r^2\dot{\theta}^2\right) - \frac{e^2}{r} \qquad \text{(for hydrogen atom)}$$

$$= \frac{1}{2m}\left(m^2\dot{r}^2 + m^2r^2\dot{\theta}^2\right) - \frac{e^2}{r}$$

$$= \frac{1}{2m}\left(P_r^{\,2} + \frac{m^2r^4\dot{\theta}^2}{r^2}\right) - \frac{c^2}{r} \qquad [\because P_r = m\dot{r}\,]$$

$$= \frac{1}{2m}\left\{P_r^{\,2} + \left(\frac{mr^2\dot{\theta}}{r}\right)^2\right\} - \frac{e^2}{r}$$

$$= \frac{1}{2m}\left\{P_r^{\,2} + \left(\frac{P_\theta}{r}\right)^2\right\} - \frac{e^2}{r} \qquad \left[P_\theta = mr^2\dot{\theta}\,\right]$$

$$= \frac{1}{2m}\left\{\frac{P_\theta^{\,2}}{r^4} \times \left(\frac{dr}{d\theta}\right)^2 + \frac{P_\theta^{\,2}}{r^2}\right\} - \frac{e^2}{r} \qquad \left[\because P_r = \frac{P_\theta}{r^2}\left(\frac{dr}{d\theta}\right)\right]$$

$$= \frac{1}{2m}\left[\frac{P_\theta^2}{r^2}\left\{\frac{1}{r^2}\left(\frac{dr}{d\theta}\right)^2+1\right\}\right]-\frac{e^2}{r}$$

$$= \frac{P_\theta^2}{2mr^2}\left\{\frac{\varepsilon^2\sin^2\theta}{(1+\varepsilon\cos\theta)^2}+1\right\}-\frac{e^2}{r} \qquad \left[\text{as } r = a\left(\frac{1-\varepsilon^2}{1+\varepsilon\cos\theta}\right) \text{ by (1.40)}\right]$$

$$= \frac{P_\theta^2(1+\varepsilon\cos\theta)^2}{2ma^2(1-\varepsilon^2)^2}\times\left\{\frac{\varepsilon^2\sin^2\theta}{(1+\varepsilon\cos\theta)^2}+1\right\}-\frac{e^2}{r} \qquad \text{[by substituting } r\text{]}$$

$$= \frac{P_\theta^2}{2ma^2(1-\varepsilon^2)^2}\left\{\varepsilon^2\sin^2\theta+(1+\varepsilon\cos\theta)^2\right\}-\frac{e^2}{r}$$

$$= \frac{P_\theta^2}{2ma^2(1-\varepsilon^2)^2}\left\{\varepsilon^2(\sin^2\theta+\cos^2\theta)+1+2\varepsilon\cos\theta\right\}-\frac{e^2}{r}$$

$$= \frac{P_\theta^2}{2ma^2(1-\varepsilon^2)^2}\left\{\varepsilon^2+1+2\varepsilon\cos\theta\right\}-\frac{e^2}{r}$$

$$= \frac{P_\theta^2}{ma^2(1-\varepsilon^2)^2}\left\{\frac{\varepsilon^2+1}{2}+\varepsilon\cos\theta\right\}-\frac{e^2}{a}\times\frac{(1+\varepsilon\cos\theta)}{(1-\varepsilon^2)} \qquad \text{[substituting } r\text{]} \tag{1.44}$$

As the energy of an electron in an orbit is constant, it does not depend upon θ, so the coefficient of $\cos\theta$ must be zero. Now the coefficient of $\cos\theta$ is

$$\frac{P_\theta^2}{ma^2(1-\varepsilon^2)^2}-\frac{e^2}{a(1-\varepsilon^2)}=0$$

or, $$\frac{P_\theta^2}{ma^2(1-\varepsilon^2)^2}=\frac{e^2}{a(1-\varepsilon^2)}$$

or, $$e^2=\frac{P_\theta^2}{ma(1-\varepsilon^2)}$$

or, $$a=\frac{P_\theta^2}{me^2(1-\varepsilon^2)}=\frac{k^2h^2n^2}{4\pi^2me^2k^2}=\frac{n^2h^2}{4\pi^2me^2}$$

and $b=\dfrac{nkh^2}{4\pi^2me^2}$.

Thus we see that the equation for semi-major axis of the ellipse is the same as that of Bohr's circular orbit. The remaining part of the equation (1.44) will give energy of the electron. So,

$$E = \frac{P_\theta^2}{ma^2(1-\varepsilon^2)^2} \times \frac{\varepsilon^2+1}{2} - \frac{e^2}{a(1-\varepsilon^2)}$$

$$= \frac{e^2}{a(1-\varepsilon^2)} \times \frac{\varepsilon^2+1}{2} - \frac{e^2}{a(1-\varepsilon^2)} \quad \left[\text{as } e^2 = \frac{P_\theta^2}{ma(1-\varepsilon^2)^2}\right]$$

$$= \frac{e^2(\varepsilon^2+1)-2e^2}{2a(1-\varepsilon^2)} = \frac{e^2\varepsilon^2-e^2}{2a(1-\varepsilon^2)} = \frac{e^2(\varepsilon^2-1)}{2a(1-\varepsilon^2)}$$

$$= -\frac{e^2}{2a} = -\frac{e^2}{2} \times \frac{4\pi^2me^2}{n^2h^2} = -\frac{2\pi^2e^4m}{n^2h^2}.$$

Thus the energy of electron in elliptical orbit is determined by $n = r + k$, i.e., sum of azimuthal and radial quantum number and not by their individual contribution to the total. The energy equation for electron in elliptical path is the same as that in Bohr's circular orbit. The frequency of transition $\bar{\nu}$ will be

$$\bar{\nu} = \frac{2\pi^2e^4m}{h^3c}\left\{\frac{1}{(r_1+k_1)^2} - \frac{1}{(r_2+k_2)^2}\right\} = \frac{2\pi^2e^4m}{h^3c}\left(\frac{1}{n_1^2} - \frac{1}{n_2^2}\right).$$

The frequency of transition is the same in both cases. So fine structure in the hydrogen spectrum remains unexplained although Sommerfeld's elliptical orbit indicates that transition may occur in a number of ways. Bohr and Sommerfeld's concepts are as given below:

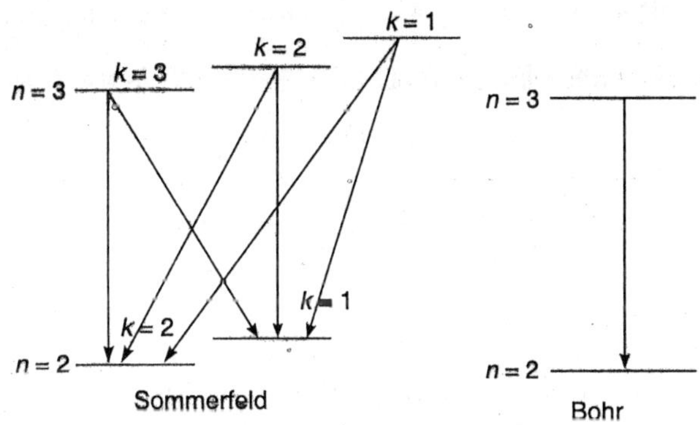

Fig. 1.21 Comparison of Bohr and Sommerfeld schemes.

As the sub-levels are degenerate in elliptical orbit, energy change will be the same. To explain the fine structures and also to remove the degeneracy of orbital, Sommerfeld introduced his second modification by considering the relative velocity and hence relative mass of the orbiting electron round the nucleus at the one focus of the ellipse. The orbiting electron will have different velocity at different position in the ellipse.

When it is closest to the nucleus (perihelion), electrostatic attraction being maximum, velocity will be the highest and when it is farthest (aphelion) from the nucleus, velocity will be minimum. This variation in velocity causes the relative mass of the electron to be changed at different positions, and causes perturbation of balancing force. The relative mass with velocity is given by the equation:

$$m = \frac{m_0}{\sqrt{1-\frac{v^2}{c^2}}}$$

This perturbation causes the electron to be slightly dragged from its original position in the elliptical path and then it travels along the new path. So the electronic path will not be a closed ellipse but an open ellipse and the ellipse itself will rotate in the same plane. This is called the *precessional motion* or rosette as shown in the figure, the major axis will rotate in plane and the perihelion and aphelion will also shift at every time. Considering this rotation Sommerfeld gave the energy value to be

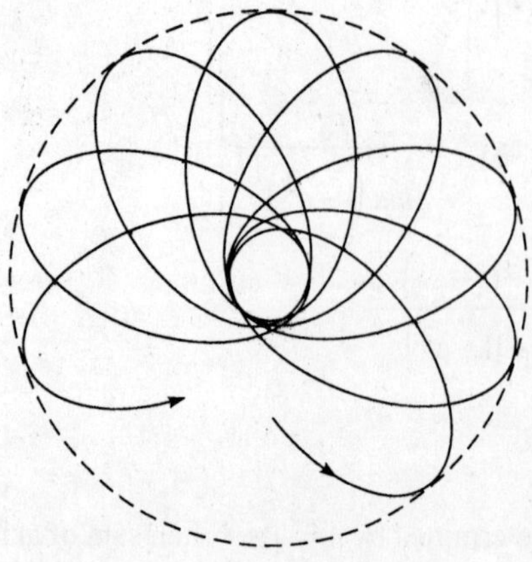

Fig. 1.22 Precession of an elliptic orbit according to the Sommerfeld theory.

$$E = -\frac{2\pi^2 Z^2 e^4 m_0}{n^2 h^2}\left[1+\frac{\mu^2}{n^2}\left(\frac{n}{k}-\frac{3}{4}\right)\right],$$

where μ is called the *fine structure constant* and its value is $\mu = \frac{2\pi e^2}{ch}$ and m_0 is the rest mass of electron. Unlike before, now the energy depends on both n and k, and so in the same quantum state slight variation in energy depending on k will result and fine structures in the hydrogen spectrum will be obtained. Thus for transition of electron from $n = 3$ to $n = 2$ gives a total of six transitions with slight varying energy. Actually, all such transitions are not allowed due to the restriction of the selection rule $\Delta k = \pm 1$.

Sommerfeld theory is thus quite satisfactory to calculate the energy of an electron, to remove the degeneracy of orbitals and to calculate the ionization potential and resonance potential.

1.18 THE MERITS AND DEMERITS OF SOMMERFELD THEORY

1. The achievement of the theory lies in the subdivision of Bohr's original stationary orbit to sub-orbits with slightly varying energy depending upon their shapes.
2. It introduces the idea of penetrating effect of the orbits. The orbit with which the electron approaches more close to the nucleus, becomes more penetrating. Thus, penetrating power changes in this way $s > p > d > f$ and hence, chemical properties also changes accordingly.
3. The demerit of the theory is that it is not applicable to multi-electron system. Say for alkali metals the energy of an electron, we know is

$$E = -\frac{2\pi^2 Z^2 c^4 m}{n^2 h^2}$$

where n, the principal quantum number is found to be non-integral which goes beyond the principle of Bohr-Sommerfeld. If we denote non-integral quantum number by n^*, then we can write $n^* = n - \sigma$, where σ is the deviation of quantum number from integral value (called *quantum defect*). The value of σ depends on the penetrating power of the orbit. Quantum defect of s-electron is maximum and minimum for f-electron. The interpenetration of orbits gives rise to precession of the path.

4. The major fault lies with the value of azimuthal quantum number k. According to Sommerfeld k has values 1 to n, but later developments replaces k by l ; the orbital angular momentum quantum number, which has values 0 to n–1. Hence they are related by $l = k-1$.
5. According to Sommerfeld the orbits are planar, but later developments show they are three-dimensional.

1.19 THE MAGNETIC QUANTUM NUMBER OR ZEEMAN EFFECT

When the atomic spectra are taken by placing the source in strong magnetic or electric field (called *Zeeman* or *Stark effect*), the spectral lines of the atoms are found to split up into several components, called *hyperfine structures*. Yellow doublet of sodium gives ten such lines under 10 000 gauss magnetic field. If the magnetic field is weak, then complex nature of large number of lines is produced, this is called anomalous *Zeeman effect*, which is caused due to combined effect of orbital and spin angular momentum. The experimental set-up is as follows:

Ne	- neon tube	P	- high resolving power spectroscope with a constant deviation prism
C	- collimator	T	- telescope
L	- converging lens	N. S.	- powerful electromagnet
L.P.	- lummer plate		

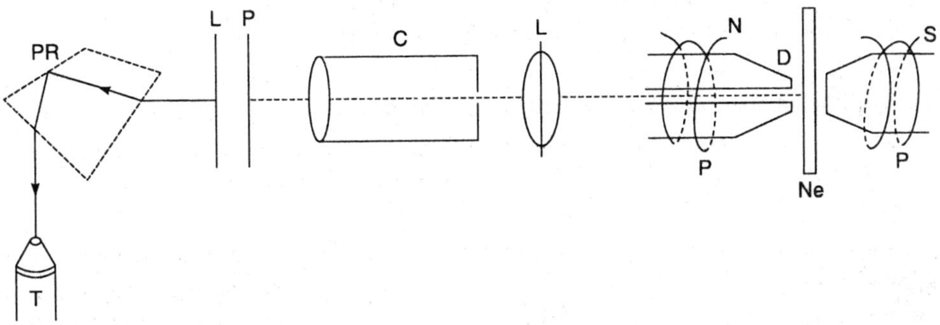

Fig. 1.23 Experimental set up for studies with Zeeman effect.

Light from neon source giving fine line of great intensity passes through the hole in the powerful electromagnet, through lens, collimator and then high resolving power spectroscope with a constant deviation prism, the resolution is then viewed through telescope.

When the spectral lines are viewed parallel to the direction of the applied field (*longitudinal Zeeman effect*), the original single line observed before the application of the field. But after the application of the field the single line is split up into two components—one having slightly greater and the other having slightly lower frequency than the original line, which is missing. But, if the direction of observation is perpendicular to the applied field (called *transverse Zeeman effect*), then the original line together with the other two as before, i.e., total three lines are observed. This indicates each energy state which under the influence of the magnetic field splits up to different other states having slightly different energy. Let us consider how the splitting occurs:

Each orbiting electron in a closed circular or elliptical orbit induces a magnetic field at the centre and hence, each orbit with moving electron may be considered as a tiny magnet placed at the centre of the orbit. In absence of externally applied field the moments of these magnets are not felt due random arrangement of the orbital and each moment will cancel each other out. When external field is applied, these atomic magnets will no longer cancel each other but will tend to orient themselves in space in the direction of the applied field and this orientative nature of the orbital will be guided

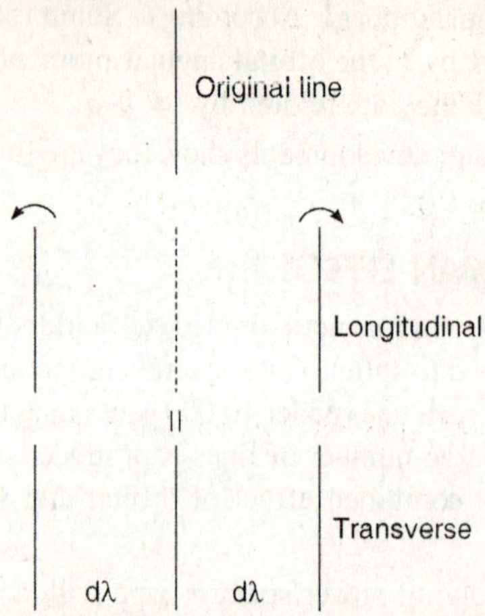

Fig. 1.24 Normal Zeeman effect.

by mechanical moment of the magnets. Thus, the electrons are moving not in a plane but in space of three dimensions, i.e., circular orbit change to a sphere and elliptical orbit changes to an ellipsoid. This change to three-dimensional figure can also be understood as follows: During revolution when the electron approaches perihelion (closest), the electron is dragged a little towards the nucleus and at the same time plane of the orbit will also be shifted and the electron will now describe the new elliptical path in a new plane.

In this way every time the elliptical orbit will be shifted a little in space and thus becoming an ellipsoid—just as a moving car in a sharp-bent moves in a new plane having outer track at a higher level, otherwise the car will skid out. The motion of electron in the ellipsoid necessitates the introduction of another new quantum condition besides n and K.

Now, let us calculate the mechanical moment of an electron. The frequency of an electron round the orbit is given by $\nu = \frac{\omega}{2\pi}$ and the current $i = \frac{\omega e}{2\pi}$.

Let us take the circular orbit for simplicity. So moment

$$M = \frac{\omega e}{2\pi} \times \pi r^2 = \frac{\omega e r^2}{2}(\text{esu}) = \frac{\omega e r^2}{2c}\text{ emu}$$

But we know angular momentum

$$P_\theta = mr^2\omega, \quad \text{so } \omega = \frac{P_\theta}{mr^2}$$

and $M = \frac{P_\theta}{mr^2} \times \frac{er^2}{2c} = \frac{P_\theta e}{2mc} = \frac{khe}{4\pi mc}$.

As P_θ is quantised, magnetic moment M is quantised and is an integral multiple (k) of the quantity $\frac{eh}{4\pi mc}$, and this unit of moment is called *Bohr magneton* whose value is $9{\cdot}1 \times 10^{-21}$ emu.

As the planes of the orbit are inclined to the applied field at different angles, and ϕ be the angle between the magnetic axis of orbit and the applied field, then corresponding momentum $P_\phi = P_\theta \cos\phi$ and the quantum condition will be $P_\phi = \frac{m_l h}{2\pi}$, where m is an integer called *magnetic quantum number* [more correctly $P_\phi = \frac{\sqrt{m_l(m_l+1)}h}{2\pi}$].

$$\text{So, } \cos\phi = P_\phi / P_\theta = \frac{\frac{m_l h}{2\pi}}{\frac{kh}{2\pi}} = \frac{m_l}{k}.$$

As ϕ may change from 0^0 to π, i.e., $\cos\phi = \pm 1$, So $m_l = \pm k$ or $m_l = \pm l$.

Thus for any k or l value, total $(2k + 1)$ or $(2l + 1)$ orientations are possible.

For $n = 1$, $k \approx l = 0$ so $m_l = 0$ singlet, 1s

For $n = 2$, $k \approx l = 0$ $k \approx l = 1$ so $m_l = 0, \pm 1$, i.e., triplet state

$m_l = 0$ for $m_l = 1$, it is p_x

2s for $m_l = -1$, it is p_y

for $m_l = 0$, it is p_z

So p orbitals are oriented in three coordinate axes mutually perpendicular to each other.

For $n = 3$, $k \approx l = 0, 1, 2$.

when $l = 0$, it is 3s and singlet state

when $l = 1$, it is 3p and triplet state as 2p

when $l = 2$, it is 3d and divided to five states

depending on value of $m_l = +2, +1, 0, -1, -2$, the orbitals are named as $d_{x^2-y^2}$, d_{xz}, d_{z^2}, d_{yz} and d_{xy} respectively.

This can be diagrammatically represented in Fig. 1.25 and 1.25a.

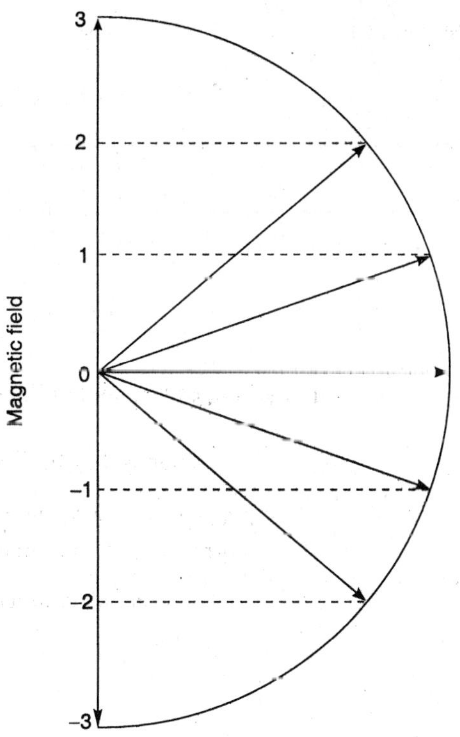

Fig. 1.25 Space quantization in a magnetic field.

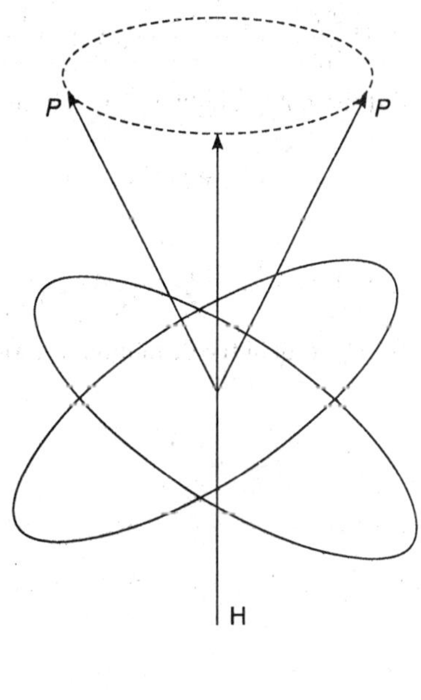

Fig. 1.25a Quantisation of P_ϕ when orbital magnetic axis P_θ is inclined to the applied field H by an angle ϕ.

The shapes of these orbitals are drawn from the solution of wave equation as in figure 1.26.

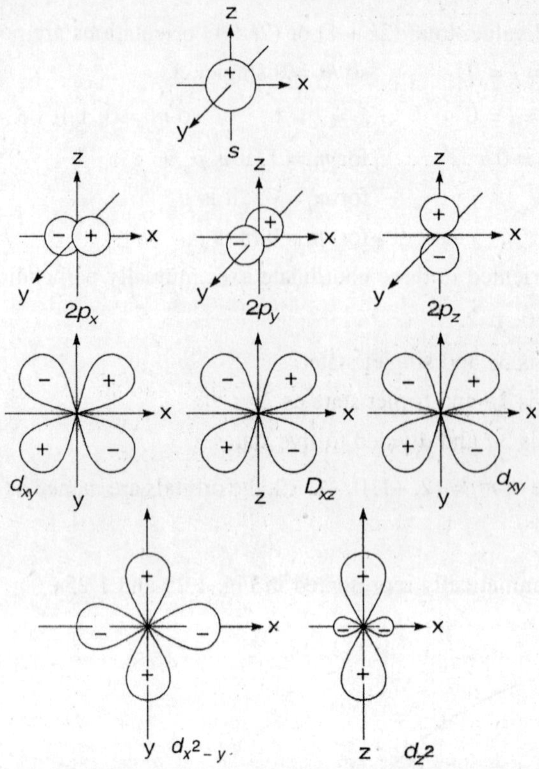

Fig. 1.26 Boundary surface of the angular part of wave function for 1*s*, 2*p* and 3*s* orbitals of hydrogen atom shown is polar diagrams.

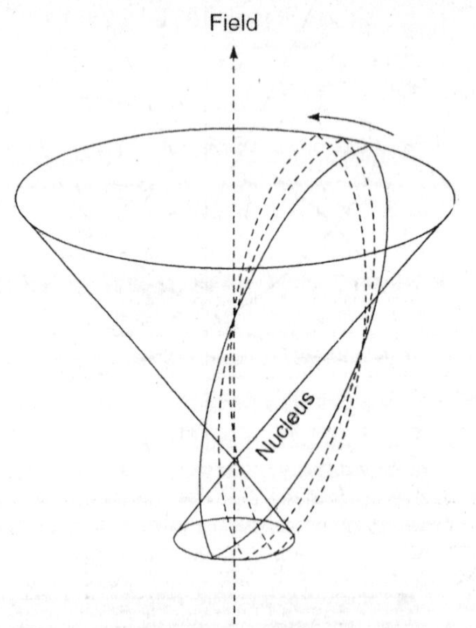

Fig. 1.27 Larmor precession of an electronic orbit in a magnetic field.

By the influence of the external field, the elliptical orbit will be subjected to a torque and undergoes Larmor precession describing the surface of a double cone joined head to head or vertex to vertex, as shown in the figure 1.27. It is just like the motion of a spinning top under small disturbing force.

The additional energy change of the electron due to interaction between the magnetic field and its own magnet is given by

$$E_m = -MH\cos\phi$$

$$= -\frac{khe}{4\pi mc} \times H \times \frac{m_l}{k}$$

$$= -\frac{eH \times m_l h}{4\pi mc}$$

As m_l has either positive or negative value, the energy will either increase or decrease in presence of field. So total energy of the electron is

$$E_{n k l m l} = -\frac{2\pi^2 Z^2 e^4 \mu}{h^2 (r+k)^2} + \frac{eH}{4\pi mc} \times m_l h$$

and the frequency of transition will be

$$\nu = \frac{2\pi^2 Z^2 e^4 \mu}{h^3} \left[\frac{1}{(r_1 + k_1)^2} - \frac{1}{(r_2 + k_2)^2} \right] + \frac{eH}{4\pi mc} (m_{l_1} - m_{l_2}).$$

This shows that together with original line other lines on both sides will appear depending on the values of m_{l_1} and m_{l_2}, which are either positive or negative.

1.20 SPIN QUANTUM NUMBER

Alkali metals have only one electron in the valence shell and the nucleus is well screened by the inner orbital electrons and the orbital angular momentums of these inner electrons cancel each other. So alkali metals can be considered like that of a hydrogen atom.

Arthur Compton in 1921 first gave idea that electron might have angular momentum due to spinning on its own axis, later it was proved by observing single yellow D-line of Na under high resolving power spectroscope, which appears to be a doublet D^1 and D^2, with gap of 6 Å. Again, under magnetic field of moderate strength, these lines are divided into four or six more lines. This was explained by assuming spinning of electron on its own axis together with its orbital motion round the nucleus. Thus, electron has both orbital angular momentum and spin angular momentum.

From quantum theory, this momentum will also be quantized and hence it is necessary to introduce another quantum number, the spin quantum number m_s and spin angular momentum s is

$$s = mvr = \sqrt{m_s(m_s+1)}.\frac{h}{2\pi}$$

Here, m_s or s will have two values $+\frac{1}{2}$ and $-\frac{1}{2}$, i.e., either clockwise or anti-clockwise rotation adds to, or opposes or decreases the orbital momentum as shown in the figure 1.28. Like azimuthal quantum number having $2k + 1$ settings, spin will have $(2s + 1)$ settings. As it is a doublet, $2s + 1 = 2$ or $s = \frac{1}{2}$, for clockwise or anti-clockwise, s will be $+\frac{1}{2}$ and $-\frac{1}{2}$.

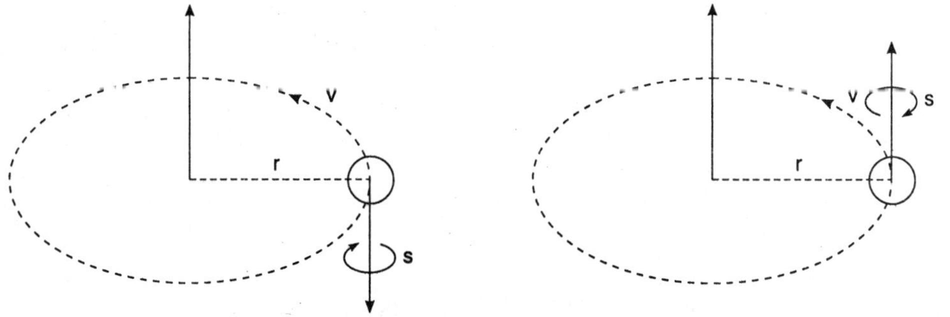

Fig. 1.28 Orientation of spin momentum with reference to orbital angular momentum.

1.21 SIGNIFICANCE OF QUANTUM NUMBERS

From the above discussions it is clear that four quantum numbers are necessary to identify an electron in an atom, just like to identify a man we require his specifications. These four quantum numbers n, $l(k)$, m_l and m_s were necessary to explain the experimental facts in the atomic spectra. These were quite arbitrarily selected. But in modern wave mechanics we consider electron as a cloud. The use of Schrödinger wave equation for hydrogen atom gives the allowed energy states, which requires the quantum numbers. Thus, the ideas of quantum numbers were quite arbitrary in old concept, but they are the natural outcome in modern wave mechanics.

1. **The principal quantum number (*n*):**

   This quantum number indicates the postulated energy states and with increasing value of *n*, the energy states gradually recedes from the nucleus and having negative sign energy value increases. From Bohr's calculation, we know $r_1 = 0.53$ Å, $r_2 = 2.12$ Å and $r_3 = 4.77$ Å and so on. Thus, with increasing value of n_1, we see the electron recedes from the nucleus, radius of the circular orbit increases. But from uncertainty principle we see such calculations are meaningless. The electron clouds are distributed in space within the atom and their density is maximum at the location of the stationary states, and density is less at closest and at furthest from the nucleus. It is evident from the plot of ψ^2 against *r*. Electron transition from one energy level to the other will give rise to atomic spectra, not the electron jump from one orbit to the other as considered here.

2. **Azimuthal quantum number (*l*):**

   This quantum number describes the splitting of each energy state defined by *n*, into other sub-state of slightly varying energy to explain the fine structure of the hydrogen spectrum. Depending on the value of *n*, it has $k = 1, 2, 3, \ldots, n$ $[l = 0, 1, \ldots, n-1]$ and depending on the value of *k* or *l*, eccentricity of the elliptical orbits are determined. In wave mechanics the plots of $\psi^2 4\pi r^2 dr$ against *r* for different values of *n* and *l* give the curves that demonstrate the elliptical orbits to show the electron probability distribution at various distances.

3. **Magnetic quantum number (m_l):**

   This quantum number indicates the orientation of orbitals in space and gives three-dimensional figure of orbitals instead of two-dimensional one. The directional nature of orbitals in space is limited by the value of *l* or *k* and total orientations are $+k$ through 0 to $-k$, totaling $(2k+1)$ orientations. The angular momentum of the electron about the axis ranges by integral steps from $+\left(l \times \frac{h}{2\pi}\right)$ through 0 to $-\left(l \times \frac{h}{2\pi}\right)$, inclined at various discrete angles to the axis. From wave mechanics although this value is $\sqrt{l(l+1)} \times \frac{h}{2\pi}$.

4. **Spin quantum number (m_s):**

   This quantum number was proposed to explain the doublet state of yellow *D*-line assuming electron rotating on its axis like top.

   Dirac (1928) gave a new wave equation for electron (unlike given by Schrödinger) which has four wave functions, and their solutions gave automatically the result equivalent to those obtained considering electron spin. It indicates that the wave packet that is formed and behaves as an electron is something like a whirlpool of closed current that can be considered as spin motion. Thus, wave mechanics proves the assumption of rotation of electron on its own axis.

1.22 ELECTRONIC CONFIGURATION OF ATOMS

From the discussions before, we have seen that the energy levels close to the nucleus have lower energy and it increases as the orbits are away from the nucleus. Again in the same quantum level where there are *s*, *p*, *d* and *f* orbitals present, the energy of the orbitals changes with penetrating power which is of the order $s > p > d > f$. Consequently, the energy is $s < p < d < f$. This is the general consequences and followed up to $n = 3$. But for elements with more electrons the sequence of energy levels changes with more and more addition of protons and neutrons in the nucleus and electrons in the outer orbitals. There we have to rely on spectroscopic data to ascertain the energy levels of gradual increasing order. The energy of the orbital electron depends on (a) effective nuclear charge depending on shielding effect, (b) penetration effect of orbitals, (c) inter-electron repulsion and (d) spin-correlation between electrons in an orbital.

(a) Effective nuclear charge: It is defined as the effective force of attraction of the nucleus on the valence electron as if no other electron is present in the inner orbital of the atom. It is expressed as $Z_{eff} = Z_{actual} - S$ (where S is the screening or shielding constant) that depends on inter-electron repulsion.

The Slater's rules for calculating the screening constant are:

1. Electrons outside the valence electron will contribute zero to the shielding.
2. All additional electrons other than the electron, considered in the nth level contribute 0.35 and if it is 1s, then 0.30.
3. All electrons in the (n–1)th shell contribute 0.85 each.
4. All electrons in the (n–2)th shell contribute 1.00 each.
5. If the electron considered is in d or f shell, then each additional electron in the same level will contribute 0.35 and all other electrons in the inner shell contribute 1.00.

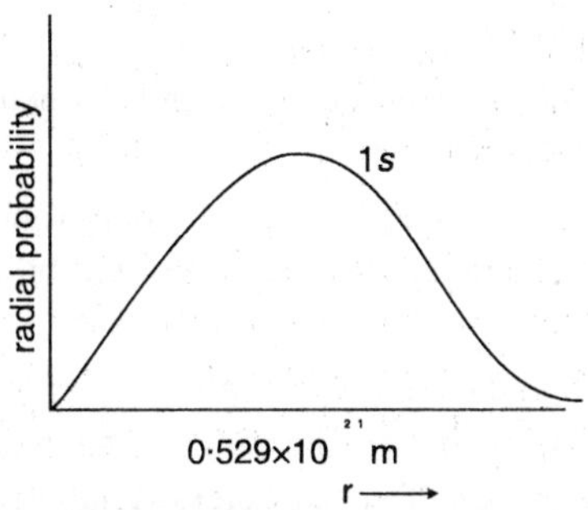

Fig. 1.28(a) The radial distribution function of 1s orbital of hydrogen.

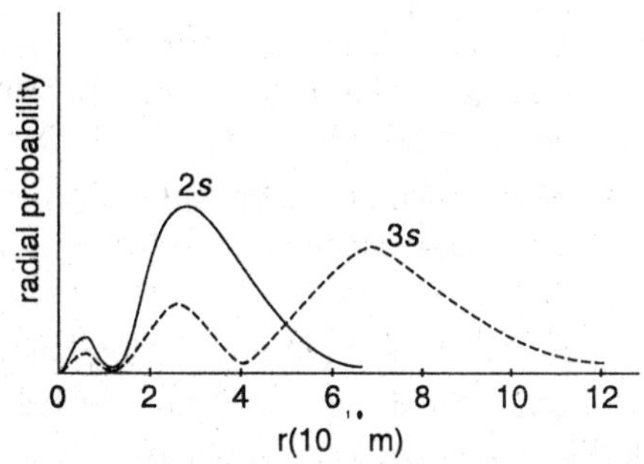

Fig. 1.28(b) The radial distribution function of 2s and 3s orbitals of hydrogen atom.

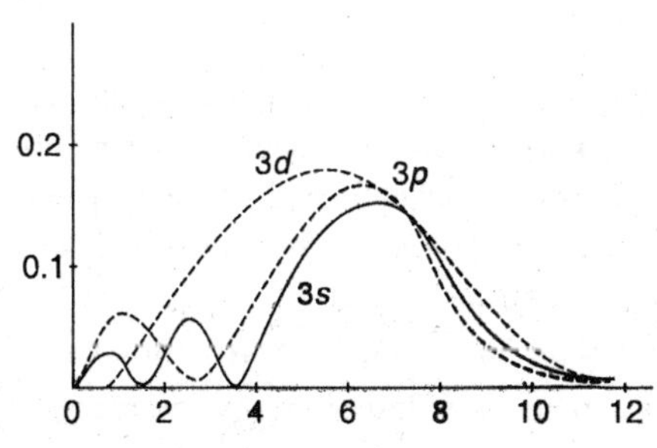

Fig. 1.28(c) The radial distribution function of 3s, 3p and 3d orbitals of hydrogen atom.

For clear understanding, let us calculate the effective nuclear change on the valance electron in zinc. The electronic configuration of Zn is $1s^2\, 2s^2\, 2p^6\, 3s^2\, 3p^6\, 3d^{10}\, 4s^2$.

Screening constant, $S = (1 \times 0.35) + (18 \times 0.85) + (10 \times 1.00) = 25.65$

$$Z_{eff}\,(\text{Zn}) = 30 - 25.65 = 4.35 \text{ for valence electrons}$$

and for $3d$ electron in Zn it is as follows:

$$S = (9 \times 0.35) + (18 \times 1.00) = 21.15$$

So, Z_{eff} (Zn) = 30–21.15 = 8.85 for the $3d$ electron.

What does this value signify? Suppose no electron is present in the atom except the electron considered $4s^1$ in the former and $3d^9$ in the later. Force of attraction of this electron by the nucleus will be as if 4.35 number of proton is present in the nucleus of Zn for former one and 8.85 number of protons in the latter one.

(b) By penetration effect of orbital we mean how close the orbital electron approaches the nucleus and hence the outer electrons are not well screened by the inner electrons, the effective nuclear change of the outer electrons will be higher. This will be clear considering the nature of radial probability functions for $1s$, $2s$, $2p$, $3s$, $3p$ and $3d$ orbitals as in the figure 1.28(a), 1.28(b) and 1.28(c). The radial density of $2s$ orbital spreads to penetrate into the curve for $1s$; $3s$ orbital penetrates into the $1s$ and $2s$ orbital and so on. If the radial distribution of $2s$ electron does not enter into $1s$ orbital, the $2s$ electron will be completely screened by $1s$ electron.

Thus less screening will increase the effective nuclear charge of outer electron. The extent of penetration can be as $s > p > d > f$.

Thus the sub-group A elements of the periodic table have less effective nuclear charge and sub-group B have higher value as a result, their ionisation potential, electronegetivity, polarisability are different.

Slater's rule considers only the inter-electron repulsion and does not consider the penetrating effect of orbitals and spin-correlation between electrons in the orbital and hence gives approximate result. More accurate result is obtained by considering penetration effect of outer orbital electrons given by Clementi and Raimondi, which is more complex one with number of parameters and beyond the scope of this book.

Still more accurate result of screening constant is obtained by self-consistent field method where the potential field of each electron is calculated from wave functions and then averaging the field of others.

1.23 AUFBAU PRINCIPLE

This German term means "building up". This rule states that atoms of multi-electron system can be achieved by adding more and more protons and neutrons to the nucleus and consequently, adding more and more electrons to the valence shell with gradually increasing energy, i.e., gradually increasing $(n+l)$ value. When for two orbitals $(n+l)$ is equal, then the orbital with lower n value will be occupied first. The sequence of energy level is as follows:

$$1s < 2s < 2p < 3s < 3p < 4s < 3d < 4p < 5s < 4d < 5p < 6s \approx 5d \approx 4f < 6p < 7s \approx 6d \approx 5f$$

After Ca (20), the $3d$ orbital becomes lower energy and for Sc (21), the electron will go to inner $3d$ orbital. In general this sequence is followed with very minor exception. In Cr (24), the configuration becomes $3d^5 4s^1$ to maintain symmetry of electron distribution in $3d$ and similar case occurs in case of Cu (29) where it is $3d^{10} 4s^1$. Same case appears for Nb (41) $4d^4 5s^1$ and Mo (42) $4d^5 5s^1$and so on. For La (57) $5d^1 6s^2$, Ce (58) $4f^1 5d^1 6s^2$ and Pr (59) $4f^3\, 6s^2$. At the higher atomic weight region the energy of orbitals are very close to each other.

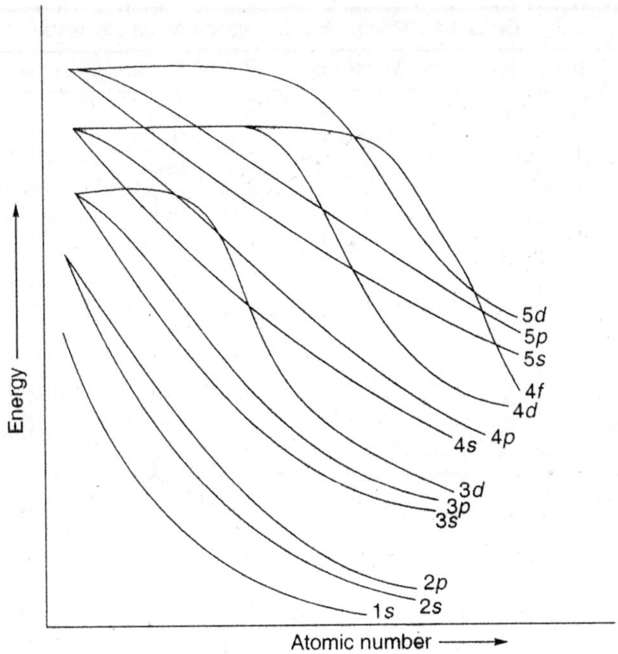

Fig. 1.29 Nature of variation of orbital energy with atomic number.

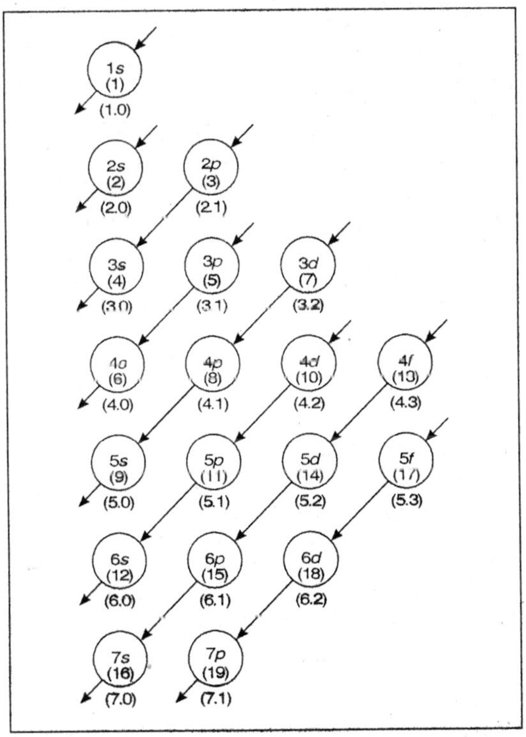

Fig. 1.30 Mnemonic scheme showing the order of occupancy of atomic orbitals (the 7*p* orbitals are not occupied in any of the elements occurring in nature).

Table 1.1 : Electronic configuration of the elements

Elements/Symbol	1s	2s	2p	3s	3p	3d	4s	4p	4d	4f	5s	5p	5d	5f	5g	6s	6p	6d	6f	6g	6h	7s
1. Hydrogen (H)	1																					
2. Helium (He)	2																					
3. Lithium (Li)	2	1																				
4. Beryllium (Be)	2	2																				
5. Boron (B)	2	2	1																			
6. Carbon (C)	2	2	2																			
7. Nitrogen (N)	2	2	3																			
8. Oxygen (O)	2	2	4																			
9. Fluorine (F)	2	2	5																			
10. Neon (Ne)	2	2	6																			
11. Sodium (Na)	2	2	6	1																		
12. Magnesium (Mg)	2	2	6	2																		
13. Aluminium (Al)	2	2	6	2	1																	
14. Silicon (Si)	2	2	6	2	2																	
15. Phosphorus (P)	2	2	6	2	3																	
16. Sulphur (S)	2	2	6	2	4																	
17. Chlorine (Cl)	2	2	6	2	5																	
18. Argon (Ar)	2	2	6	2	6																	
19. Potassium (K)	2	2	6	2	6	–	1															
20. Calcium (Ca)	2	2	6	2	6	–	2															
21. Scandium (Sc)	2	2	6	2	6	1	2															
22. Titanium (Ti)	2	2	6	2	6	2	2															
23. Vanadium (V)	2	2	6	2	6	3	2															
24. Chromium (Cr)	2	2	6	2	6	5	1															
25. Manganese (Mn)	2	2	6	2	6	5	2															
26. Iron (Fe)	2	2	6	2	6	6	2															
27. Cobalt (Co)	2	2	6	2	6	7	2															
28. Nickel (Ni)	2	2	6	2	6	8	2															
29. Copper (Cu)	2	2	6	2	6	10	1															
30. Zinc (Zn)	2	2	6	2	6	10	2															
31. Gallium (Ga)	2	2	6	2	6	10	2	1														
32. Germanium (Ge)	2	2	6	2	6	10	2	2														
33. Arsenic (As)	2	2	6	2	6	10	2	3														
34. Selenium (Se)	2	2	6	2	6	10	2	4														
35. Bromine (Br)	2	2	6	2	6	19	2	5														
36. Krypton (Kr)	2	2	6	2	6	10	2	6														
37. Rubidium (Rb)	2	2	6	2	6	10	2	6	–	–	1											
38. Strontium (Sr)	2	2	6	2	6	10	2	6	–	–	2											
39. Yttrium (Y)	2	2	6	2	6	10	2	6	1	–	2											
40. Zirconium (Zr)	2	2	6	2	6	10	2	6	2	–	2											
41. Niobium (Nb)	2	2	6	2	6	10	2	6	4	–	1											
42. Molybdenum (Mo)	2	2	6	2	6	10	2	6	5	–	1											
43. Technetium (Tc)	2	2	6	2	6	10	2	2	6	–	1											
44. Ruthenium (Ru)	2	2	6	2	6	10	2	6	7	–	1											
45. Rhodium (Rh)	2	2	6	2	6	10	2	6	8	–	1											
46. Palladium (Pd)	2	2	6	2	6	10	2	6	10	–	–											
47. Silver (Ag)	2	2	6	2	6	10	2	6	10	–	1											
48. Cadmium (Cd)	2	2	6	2	6	10	2	6	10	–	2											
49. Indium (In)	2	2	6	2	6	10	2	6	10	–	2	1										
50. Tin (Sn)	2	2	6	2	6	10	2	6	10	–	2	2										

Contd.

Elements/Symbol	1s	2s	2p	3s	3p	3d	4s	4p	4d	4f	5s	5p	5d	5f	5g	6s	6p	6d	6f	6g	6h	7s
51. Antimony (Sb)	2	2	6	2	6	10	2	6	10	–	2	3										
52. Tellurium (Te)	2	2	6	2	6	10	2	6	10	–	2	4										
53. Iodine (I)	2	2	6	2	6	10	2	6	10	–	2	5										
54. Xenon (Xe)	2	2	6	2	6	10	2	6	10	–	2	6										
55. Caesium (Cs)	2	2	6	2	6	10	2	6	10	–	2	6	–									
56. Barium (Ba)	2	2	6	2	6	10	2	6	10	–	2	6	–									
57. Lanthanum (La)	2	2	6	2	6	10	2	6	10	–	2	6	1									
58. Cerium (Ce)	2	2	6	2	6	10	2	6	10	2	2	6	–									
59. Praseodymium (Pr)	2	2	6	2	6	10	2	6	10	3	2	6	–									
60. Neodymium (Nd)	2	2	6	2	6	10	2	6	10	4	2	6	–									
61. Promethium (Pm)	2	2	6	2	6	10	2	6	10	5	2	6	–									
62. Samarium (Sm)	2	2	6	2	6	10	2	6	10	6	2	6	–									
63. Europium (Eu)	2	2	6	2	6	10	2	6	10	7	2	6	–									
64. Gadolinium (Gd)	2	2	6	2	6	10	2	6	10	7	2	6	1									
65. Terbium (Tb)	2	2	6	2	6	10	2	6	10	9	2	6	–									
66. Dysprosium (Dy)	2	2	6	2	6	10	2	6	10	10	2	6	–									
67. Holmium (Ho)	2	2	6	2	6	10	2	6	10	11	2	6	–									
68. Erbium (Er)	2	2	6	2	6	10	2	6	10	12	2	6	–									
69. Thullium (Tm)	2	2	6	2	6	10	2	6	10	13	2	6	–									
70. Ytterbium (Yb)	2	2	6	2	6	10	2	6	10	14	2	6	–									
71. Lutecium (Lu)	2	2	6	2	6	10	2	6	10	14	2	6	1									
72. Hafnium (Hf)	2	2	6	2	6	10	2	6	10	14	2	6	2									
73. Tantalum (Ta)	2	2	6	2	6	10	2	6	10	14	2	6	3									
74. Tungsten (W)	2	2	6	2	6	10	2	6	10	14	2	6	4									
75. Rhenium (Re)	2	2	6	2	6	10	2	6	10	14	2	6	5									
76. Osmium (Os)	2	2	6	2	6	10	2	6	10	14	2	6	6									
77. Iridium (Ir)	2	2	6	2	6	10	2	6	10	14	2	6	7	–	–	2						
78. Platinum (Pt)	2	2	6	2	6	10	2	6	10	14	2	6	9	–	–	1						
79. Gold (Au)	2	2	6	2	6	10	2	6	10	14	2	6	10	–	–	1						
80. Mercury (Hg)	2	2	6	2	6	10	2	6	10	14	2	6	10	–	–	2						
81. Thallium (Tl)	2	2	6	2	6	10	2	6	10	14	2	6	10	–	–	2	1					
82. Lead (Pb)	2	2	6	2	6	10	2	6	10	14	2	6	10	–	–	2	2					
83. Bismuth (Bi)	2	2	6	2	6	10	2	6	10	14	2	6	10	–	–	2	3					
84. Polonium (Po)	2	2	6	2	6	10	2	6	10	14	2	6	10	–	–	2	4					
85. Astatine (At)	2	2	6	2	6	10	2	6	10	14	2	6	10	–	–	2	5					
86. Radon (Rn)	2	2	6	2	6	10	2	6	10	14	2	6	10	–	–	2	6					
87. Francium (Fr)	2	2	6	2	6	10	2	6	10	14	2	6	10	–	–	2	6	–	–	–	–	1
88. Radium (Ra)	2	2	6	2	6	10	2	6	10	14	2	6	10	–	–	2	6	–	–	–	–	2
89. Actinium (Ac)	2	2	6	2	6	10	2	6	10	14	2	6	10	–	–	2	6	1	–	–	–	2
90. Thorium (Th)	2	2	6	2	6	10	2	6	10	14	2	6	10	–	–	2	6	2	–	–	–	2
91. Protactinium (Pa)	2	2	6	2	6	10	2	6	10	14	2	6	10	2	–	2	6	1	–	–	–	2
92. Uranium (U)	2	2	6	2	6	10	2	6	10	14	2	6	10	3	–	2	6	1	–	–	–	2
93. Neptunium (Np)	2	2	6	2	6	10	2	6	10	14	2	6	10	5	–	2	6	–	–	–	–	2
94. Plutonium (Pu)	2	2	6	2	6	10	2	6	10	14	2	6	10	6	–	2	6	–	–	–	–	2
95. Americium (Am)	2	2	6	2	6	10	2	6	10	14	2	6	10	7	–	2	6	–	–	–	–	2
96. Curium (Cm)	2	2	6	2	6	10	2	6	10	14	2	6	10	7	–	2	6	1	–	–	–	2
97. Berkelium (Bk)	2	2	6	2	6	10	2	6	10	14	2	6	10	8	–	2	6	1	–	–	–	2
98. Californium (Cf)	2	2	6	2	6	10	2	6	10	14	2	6	10	10	–	2	6	–	–	–	–	2
99. Einsteinium (Es)	2	2	6	2	6	10	2	6	10	14	2	6	10	11	–	2	6	–	–	–	–	2
100. Fermium (Fm)	2	2	6	2	6	10	2	6	10	14	2	6	10	12	–	2	6	–	–	–	–	2
101. Mendelevium (Md)	2	2	6	2	6	10	2	6	10	14	2	6	10	13	–	2	6	–	–	–	–	2
102. Nobelium (No)	2	2	6	2	6	10	2	6	10	14	2	6	10	14	–	2	6	–	–	–	–	2
103. Lawrencium (Lr)	2	2	6	2	6	10	2	6	10	14	2	6	10	14	–	2	6	1	–	–	–	2

1.24 PAULI'S EXCLUSION PRINCIPLE OR EQUIVALENCE PRINCIPLE

The principle states that every completely defined quantum state having same values of quantum numbers can have only one electron i.e. in an atom no two electrons will have the four quantum numbers n, l, m_l and m_s identical. If happens, then one of the two will be excluded from entering into the constitution of the atom, hence named as exclusion principle. The electrons in an atom having same n and l i.e. same sub-group are called equivalent electrons. For $l = 0$, sub-group is s-electrons, for $l = 1$ the p-electrons and so on. Suppose for valence electrons in N ($Z = 7$) are $2s^2$, $2p^3$ are equivalent electrons. They can be written with their quantum numbers as follows:

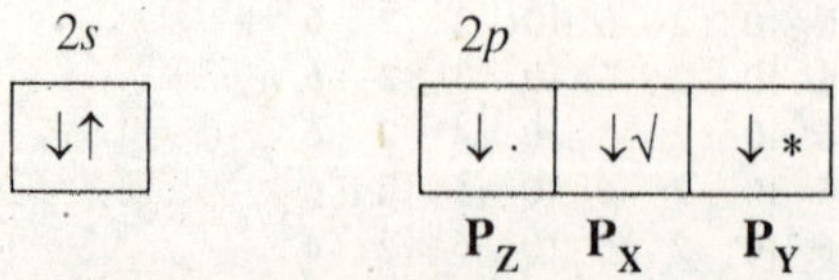

Fig. 1.31 Equavalence electrons with different quantum number.

The electrons marked ↓· has quantum numbers 2, 1, 0, + ½ and ↓√ has 2, 1, +1, + ½ and ↓* has 2, 1, –1, + ½. Following this principle maximum number of electrons that can be accommodated in a shell can be determined. For $n = 3$,

$l = 0$	1	2
$m_l = 0$	+1 0, –1	+2, +1, 0, –1, –2
$m_S = \pm ½$	±½, ±½, ±½	±½, ±½, ±½, ±½, ±½
= 2	= 6	=10

Total 2 + 6 + 10 = 18 and the shell is closed.

For fourth quantum states $n = 4$, for the values up to $l = 2$, electrons will be as before and 18 electrons are there, for $l = 3$ and $m_l = +3$ to –3 total 7 m_l values and added electrons will be 7 × 2 = 14.

Table 1.2 : Arrangement of electrons in quantum states up to $n = 4$

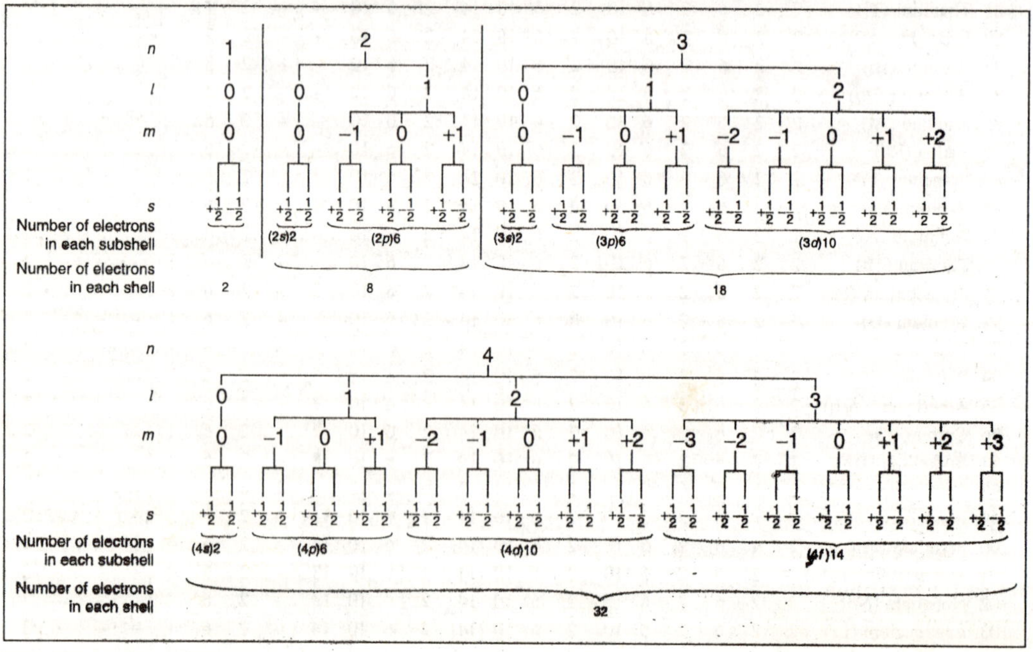

So 4th quantum state will hold 18 + 14 = 32 electrons and the shell becomes closed. A direct consequence of this principle is that every shell or sub-shell when closed, its angular momentum and spin momentum become balanced and do not contribute to the total angular momentum, that is why, the chemical and physical properties of elements depend only upon electrons outside the closed shell.

1.25 HUND'S RULE OF MAXIMUM MULTIPLICITY

We know electron will always occupy the lowest energy orbital and that will be the ground state configuration of an atom. If the atom has degenerate orbitals (i.e., same energy), then occupancy of the orbital by electron is guided by Hund's rule. The energy of the electrons also depends on the exchange energy besides the electrostatic interaction between electrons and protons. The more is the exchange energy, the more stable will be the configuration. The exchange energy, $E_{ex} \propto \frac{n(n-1)}{2}$ where n is the number of electrons of parallel spin.

For p^1 system $E_{ex} \propto \frac{1(1-1)}{2} = 0$

For p^2, two configurations are possible:

when spin is parallel (↑↑), its $E_{ex} \propto \frac{2(2-1)}{2} = 0\ 1$

when spin is paired (↑↓), its $E_{ex} \propto \frac{1(1-1)}{2} + \frac{1(1-1)}{2} = 0 + 0 = 0$

For p^3 system, two arrangements are possible:

when ↑↑↑ spins, its $E_{ex} \propto \frac{3(3-1)}{2} = 3$; when it is ↓↑↑, $E_{ex} \propto \frac{2(2-1)}{2} + \frac{1(1-1)}{2} = 1 + 0$

For p^4 system, the arrangements are:

when ↑↑↑↑, it is $E_{ex} \propto \frac{4(4-1)}{2} = 6$; when it is ↑↓↑↑, its $E_{ex} \propto \frac{3(3-1)}{2} + \frac{1(1-1)}{2} = 3 + 0$

when it is ↑↓ ↑↓, it is $E_{ex} \propto \frac{2(2-1)}{2} + \frac{2(2-1)}{2} = 1 + 1$ and so on.

Thus, we see that the exchange energy is maximum for parallel spin and that will be the ground state of the atom. So, Hund's rule can be stated in this way:

The ground state of an atom will be determined by having maximum number of unpaired electrons in the same sub-shell, i.e., minimum electrostatic repulsion. Single electron will preferentially fill up the degenerate states. Explanations of atomic spectra support this rule. For Cr the electronic configuration of the valence shell is $3d^5 4s^1$, because exchange energy for this is proportional to 10 and for $3d^4 5s^2$ it is ≈ 6. This difference is sufficient to compensate the loss in energy for moving the electron from 4s to 3d. Likewise, calculation can show why Cu has $3d^{10} 4s^1$. Thus, it can be said that the "half-filled and full-filled orbitals are more stable than other configuration." This is reflected in the evaluation of ionization potential of elements.

1.26 CLASSIFICATION OF ELEMENTS ON THE BASIS OF ELECTRONIC CONFIGURATION, THEIR POSITION IN THE PERIODIC TABLE

When inner orbitals of an atom is closed, their total orbital angular momentum and spin angular momentum balance each other and the chemical properties depend only upon the nature of the valence electrons. Now properties of electrons are dependent upon the nature of orbital to which it belongs, because *s*, *p*, *d*, *f* have different penetrating powers. Accordingly, elements are broadly classified as

s-block	having valence orbital	ns^{1-2}
p-block	... ...	ns^2np^{1-6}
d-block	... ...	$(n-1)d^{1-10}ns^2$
f-block	... ...	$(n-2)f^{1-14}(n-1)d^{0/1}ns^{2/1}$

s and *p* block elements are called representative elements as they represent the groups of the periodic table. The *d*-block elements are called the transitional elements as their characteristics gradually change from basic or alkaline nature to the acidic nature. They are Sc (21) to Zn (30) having $3d^{1-10}\,4s^{1\text{ or }2}$ configurations, Y (39) to Cd (48) having $4d^{1-10}\,5s^{1\text{ or }2}$, La (57), Hf (72) to Hg (80) $5d^{1-10}\,6s^{1\text{ or }2}$. Although last but one member of each series have $(n-1)\,d^{10}ns^1$, i.e., the penultimate shell is complete and hence behave as a representative element, the involvement of inner *d* shell electron take part in chemical combinations and thus the *d* shell has incomplete configuration and behaves as a transitional element.

The *f*-block elements are called the inner transition elements, i.e., a group of elements within the elements of transition series. There are two such series, the first one starts with Lanthanum La (57) having $4f^{1-14}$ called *Lanthanides* and the other starts with Actinium (89) having $5f^{1-14}$ called *Actinides*. The elements of these two groups are chemically indistinguishable and very rare in nature and hence called *rare earths*. Within Actinide series naturally occurring element is up to Uranium (92) and beyond Uranium all are artificial or synthetic elements, called *Trans-uranic elements*.

Another group of elements ranging Z = 104–121 was supposed to exist analogues to element Hf (72) to Ac (89), these were supposed to be named as *super-Actinide series*, also termed as *super-heavy elements*.

There are four elements—Tc (43), Pm (61), At (85) and Fr (87); these are not found in nature and so called *missing elements*. In the beginning of the earth they were present, the $T_{1/2}$ of these elements are very short and they have decayed totally. Another very rare element is Re (75) with abundance 0.0007 ppm, i.e., very trace in nature.

The position of representative elements and transition elements in the periodic table can be ascertained easily knowing their electronic configurations. For representative elements, the last quantum state determines their period, and number of electrons in the valence shell determines their group. For transition elements, last quantum state (*n*) determines their period and number of electrons in $(n-1)$ *d* shell and *ns* shell will be their group number in the periodic table. These elements are in general placed with the representative elements in the periodic table. The more basic elements in sub-group A and the less basic elements in sub-group B are placed. But in modern tables, the groups has been extended, the *s*-block and *d*-block in sub-group A and *p*-block in sub-group B, the *d*-block has been placed in between group IIA and group IIIB, total 18 groups. The *f*-block elements are placed in group IIIB with La (57) and Ac(89) and these are shown in a separate caption below in the periodic table. The new elements yet to be discovered are Unq (104), Unp (105), Unh (106), Uns (107), Uno(108), Une (109), as IUPAC named and symbolized them respectively will be placed in group IV to group IX respectively in the extended form of the periodic table with regard to their valence electrons.

Table 1.3 : A modern form of the Periodic Table (valence shell electron configurations indicated)

	s^1	s^2	$d^2s^2(n = 1–10; x = 0, 1 \text{ or } 2)$										s^2p^1	s^2p^2	s^2p^3	s^2p^4	s^2p^5	s^2p^6
1*s*							1 H	2 He										
2*s*2*p*	3 Li	4 Be											5 B	6 C	7 N	8 O	9 F	10 Ne
3*s*3*p*	11 Na	12 Mg											13 Al	14 Si	15 P	16 S	17 Cl	18 Ar
4*s*4*d*4*p*	19 K	20 Ca	21 Sc	22 Ti	23 V	24 Cr	25 Mn	26 Fe	27 Co	28 Ni	29 Cu	30 Zn	31 Ga	32 Ge	33 As	34 Se	35 Br	36 Kr
5*s*4*d*5*p*	37 Rb	38 Sr	39 Y	40 Zr	41 Nb	42 Mo	43 Tc	44 Ru	45 Rh	46 Pd	47 Ag	48 Cd	49 In	50 Sn	51 Sb	52 Te	53 I	54 Xe
6*s*(4*f*) 5*d* 6*p*	55 Cs	56 Ba	57 La	72 Hf	73 Ta	74 W	75 Re	76 Os	77 Ir	78 Pt	79 Au	80 Hg	81 Tl	82 Pb	83 Bi	84 Po	85 At	86 Rn
7*s*(5*f*) 6*d*	87 Fr	88 Ra	89 Ac	104 Unq	105 Unp	106 Unh	107 Uns	108 Uno	109 Une									

	$f^md^ns^2$ (m = 1–14, n = 0 or 1 (2 for Th))													
* Lanthanide series (4*f*)	58 Ce	59 Pr	60 Nd	61 Pm	62 Sm	63 Eu	64 Gd	65 Tb	66 Dy	67 Ho	68 Er	69 Tm	70 Yb	71 Lu
** Actinide series (5*f*)	90 Th	91 Pa	92 U	93 Np	94 Pu	95 Am	96 Cm	97 Bk	98 Cf	99 Es	100 Fm	101 Md	102 No	103 Lr

1.27 WAVE PROPERTIES OF PARTICLES

In 1905 Einstein and in 1924 Compton established the particle nature of electromagnetic waves to explain the different experimental results. In 1923 Louis de Broglie first gave the idea that matter may have the wave character and was proved first by C. J. Davisson and L. H. Germer in 1927. Accelerated electron beam is allowed to fall on nickel crystal and studied the reflection and scattering of electron. The regular arrangement of atoms in different planes in solid crystal provides a suitably spaced three-dimensional grating for the diffraction, and the diffraction pattern was detected by using a photographic plate. The similar experiment using fast moving electron beam and very thin metal sheet was carried out by G. P. Thomson in 1927 and get the same pattern consisting of a series of concentric circles indicating the wave properties of electron. The patterns can be deformed by using magnetic field showing that it is produced by electrons not by any extraneous radiation.

Similar diffraction pattern was obtained by using accelerated stream of hydrogen, helium ions, neutrons, etc. The diffracted electron beam passing through crystals meets the various sets of atoms or ions in plane and the relation can be obtained from Bragg's equation $n\lambda = 2d \sin \theta$, where λ is the wavelength of beam, d is the distance between two planes and θ is the angle of incidence of the beam on the plane, n is an integer.

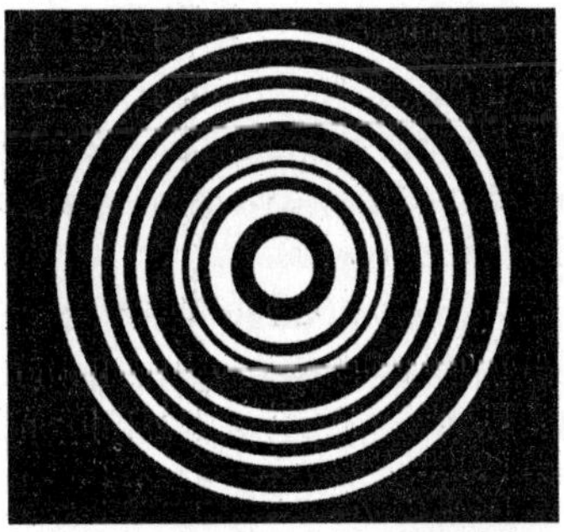

Fig. 1.32 Electron beam diffraction pattern.

From different experimental results we see that electron in some cases behaves as a particle and in some cases as a wave. Normally, we cannot think of how a particle simultaneously can behave like a wave. This can be understood from the following experiment in which in some part of the experiment it behaves like a wave causing diffraction and in the later part as a particle emitting photoelectron.

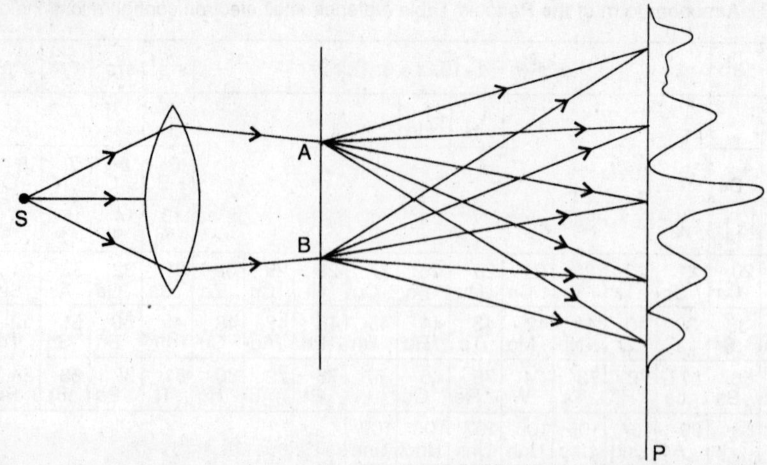

Fig. 1.33 Experiment proving dual nature of electron.

The monochromatic light from point source S passes through converging lens L, then through opaque slits A and B and ultimately reaches the screen P that consists of many adjacent photoelectric cells. The monochromatic light here on the screen behaves as a particle and amount of current produced at the point occurs where interference of a pair of coherent wave-trains takes place. Interference fringes in the screen can be understood if we consider light as a wave from source to the screen. We may think the fringes may occur due to interference of photons passing through two slits to explain all in the light of a particle. But when intensity of light is decreased to such an extent that only one photon enters the apparatus at a time through the slit, the diffraction pattern is still obtained. Again, how can a photon interfere with itself? If we consider light as a wave, which are spread out in space, then interference is possible. Thus, the dual nature of electron is established in the same experiment.

Louis de Broglie considering the dual nature of electron gave his hypothesis. According to him, "the relation between wavelength and momentum of photon is completely general and applies equally to both photons and material particles." He co-relates these properties considering Planck's quantum theory and theory of relativity of Einstein. From Planck's theory $E = h\nu$ and Einstein equation of mass and energy of particle $E = mc^2$ so we may write

$$h\nu = mc^2 \quad \text{or,} \quad h\times\frac{c}{\lambda} = mc^2 \quad \text{or,} \quad \frac{h}{\lambda} = mc = p \text{ (momentum)} \quad \text{or,} \quad \lambda = \frac{h}{p}$$

Thus, moving particle of matter has always got a wave associated with it. As an example, the electrons are particles, their charge, mass, energy, etc., are known when we consider in particle form, but when we want to know their path, their reflections, scatterings, etc., we treat them as wave, called *matter wave*.

Velocity of de Broglie's wave (u):

The velocity of de Broglie's wave $(u) = \lambda \times \nu$

$$\text{or, } u = \frac{h}{mv}\times\frac{mc^2}{h} \left[E = h\nu = mc^2, \quad \therefore \quad \nu = \frac{mc^2}{h}\right]$$

$= \frac{c^2}{v}$, where c is the velocity of light and v is the velocity of material particle.

As $c > v$, therefore $u >> c$. This u is called the *phase velocity*. So, it is evident that every moving material particle has two types of velocity: (a) the mechanical motion of the particle represented by v and the other is (b) the velocity of propagation associated with wave c, i.e., speed of light is the limiting value of mass-energy relation but u is not of this kind.

Another concept can be obtained from this phase velocity. If a number of such waves travel in the same medium, then superposition of these waves may take place forming a wave-packet and this wave-packet can be considered as particle. It can be shown that velocity of this wave-packet, called *group velocity*, is equal to the velocity of the particle. The difficulty in this concept is that wave-packet formed very soon dissipates.

The concept of matter wave is negligible and insignificant in case of macro-bodies, because λ associated with them is very low in comparison to the dimension of the matter but in case of micro-world e^{-1}, ${}_{1}^{1}p$, ${}_{0}^{1}n$, etc., λ is within the dimension of the atomic system and hence cannot be neglected. λ for an e^{-1} moving with $\frac{c}{10}$ is

$$\lambda = \frac{6.67 \times 10^{-27}\ \text{erg sec}}{9.1 \times 10^{-28} \times 3 \times 10^{9}} = 0.24 \times 10^{-8}\ \text{cm}.$$

1.28 HEISENBERG'S UNCERTAINTY PRINCIPLE

A particle has definite momentum, position and energy at a particular time and for a macroscopic substance these are determined by Newtonian mechanics. A wave has its amplitude, wavelength, etc. A submicroscopic particle has both particle and wave properties. Heisenberg in1927 first stated that it is not possible to determine simultaneously both the position and momentum of a submicroscopic particle accurately, e.g., we want to locate an electron in an atom. Light of comparable wavelength of the dimension of electron to be used, i.e., X-ray by scattering, when X-ray collides with electron, the electron will be located but in doing so, collision will change the momentum of the electron, and cannot be determined accurately. So determination of position and momentum becomes uncertain.

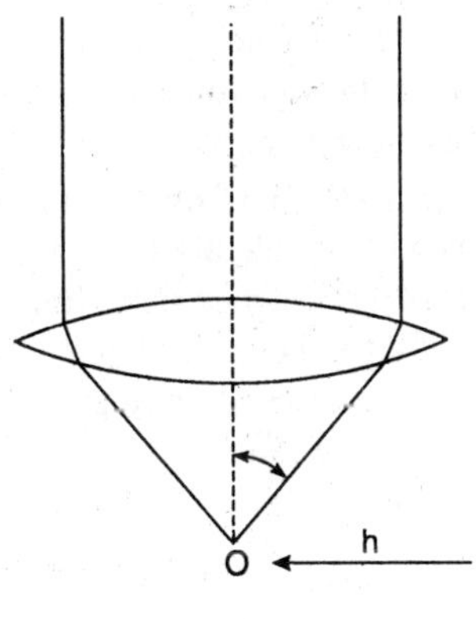

Fig. 1.34 Diagrammatic operation of a gamma ray microscope.

If we want to know the wavelength of electron wave from diffraction pattern, and also momentum from de Broglie's relation, it can be done, but position of the electron will be undetermined as scattering angle is uncertain. According to wave concept, the wave packet may be considered as a particle, extending over a small region. So electron as a particle can be in any position within the wave packet, so its position becomes uncertain at any instant. Thus there is no single experiment by which the particle properties and wave properties of any system can be determined accurately. This can be illustrated by the following experiment:

An electron beam of momentum P passing through the slit of width ΔQ hits the screen at AA' and produces patterns. The angle of deviation is α from the original direction. The gain in momentum in vertical direction is

$$\Delta P = p \sin\alpha \text{ and } \alpha = \frac{\lambda}{\Delta q}.$$

$$\text{So, } \Delta P = p \times \sin\frac{\lambda}{\Delta q} \simeq P \times \frac{\lambda}{\Delta q} \text{ as } \alpha \text{ is small,}$$

or, $\Delta P = \frac{h}{\lambda} \times \frac{\lambda}{\Delta q} = \frac{h}{\Delta q}$; so, $\Delta P \times \Delta q \simeq h$ [for momentum and position].

Similar equations are obtained $\Delta E \times \Delta t \geq \frac{h}{4\pi}$ for energy and time, and $\Delta v \times \Delta x \simeq \frac{h}{m}$ for velocity and position.

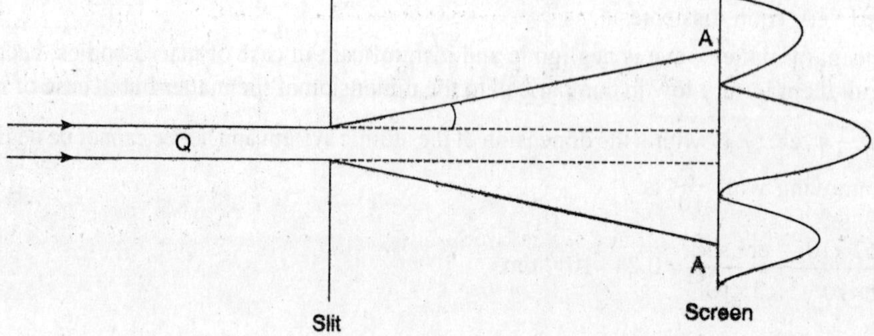

Fig 1.35 Experiment proving Heisenberg's uncertainty principle.

In Bohr's theory electron is revolving round the nucleus with constant speed, but this principle considers the electron to be moving from point to point with varying velocity and considers probability of finding electron at a point.

In 1925, Heisenberg and Schrödinger, taking the idea of matter wave of de Broglie, tried to explain independently spectral anomalies of atoms from two different angles. Heisenberg developed quantum mechanics using matrix calculus, which is very complicated branch of mathematics. Schrödinger developed wave mechanics, which is simple and gives pictorial representation. Schrödinger later showed that Heisenberg's mechanics and his own were mathematically equivalent and gives the same result.

CHAPTER

2

WAVE MECHANICS

2.01 WAVE NATURE OF ELECTRON

An electron although a particle of definite mass and charge behaves like a wave in many situations (e.g., diffraction) and so its motion is to be represented by a wave equation that involves a second order differential equation as

$$\frac{\partial^2 \phi}{\partial x^2} = \frac{1}{c^2} \times \frac{\partial^2 \phi}{dt^2} \qquad \text{(for one dimension)}$$

$$\frac{\partial^2 \phi}{\partial x^2} + \frac{\partial^2 \phi}{\partial y^2} + \frac{\partial^2 \phi}{\partial z^2} = \frac{1}{c^2} \times \frac{\partial^2 \phi}{\partial t^2} \text{ or, } \nabla^2 \phi = \frac{1}{c^2} \times \frac{\partial^2 \phi}{\partial t^2} \qquad \text{(for three dimensions)}$$

where c is the velocity of propagation of wave and if it were an electromagnetic wave, then c will be the velocity of light. The function ϕ is the displacement of the wave as a function of coordinates x, y, z and time t.

Here, ∇^2 is called Laplacian operator. The solution of the equation (one dimension) may be of the form

$$\phi = A \sin \frac{2\pi}{\lambda}(x - ct) \quad \text{or,} \quad \phi = Ae^{i(k.x - \omega t)}$$

i.e., ϕ is the amplitude function of wave which is the function of space co-ordinates and time. As the electron is bound within the atom with attractive force, it behaves like a standing wave and so will be time-independent. When the electron is free, then wave function will be time-dependent as in case of emission/absorption of radiation. So, ψ for an electron within the atom can be written as $\psi(x, y, z, t) = \psi(x, y, z)\,\phi(t)$, where $\psi(x, y, z)$, is the function of space co-ordinates and $\phi(t)$ is the function of t; $\phi(t)$ may be of the form $\sin 2\pi\nu t$ or $e^{2\pi i\nu t}$.

To derive the Schrödinger wave equation two basic postulates are:

1. The equation must be linear so that its solution can be super-imposed to produce interference and thus may form the wave packet.
2. The coefficient of the equation must be a constant like charge, mass, etc., of electron but not the parameters like momentum, energy, etc. This is because, we leave open the possibility of superimposing solutions that belong to different values of parameters, i.e., more general expression cannot be a solution.

Remembering these two points we can write wave equation as under:

$$\nabla^2\psi(x,y,z)\,e^{2\pi i\nu t} = \frac{1}{c^2}\times\frac{\partial^2}{\partial t^2}\,\psi(x,y,z)e^{2\pi i\nu t}$$

and for atomic system ψ will be time-independent and hence constant, so

$$e^{2\pi i\nu t}\nabla^2\psi(x,y,z) = \frac{1}{c^2}\times\frac{\partial^2}{\partial t^2}\,\psi(x,y,z)e^{2\pi i\nu t} \quad = \frac{1}{c^2}\psi(x,y,z)\frac{\partial^2}{\partial t^2}\,e^{2\pi i\nu t}$$

or, $$e^{2\pi i\nu t}\nabla^2\psi(x,y,z) = \frac{1}{c^2}\psi(x,y,z)\times\frac{\partial\,(2\pi i\nu e^{2\pi i\nu t})}{\partial t}$$

$$= \frac{1}{c^2}\psi(x,y,z)\times\left(-4\pi^2\nu^2\right)e^{2\pi i\nu t}$$

So, $$\nabla^2\psi(x,y,z) = \frac{1}{c^2}\times\psi(x,y,z)\times\left(-4\pi^2\nu^2\right) = -\frac{4\pi^2\nu^2}{c^2}\psi(x,y,z).$$

This is the wave equation for electron that contains no time-dependent part but only space co-ordinate system. Introducing the particle character to the equation, we get

$$\nabla^2\psi(x,y,z) = \frac{-4\pi^2p^2}{h^2}\psi(x,y,z) \qquad [\text{as } c=\lambda\nu,\ c^2=\lambda^2\nu^2 \text{ and } \lambda=\frac{h}{p}\ ]$$

$$= -4\pi^2\times\frac{2Tm}{h^2}\times\psi(x,y,z) \qquad [\ T=\frac{1}{2}m\nu^2=\frac{p^2}{2m}\ ]$$

$$= -\frac{8\pi^2m}{h^2}(E-V)\times\psi(x,y,z) \qquad [E=T+V]$$

or, $$\nabla^2\psi(x,y,z)+\frac{8\pi^2m}{h^2}(E-V)\times\psi(x,y,z)=0$$

This is the Schrödinger's time-independent wave equation.

The Schrödinger equation can also be written in another form. Multiplying with $\frac{h^2}{8\pi^2m}$ or $\frac{\hbar^2}{2m}$,

$$\frac{h^2}{8\pi^2m}\times\nabla^2\psi(x,y,z)-V\psi(x,y,z)+E\psi(x,y,z)=0$$

or, $$\left(-\frac{h^2}{8\pi^2m}\times\nabla^2+V\right)\psi(x,y,z)=E\psi(x,y,z).$$

Now, the L.H.S. contains the mathematical tool $-\frac{h^2}{8\pi^2 m} \times \nabla^2 + V$ that when acting on the $\psi(x, y, z)$ gives the eigenvalue E, that corresponds to the total energy of the system. This operator is called *Hamiltonian operator* H and the equation becomes

$$H\psi(x, y, z) = E\psi(x, y, z).$$

In the above equation when $V = 0$, i.e., when the electron will have no potential energy, $\psi(x, y, z)$ will represent progressive wave with kinetic energy only, the electron will behave as a free electron. In atom the electron bound to the nucleus cannot leave without external energy, hence the potential energy is negative. So electron in atom behaves as a stationary wave and the value of ψ is zero at infinite distance.

2.02 PHYSICAL INTERPRETATION OF WAVE FUNCTION ψ

(a) According to Millikan ψ is merely an "artifice of calculation", i.e., an auxiliary mathematical quantity employed to facilitate computations relative to experimental results. This is true to some extent in certain cases.

(b) Schrödinger himself interpreted ψ in terms of charge density. An electromagnetic wave of amplitude A has the energy density, i.e., energy per unit volume A^2. Each photon has energy $h\nu$, so number of photon per unit volume, i.e., photon density is $\frac{A^2}{h\nu}$. So, photon density $\propto A^2/$.

Similarly, if ψ is the amplitude of matter wave at any point in space ψ^2 or absolute value of $|\psi^2|$ will be the measure of particle density. When ψ^2 is multiplied by charge of electron e, then we get charge density. Thus ψ^2 is a measure of charge density also.

This consideration faced with the following difficulties:

(i) Charge density is maximum in wave packets, which is identically considered to be a particle. In nature every kind of wave spread through space in course of time. If so, it cannot represent a particle as particle has definite identity but not the wave packet when it dissipates.

(ii) When two electrons or in other words wave packets collide with each other, their waves propagate in space in three dimensions. When number of such wave packets, i.e., particles collide, their waves will propagate in dimensions three times the number of colliding particles which is purely a hypothetical mathematical concept, so wave is also an auxiliary mathematical concept.

(c) In 1926, Max Born considered $|\psi^2|$ as the probable density of the particle, i.e., probability of finding the particle at that point at a certain instant where we consider the probability in volume $dx \times dy \times dz \approx dv$. Heisenberg, Dirac and Bohr supported this

$P = \int_{-\infty}^{\infty} \psi^2 dv$ or $P = \int \psi \psi^* dv$ will be unity, i.e., $v = 1$. So ψ is called the *probability amplitude* and ψ is also called the *normalized wave function*. Generally, ψ is a complex quantity and ψ^* is the conjugate complex of ψ. When $P = 0$, the particle does not exist. When $P = \infty$, the particle is everywhere simultaneous. $|\psi^2| = P$ cannot be negative or complex because ψ is well-defined single-valued function.

2.03 NORMALIZED AND ORTHOGONAL WAVE FUNCTION

We know ψ^2 represents the probability of finding the electron in a volume element $dx \times dy \times dz \approx dv$ and $\int \psi^2 dv = 1$, where integration extends over whole space. In this case, the wave function is said to be normalized. Under certain conditions the wave function may be normalized by multiplying or dividing by a constant and in that case, $A\psi$ will be the normalized wave function (solution of the wave equation).

In that case, $\int A\psi \times A\psi \, dv = 1$ or, $A^2 \int \psi^2 dv = 1$

or, $\int \psi^2 \, dv = \frac{1}{A^2}$.

Thus, A is called *normalizing factor* and the normalized wave function is $A\psi$. When wave functions are same in a system, then $\int \psi_1 \psi_2 dv = 1, \psi$ is called ***normalized function*** and when they are different $\int \psi_1 \psi_2 dv = 0$, i.e., vanishes over entire space, the wave functions are said to be *orthogonal*. The above two concepts may be combined and functions which satisfy both the conditions are known as orthonormal.

$$\int_{-\infty}^{\infty} \psi_i \psi_j dz = \delta_{ij}$$

$$\delta_{ij} = 1 \text{ when } i = j$$

$$= 0 \text{ when } i \neq j$$

Eigenvalue and Eigenfunction:

We know the total energy operator in Schrödinger equation, $H = -\frac{h^2}{8\pi^2 m} \times \nabla^2 + V$ and H is called the *Hamiltonian* that when operated on a wave function of system gives the energy E for that system. Any equation of this form, where an operator acting on the function gives the function itself multiplied by some constant is called an eigenvalue equation. The constant is called ***eigenvalue*** and the corresponding wave function is called *eigenfunction*. These terms came from German *eigen* 'own' meaning proper or characteristic.

So in the equation, $H\psi = E\psi$, ψ is the *eigenfunction* and **E is the *eigenvalue*** of the operator H.

The characteristics of *eigenfunctions* are:

(a) ψ must be single-valued as the probability of finding the electron at any point has only one value.

(b) ψ must be continuous for meaningful solution.

(c) ψ must be finite as wave nature of the electron is considered.

(d) ψ must be normalized as the probability of finding the electron in the whole space will be unity, i.e., $\int \psi^2 dv = 1$.

2.04 APPLICATION TO HYDROGEN ATOM

Hydrogen atom consists of two interacting point particles, the nucleus of mass M and charge Ze^+ (Z is the atomic number, for Hydrogen $Z = 1$) and the electron of mass m and charge e^- and there is coulombic attraction $-\frac{Ze^2}{r}$ between the two. Let us take the co-ordinates of the nucleus as

(x_1, y_1, z_1) and that of electron as (x_2, y_2, z_2) with reference to some arbitrary axes. [Here kinetic energy of the atom is the kinetic energy of electron and the same for the nucleus as both are revolving about the common centre of mass G.]

So the Schrödinger wave equation for the system will be

$$-\frac{h^2}{8\pi^2 M}\left(\frac{\partial^2}{\partial x_1^2}+\frac{\partial^2}{\partial y_1^2}+\frac{\partial^2}{\partial z_1^2}\right)\psi-\frac{h^2}{8\pi^2 m}\left(\frac{\partial^2}{\partial x_2^2}+\frac{\partial^2}{\partial y_2^2}+\frac{\partial^2}{\partial z_2^2}\right)\psi+V\psi=E\psi .$$

Let the origin of the co-ordinate system be orthogonally transformed such that the origin of the system is now at M, the position of the nucleus. The new co-ordinates of the electron are now (x, y, z).

Therefore, $x = x_2 - x_1$, $y = y_2 - y_1$ and $z = z_2 - z_1$. Let the co-ordinates of the center of mass be (X, Y, Z). Considering the moment of the particles about G, we have,

$Mx_1 = + mx_2 = (m+M)X$

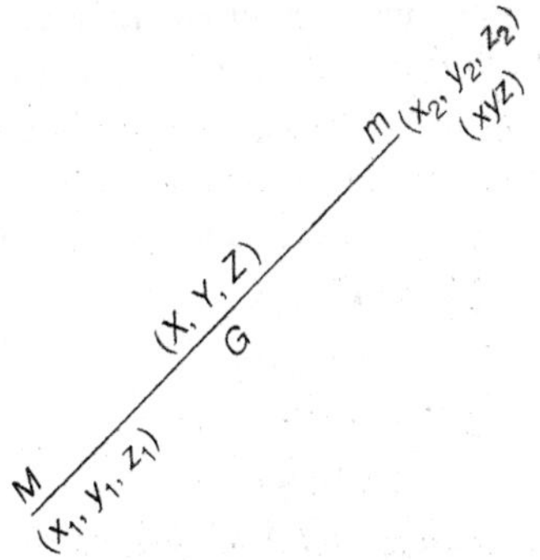

Fig 2.1 Showing hydrogen atom system with centre of mass.

$$\text{or, } X=\frac{M}{m+M}\times x_1+\frac{m}{m+M}\times x_2=\frac{\mu}{m}\times x_1+\frac{\mu}{M}\times x_2$$

$$=\frac{\mu}{m}\times x_1+\frac{\mu}{M}(x+x_1)$$

$$=\mu\left(\frac{1}{m}+\frac{1}{M}\right)x_1+\frac{\mu}{M}x=x_1+\frac{\mu}{M}x$$

[μ = reduced mass of the system = $mM/(m + M)$]

$$\text{or, } x_1=X-\frac{\mu}{M}x,\ y_1=Y-\frac{\mu}{M}y \text{ and } z_1=Z-\frac{\mu}{M}z .$$

Similarly,

$$x_2=X+\frac{\mu}{m}x,\ y_2=Y+\frac{\mu}{m}y \text{ and } z_2=Z+\frac{\mu}{m}z$$

Thus, the co-ordinates of the nucleus and the electron are expressed in terms of μ, m and M. Now the kinetic energy of the atom is

$$\text{K.E.} =\frac{1}{2}MV^2+\frac{1}{2}mv^2=\frac{1}{2}M\left(\dot{x}_1^2+\dot{y}_1^2+\dot{z}_1^2\right)+\frac{1}{2}m\left(\dot{x}_2^2+\dot{y}_2^2+\dot{z}_2^2\right)$$

$$=\frac{1}{2}\left[M\dot{x}_1^2+M\dot{y}_1^2+M\dot{z}_1^2\right]+\frac{1}{2}\left[m\dot{x}_2^2+m\dot{y}_2^2+m\dot{z}_2^2\right]$$

$$=\frac{1}{2}\left[\left(M\dot{x}_1^2+m\dot{x}_2^2\right)+\left(M\dot{y}_1^2+m\dot{y}_2^2\right)+\left(M\dot{z}_1^2+m\dot{z}_2^2\right)\right]$$

Taking one part,

$$M\dot{x}_1^2+m\dot{x}_2^2=M\left(\dot{X}-\frac{\mu}{M}\dot{x}\right)^2+m\left(\dot{X}+\frac{\mu}{m}\dot{x}\right)^2$$

$$=M\left(\dot{X}^2-\frac{2\mu}{M}\dot{x}\dot{X}+\frac{\mu^2}{M^2}\dot{x}^2\right)+m\left(\dot{X}^2+\frac{2\mu}{m}\dot{x}\dot{X}+\frac{\mu^2}{m^2}\dot{x}^2\right)$$

$$= \dot{X}^2(M+m)+\mu^2\dot{x}^2\left(\frac{1}{m}+\frac{1}{M}\right)$$

$$= (m+M)\dot{X}^2+\mu\dot{x}^2 .$$

Similarly, for other parts,

$$M\dot{y}_1{}^2+m\dot{y}_2{}^2=(m+M)\dot{Y}^2+\mu\dot{y}^2 \quad \text{and}$$

$$M\dot{z}_1{}^2+m\dot{z}_2{}^2=(m+M)\dot{Z}^2+\mu\dot{z}^2 .$$

i.e., the kinetic energy of the atom as a whole indicates one part is due to the motion of the atom as a whole round the centre of mass and the other part is due to the motion of the electron round the nucleus.

So, total K.E. $=\frac{1}{2}\left[(m+M)\left(\dot{X}^2+\dot{Y}^2+\dot{Z}^2\right)+\mu\left(\dot{x}^2+\dot{y}^2+\dot{z}^2\right)\right].$

The Schrödinger equation for the hydrogen atom will be

$$\left[-\frac{h^2}{8\pi^2(m+M)}\left(\frac{\partial^2}{\partial X^2}+\frac{\partial^2}{\partial Y^2}+\frac{\partial^2}{\partial Z^2}\right)\psi-\frac{h^2}{8\pi^2\mu}\left(\frac{\partial^2}{\partial x^2}+\frac{\partial^2}{\partial y^2}+\frac{\partial^2}{\partial z^2}\right)\psi\right]+V\psi=E\psi .$$

The two parts within the bracket will be independent of each other and will represent a part of constant energy.

So, we can write

$$-\frac{h^2}{8\pi^2(m+M)}\times\left(\frac{\partial^2}{\partial X^2}+\frac{\partial^2}{\partial Y^2}+\frac{\partial^2}{\partial Z^2}\right)\Psi_c+V_c\Psi_c=E_c\Psi_c$$

and $$-\frac{h^2}{8\pi^2\mu}\left(\frac{\partial^2}{\partial x^2}+\frac{\partial^2}{\partial y^2}+\frac{\partial^2}{\partial z^2}\right)\Psi_e+V_e\Psi_e=E_e\Psi_e$$

or, $H_c\Psi_c=E_c\Psi_c$ and $H_e\Psi_e=E_e\Psi_e$

We shall neglect the part $H_c\Psi_c=E_c\Psi_c$ for the centre of mass and shall consider that for the electronic part taking the general form

$$-\frac{h^2}{8\pi^2\mu}\left(\frac{\partial^2}{\partial x^2}+\frac{\partial^2}{\partial y^2}+\frac{\partial^2}{\partial z^2}\right)\Psi_e+V_e\Psi_e=H\Psi=E\Psi$$

or, $$-\frac{h^2}{8\pi^2\mu}\left(\frac{\partial^2}{\partial x^2}+\frac{\partial^2}{\partial y^2}+\frac{\partial^2}{\partial z^2}\right)\Psi_e-\frac{Ze^2}{\sqrt{x^2+y^2+z^2}}=E\Psi .$$

Now as the potential energy is the coulombic force between the nucleus and the electron that extends spherically symmetrical, hence spherical polar co-ordinates are to be introduced in terms of x, y and z.

$$x_2-x_1=x=r\sin\theta\cos\phi$$

$$y_2-y_1=y=r\sin\theta\sin\phi$$

$$z_2-z_1=z=r\cos\theta$$

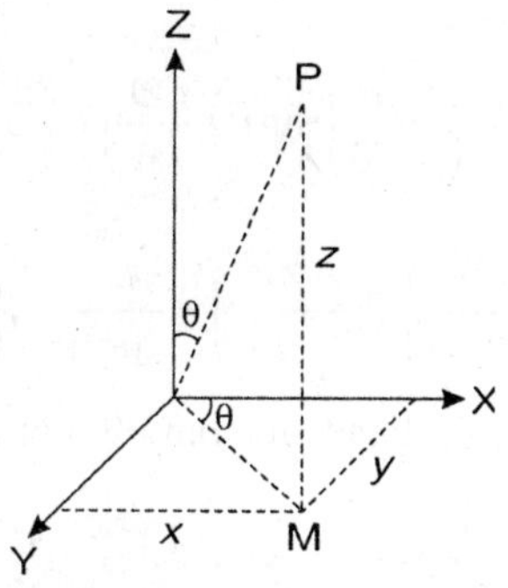

Fig 2.2 Relationship between Cartesian co-ordinates, x, y, z and the spherical polar co-ordinates r θ, ϕ ($x = r\sin\theta\cos\phi$; $y = r\sin\theta\sin\phi$; $z = r\cos\theta$).

And the Schrödinger equation in spherical polar co-ordinates takes the form as

$$\frac{1}{r^2}\times\frac{\partial}{\partial r}\left(r^2\times\frac{\partial\Psi}{\partial r}\right)+\frac{1}{r^2\sin\theta}\times\frac{\partial}{\partial\theta}\left(\sin\theta\times\frac{\partial\Psi}{\partial\theta}\right)+\frac{1}{r^2\sin^2\theta}\times\frac{\partial^2\Psi}{\partial\phi^2}+\frac{8\pi^2\mu}{h^2}\times\left(E+\frac{Ze^2}{r}\right)\Psi=0$$

Here Ψ is a function of r, θ, ϕ. To solve this equation the first step is to separate the variables. Let Ψ (r, θ and ϕ) is the product of three functions $R(r)$, Θ(θ) and Φ(Φ). So we can write

$$\Psi(r,\theta,\phi)=R(r)\times\Theta(\theta)\times\Phi(\phi)\approx R\Theta\Phi$$

Substituting this to the equation we get

$$\frac{1}{r^2}\times\frac{\partial}{\partial r}\left(\frac{r^2R\Theta\Phi}{\partial r}\right)+\frac{1}{r^2\sin\theta}\times\frac{\partial}{\partial\theta}\left(\sin\theta\times\frac{\partial R\Theta\Phi}{\partial\theta}\right)+\frac{1}{r^2\sin^2\theta}\times\frac{\partial^2}{\partial\phi^2}(R\Theta\Phi)+\frac{8\pi^2\mu}{h^2}\left(E+\frac{Ze^2}{r}\right)R\Theta\Phi=0$$

Taking out the constant terms and multiplying with $\frac{r^2\sin^2\theta}{R\Theta\Phi}$ we get

$$\frac{\sin^2\theta}{R}\times\frac{\partial}{\partial r}\left(r^2\frac{\partial R}{\partial r}\right)+\frac{\sin\theta}{\Theta}\times\frac{\partial}{\partial\theta}\left(\sin\theta\times\frac{d\Theta}{d\theta}\right)+\frac{1}{\Phi}\times\frac{\partial^2\Phi}{\partial\phi^2}+\frac{8\pi^2\mu}{h^2}\left(E+\frac{Ze^2}{r}\right)r^2\sin^2\theta=0$$

or, $$\frac{\sin^2\theta}{R}\times\frac{\partial}{\partial r}\left(r^2\times\frac{\partial R}{\partial r}\right)+\frac{\sin\theta}{\Theta}\times\frac{\partial}{\partial\theta}\left(\sin\theta\times\frac{\partial\Theta}{\partial\theta}\right)+\frac{8\pi^2\mu}{h^2}\left(E+\frac{Ze^2}{r}\right)\times r^2\sin^2\theta=-\frac{1}{\Phi}\times\left(\frac{\partial^2\Phi}{\partial\phi^2}\right)$$

The left side of this equation contains only R and θ variables and the right side contains only ϕ. As each function is independent of each other, so sum of R and Θ part is equal to Φ part. Let each side be equal to $-m^2$, then

$$-\frac{1}{\Phi}\times\frac{\partial^2\Phi}{\partial\phi^2}=m^2 \quad\text{or,}\quad \frac{\partial^2\Phi}{\partial\phi^2}+m^2\Phi=0 \tag{2.1}$$

Inserting $-m^2$ to the original equation we get

$$\frac{\sin^2\theta}{R}\times\frac{\partial}{\partial r}\left(r^2\times\frac{\partial R}{\partial r}\right)+\frac{\sin\theta}{\Theta}\times\frac{\partial}{\partial\theta}\left(\sin\theta\times\frac{\partial\Theta}{\partial\theta}\right)+\frac{8\pi^2\mu r^2\sin^2\theta}{h^2}\left(E+\frac{Ze^2}{r}\right)-m^2=0.$$

Dividing by $\sin^2\theta$ we get

$$\frac{1}{R}\times\frac{\partial}{\partial r}\left(r^2\times\frac{\partial R}{\partial r}\right)+\frac{1}{\Theta\sin\theta}\times\frac{\partial}{\partial\theta}\left(\sin\theta\times\frac{\partial\Theta}{\partial\theta}\right)+\frac{8\pi^2\mu r^2}{h^2}\left(E+\frac{Ze^2}{r}\right)-\frac{m^2}{\sin^2\theta}=0$$

or, $$\frac{1}{R}\times\frac{\partial}{\partial r}\left(r^2\times\frac{\partial R}{\partial r}\right)+\frac{8\pi^2\mu r^2}{h^2}\left(E+\frac{Ze^2}{r}\right)=\frac{m^2}{\sin^2\theta}-\frac{1}{\Theta\sin\theta}\times\frac{\partial}{\partial\theta}\left(\sin\theta\times\frac{\partial\Theta}{\partial\theta}\right)$$

The L.S. contains only R and right-hand side only θ. For satisfactory value each side must be equal to some constant β (say).

So, $$\frac{1}{R}\times\frac{\partial}{\partial r}\left(r^2\frac{\partial R}{\partial r}\right)+\frac{8\pi^2\mu r^2}{h^2}\left(E+\frac{Ze^2}{r}\right)=\beta$$

or, $$\frac{1}{r^2}\times\frac{\partial}{\partial r}\left(r^2\frac{\partial R}{\partial r}\right)+\frac{8\pi^2\mu R}{h^2}\left(E+\frac{Ze^2}{r}\right)-\frac{\beta R}{r^2}=0 \quad \left[\text{multiplying by } \frac{R}{r^2}\right] \tag{2.2}$$

and $$\frac{1}{\sin\theta}\times\frac{\partial}{\partial\theta}\left(\sin\theta\times\frac{\partial\Theta}{\partial\theta}\right)+\left(\beta-\frac{m^2}{\sin^2\theta}\right)\Theta=0. \tag{2.3}$$

Thus all variables have been separated and their solutions can be done.

2.04(a) SOLUTION OF Φ EQUATION

The ϕ equation is $\frac{\partial^2\Phi}{\partial\phi^2}+m^2\Phi=0$ which is the wave equation of a particle in a box and its solution is $\phi=Ae^{im\phi}+Be^{-im\phi}$ or, $\phi=Ae^{\pm im\phi}$ [in case of hydrogen atom], where A and B are constants. In order to be an acceptable wave function, ϕ must be a single-valued function, i.e., must have same value at $\phi=0$ and $\phi=2\pi$, i.e., ϕ will be periodic with 2π.

When $\phi=0$, then $\phi=A$ and for $\phi=2\pi$, $\phi=Ae^{\pm im2\pi}$.

Since the two functions are identical, $A=Ae^{\pm im2\pi}$

or, $$e^{+im2\pi}=e^{-im2\pi}=1$$

or, $$e^{im2\pi}+e^{-im2\pi}=2$$

or, $$\frac{e^{im2\pi}+e^{-im2\pi}}{2}=1.$$

The value of the constant A has been chosen and its value can be determined by considering normalization condition.

i.e., $$\int\phi\phi^*d\phi=1$$

or, $$\int_0^{2\pi}A\,e^{\pm im2\pi}A\,e^{\mp im2\pi}\,d\phi=1=A^2\int_0^{2\pi}d\phi=2\pi A^2$$

or, $$A=\frac{1}{\sqrt{2\pi}},\quad \text{so,}\quad \Phi=\frac{1}{\sqrt{2\pi}}e^{im\phi},\quad \text{where } m=0,\pm1,\pm2, \text{etc.}$$

Thus solution of ϕ equation introduces an integer m in its solution, which are identical to magnetic quantum numbers in Bohr-Sommerfeld model. Here the wave function contains imaginary quantity.

If we want a real solution, then we can write $\phi = C \cos m$ or $C \sin m$, where C is another normalization constant different from that considered above.

$$\therefore\ 1 = \int_0^{2\pi} \phi^2 d\phi = \frac{C^2}{2}\int_0^{2\pi} 2\cos^2 m\phi = \frac{C^2}{2}\int_0^{2\pi}(1+\cos 2m\phi)d\phi$$

$$= \frac{C^2}{2}\int_0^{2\pi} d\phi + \frac{C^2}{2}\int_0^{2\pi}\cos 2m\phi\, d\phi$$

$$= \frac{C^2}{2}\times 2\pi + \frac{C^2}{2}\left[\frac{\sin m\, 2\phi}{2m}\right]_0^{2\pi} = C^2\pi + 0 = C^2\pi$$

$$\therefore\ C^2 = \frac{1}{\pi} \quad \text{or,}\ C = \frac{1}{\sqrt{\pi}}.$$

So, $\Phi = \frac{1}{\sqrt{\pi}}\cos m\phi$, where $m = 0, \pm 1, \pm 2, \ldots$, etc.

2.04(b) SOLUTION OF θ FUNCTION

The solutions of Θ and R are not so easy. Θ function can be solved by application of Legendre's equation.

Now, $$\frac{1}{\sin\theta}\times\frac{\partial}{\partial\theta}\left(\sin\theta\times\frac{\partial\Theta}{\partial\theta}\right)+\left(\beta-\frac{m^2}{\sin^2\theta}\right)\Theta = 0.$$

Let, $\cos\theta = z$, then $\frac{d\Theta}{d\theta} = \frac{d\Theta}{dz}\times\frac{dz}{d\theta} = \frac{d\theta}{dz}\times -\sin\theta.$ [Hence $\Theta(\theta) \equiv P(z)$]

So, $$\frac{\partial}{\partial\theta} = -\sin\theta\frac{\partial}{\partial z} \quad \text{or,}\quad \sin\theta\frac{\partial}{\partial\theta} = -\sin^2\theta\frac{\partial}{\partial z}$$

Substituting this to the Θ function

$$\frac{1}{\sin\theta}\times -\sin\theta\frac{\partial}{\partial z}\left(-\sin^2\theta\frac{\partial P}{\partial z}\right)+\left(\beta-\frac{m^2}{1-z^2}\right)P = 0$$

or $$\frac{\partial}{\partial z}\left[\left(1-z^2\right)\times\frac{\partial P}{\partial z}\right]+\left(\beta-\frac{m^2}{1-z^2}\right)P = 0 \tag{2.4}$$

or $$\left(1-z^2\right)\times\frac{\partial^2 P}{\partial z^2} - 2z\times\frac{\partial P}{\partial z}+\left(\beta-\frac{m^2}{1-z^2}\right)P = 0 \tag{2.5}$$

These equations (2.4) and (2.5) both are known as Legendre's Equation. Here as $z = \cos\theta$, this has the significance only for values for z between $+1$ and -1.

Using Recursion Formula for the coefficient of power series this Legendre's equation can be solved. We get

$$\theta = \sqrt{\frac{2l+1}{2} \times \frac{(l-m)!}{(l+m)!}} \times \left(1-z^2\right)^{m/2} \frac{\partial^{l+m}}{\partial z^{l+m}} \left(z^2-1\right)^l,$$

where l is the integer having value 0, 1, 2, 3, ... and $\beta = l(l+1)$ which are equivalent to azimuthal quantum number of Bohr-Sommerfeld model of hydrogen atom and $z = \cos\theta$. Therefore, $\Psi(\theta, \phi)$ becomes

$$\Psi(\theta, \phi) = \sqrt{\frac{2l+1}{2} \times \frac{(l-m)!}{(l+m)!}} \times \left(1-z^2\right)^{m/2} \frac{\partial^{l+m}}{\partial z^{l+m}} \times \left(z^2-1\right)^l \times \frac{1}{\sqrt{2\pi}} e^{\pm im\phi}$$

2.04(c) SOLUTION FOR RADIAL EQUATION

The radial equation of the Schrödinger equation is

$$\frac{1}{r^2} \times \frac{\partial}{\partial r}\left(r^2 \frac{\partial R}{\partial r}\right) + \frac{8\pi^2 \mu R}{h^2}\left(E + \frac{Ze^2}{r}\right) - \frac{\beta R}{r^2} = 0$$

or,
$$\frac{1}{r^2} \times \frac{\partial}{\partial r}\left(r^2 \frac{\partial R}{\partial r}\right) - \frac{l(l+1)R}{r^2} + \frac{8\pi^2 \mu ER}{h^2} + \frac{8\pi^2 \mu Ze^2 R}{h^2 r} = 0$$

or,
$$\frac{\partial^2 R}{\partial r^2} + \frac{2}{r} \times \frac{\partial R}{\partial r} - \frac{l(l+1)R}{r^2} + \frac{8\pi^2 \mu ER}{h^2} + \frac{8\pi^2 \mu Ze^2 R}{h^2 r} = 0.$$

According to classical mechanics E is negative, i.e., $E < 0$, i.e., electronic orbit is elliptical and when $E > 0$, the orbit will be hyperbolic and unbounded states. In bounded case as in hydrogen atom let us put

$$\frac{8\pi^2 \mu E}{h^2} = -\alpha^2 \text{ and } \frac{4\pi^2 \mu Ze^2}{h^2} = \lambda,$$

the equation takes the form $\frac{\partial^2 R}{\partial r^2} + \frac{2}{r} \times \frac{\partial R}{\partial r} + \left[-\frac{l(l+1)}{r^2} - \alpha^2 + \frac{2\pi\alpha}{r}\right] R = 0.$

Now put another variable $\rho = 2\alpha r$.

$$\therefore \quad \frac{\partial R}{\partial r} = \frac{\partial R}{\partial \rho} \times \frac{\partial \rho}{\partial r} = 2\alpha \times \frac{\partial R}{\partial \rho}.$$

$$\therefore \quad \frac{\partial^2 R}{\partial r^2} = \frac{\partial}{\partial r} \times \left(\frac{\partial R}{\partial r}\right) = \frac{\partial}{\partial r}\left(\frac{\partial R}{\partial \rho} \times \frac{\partial \rho}{\partial r}\right) = \frac{\partial}{\partial \rho}\left(2\alpha \frac{\partial R}{\partial \rho}\right) = 4\alpha^2 \frac{\partial^2 R}{\partial \rho^2}.$$

Hence the above equation becomes (dividing by $4\alpha^2$)

$$\frac{\partial^2 R}{\partial \rho^2} + \frac{2}{\rho} \times \frac{\partial R}{\partial \rho} + \left[-\frac{l(l+1)}{\rho^2} + \frac{\lambda}{\rho} - \frac{1}{4}\right] R = 0.$$

Now ρ or r may vary from 0 to ∞ , when $\rho \to \infty$, then

$$\frac{\partial^2 R}{\partial \rho^2} - \frac{1}{4} R = 0. \quad \text{So } R = e^{\pm\frac{\rho}{2}}.$$

When ρ increases, $e^{\pm\frac{\rho}{2}}$ will increase and so wave function will be unacceptable, but when ρ or r increases R will decrease, so it is acceptable and $R(\rho) = A \times e^{-\frac{\rho}{2}}$.

Using recursion formula Laguerre Polynomial can be solved as

$$R = \sqrt{\left(\frac{2Z}{na_0}\right)^3 \frac{(n-l-1)!}{2n\{(n+l)!\}^3}}\, e^{\frac{-Zr}{na_0}} \left(\frac{2Zr}{na_0}\right)^l \times \frac{\partial^{2l+1}}{\partial \rho^{2l+1}} \left[e^{\rho} \frac{\partial^{n+l}}{\partial \rho^{\,n+l}} \left(\rho^{n+l} e^{-\rho}\right) \right]$$

[a_0 is the first Bohr orbit]

2.04(d) THE TOTAL WAVE FUNCTION $\Psi = \Phi\Theta R$.

$$\Psi = \frac{1}{\sqrt{2\pi}} e^{-m\phi i} \times \sqrt{\frac{2l+1}{2} \times \frac{(l-m)!}{(l+m)!}} \left(1-z^2\right)^{\frac{m}{2}} \frac{\partial^{l+m}}{\partial z^{l+m}} \left(z^2-1\right)^l$$

$$\times \sqrt{\left(\frac{2Z}{na_o}\right)^3 \frac{(n-l-1)!}{2n\{(n+l)!\}^3}}\, e^{\frac{-Zr}{na0}} \left(\frac{2Zr}{na_0}\right)^l \times \frac{\partial^{2l+1}}{\partial \rho^{2l+1}} \left[e^{\rho} \frac{\partial^{n+l}}{\partial \rho^{n+l}} \left(\rho^{n+l}\, e^{-\rho}\right) \right]$$

The value of ρ *from Bohr's orbit*:

We knew $\rho = 2\alpha r$.

$$\therefore \ \rho^2 = 4\left(\frac{-8\pi^2 mE}{h^2}\right) r^2 = 4 \frac{-8\pi^2 m}{h^2} \times \frac{-2\pi^2 z^2 e^4 m}{n^2 h^2} \times r^2$$

$$= \frac{64\pi^4 m^2 z^2 e^4 r^2}{n^2 h^4}, \qquad \therefore \ \rho - \frac{8\pi^2 mze^2 r}{nh^2}.$$

If a_0 is the Bohr's first orbit, then $a_0 = \dfrac{h^2}{4\pi^2 me^2}$; $\quad \therefore \ \rho = \dfrac{2Zr}{na_0}$.

Atomic energy levels and quantum numbers:

We assumed $\dfrac{8\pi^2 \mu E}{h^2} = -\alpha^2$ and $\dfrac{4\pi^2 \mu Ze^2}{\alpha h^2} = \lambda = n$.

$$\therefore \ \alpha = \frac{4\pi^2 \mu Ze^2}{h^2 n} \quad \text{or, } \alpha^2 = \frac{16\pi^4 \mu^2 z^2 e^4}{h^4 n^2}$$

$$\text{So, } \frac{16\pi^4 \mu^2 z^2 e^4}{n^2 h^4} = -\frac{8\pi^2 \mu E}{h^2} \quad \text{or, } E = -\frac{16\pi^4 \mu^2 z^2 e^4 \times h^2}{n^2 h^4 \times 8\pi^2 \mu} \quad \text{or, } E = -\frac{2\pi^2 z^2 e^4 \mu}{n^2 h^2}.$$

This result is the same as that obtained from Bohr's theory of the energy of an electron based on some postulates concerning the quantisation of angular momentum. From wave mechanics it is seen that acceptable solutions of wave equation lead to the permitted energy levels which are the same as those derived from Bohr's theory. The integral value of n, the principal quantum number of Bohr is the necessary condition for the acceptable solution.

2.05 WAVE FUNCTION FOR DIFFERENT ENERGY STATES OF AN ELECTRON

From the equation for $\Psi(\Phi, \Theta, R)$ we can arrive at the wave functions of different electrons in different orbitals (spin not considered).

$$\Psi(\Phi,\Theta,R)=\frac{1}{\sqrt{2\pi}}e^{-im\phi}\times\sqrt{\frac{2l+1}{2}\frac{(l-m)!}{(l+m)!}}\left(1-z^2\right)^{\frac{m}{2}}\frac{\partial^{l+m}}{\partial Z^{l+m}}\left(z^2-1\right)^l$$

$$\times\sqrt{\left(\frac{2Z}{na_0}\right)^3\frac{(n-l-1)!}{2n\{(n+l)!\}^3}}\,e^{\frac{-Zr}{na_0}}\left(\frac{2Zr}{na_0}\right)^l\times\frac{\partial^{2l+1}}{\partial\rho^{2l+1}}\left[e^{\rho}\frac{\partial^{n+l}}{\partial_{\rho^{n+l}}}\left(\rho^{n+l}e^{-\rho}\right)\right].$$

$$\Psi_{100}=\frac{1}{\sqrt{2\pi}}e^{-i0\phi}\sqrt{\frac{20+1}{2}\frac{(0-0)!}{(0+0)!}}\left(\sin^2\theta\right)^{0/2}\frac{\partial^{0+0}}{\partial Z^{0+0}}\left(\cos^2\theta-1\right)^0$$

$$\times\sqrt{\left(\frac{2Z}{1a_0}\right)^3\frac{(1-0-1)}{2\times1\{(1+0)!\}^3}}\,e^{\frac{-Zr}{1\times a_0}}\left(\frac{2Zr}{a_0}\right)^0\times\frac{\partial^{2\times0+1}}{\partial_{\rho^{2\cdot0+1}}}\left[e^{\rho}\frac{\partial^{1+0}}{\partial_{\rho^{1+0}}}\left(\rho^{1+0}.e^{-\rho}\right)\right]$$

or,
$$\Psi_{100}=\frac{1}{\sqrt{2\pi}}\times\frac{1}{\sqrt{2}}\times\sqrt{\left(\frac{2Z}{a_0}\right)^3}\,e^{\frac{-Zr}{a_0}}\frac{\partial}{\partial\rho}\left[e^{\rho}\times\frac{\partial}{\partial\rho}\left(\rho\times e^{-\rho}\right)\right]$$

$$=\frac{1}{\sqrt{2\pi}}\times\frac{1}{\sqrt{2}}\left(\frac{2Z}{a_0}\right)^{\frac{3}{2}}e^{\frac{-Zr}{a_0}}\frac{\partial}{\partial\rho}\left[e^{\rho}\left\{e^{-\rho}+\rho\times-\rho e^{-\rho}\right\}\right]$$

$$=\frac{1}{\sqrt{2\pi}}\times\frac{1}{\sqrt{2}}\left(\frac{2Z}{a_0}\right)^{\frac{3}{2}}e^{\frac{-Zr}{a_0}}\frac{\partial}{\partial\rho}\left[e^{\rho}\times e^{-\rho}-\rho^2e^{\rho-\rho}\right]$$

$$=\frac{1}{\sqrt{2\pi}}\times\frac{1}{\sqrt{2}}\left(\frac{2Z}{a_0}\right)^{\frac{3}{2}}e^{\frac{-Zr}{a_0}}\frac{\partial}{\partial\rho}\left(1-\rho^2\right)=\frac{1}{\sqrt{2\pi}}\times\frac{1}{\sqrt{2}}\left(\frac{2Z}{a_0}\right)^{\frac{3}{2}}e^{\frac{-Zr}{a_0}}\times-2\rho$$

$$=\frac{1}{\sqrt{\pi}}\left(\frac{2Z}{a_0}\right)^{\frac{3}{2}}\rho\times e^{\frac{-Zr}{a_0}}.$$

2.06 RADIAL DISTRIBUTION FUNCTION AND SHAPES OF ORBITALS

The wave functions for electrons of different orbitals obtained by putting the value of n, l and m are as follows:

$$\Psi_{100}=\frac{1}{\sqrt{\pi}}\left(\frac{Z}{a_0}\right)^{\frac{3}{2}}e^{\frac{-Zr}{a_0}},\quad\Psi_{200}=\frac{1}{4\sqrt{2\pi}}\left(\frac{Z}{a_0}\right)^{\frac{3}{2}}\left(2-\frac{Zr}{a_0}\right)e^{\frac{-Zr}{2a_0}},$$

$$\Psi_{210}=\frac{1}{4\sqrt{2\pi}}\left(\frac{Z}{a_0}\right)^{\frac{5}{2}}e^{\frac{-Zr}{2a_0}}r\cos\theta.$$

$$\Psi_{21+1} = \frac{1}{4\sqrt{2\pi}}\left(\frac{Z}{a_0}\right)^{\frac{5}{2}} e^{\frac{-Zr}{2a0}} r\sin\theta\cos\phi \quad \Psi_{21-1} = \frac{1}{4\sqrt{2\pi}}\left(\frac{Z}{a_0}\right)^{\frac{5}{2}} e^{\frac{-Zr}{2a0}} r\sin\theta\sin\phi$$

The radial part of these wave functions is

$$R_{100} = \frac{2}{(a_0)^{\frac{3}{2}}} e^{\frac{-Zr}{a0}}, \qquad R_{200} = \frac{1}{2\sqrt{2}(a_0)^{\frac{3}{2}}}\left(2 - \frac{r}{a_0}\right) e^{\frac{-Zr}{2a0}},$$

$$R_{210} = \frac{1}{2\sqrt{6}\, a_0^{\frac{5}{2}}} \times r \times e^{\frac{-Zr}{2a_0}}.$$

The squares of these radial functions are the probability of finding the electron in a small or unit length at a distance r from the nucleus, i.e., probability distribution function. In view of classical mechanics thus R^2 must be zero except at the point of location of electronic shell. But in wave mechanics we consider the probability of finding in all places. So instead of considering R^2 we consider $4\pi r^2 R^2$ which considers the probability in a small spherical shell at a distance r. If we plot $4\pi r^2 R^2$ against r, we get the following curves showing one maxima at the distance representing Bohr's orbit. From these plots, it is seen that probability of finding $\bar{e}$ electron at the nucleus and at large distance is zero. Comparison of $2s$ and $2p$ with increase in value of l, most probable radius decreases, and at the nodes except $1s$, $4\pi r^2 R^2$ is zero.

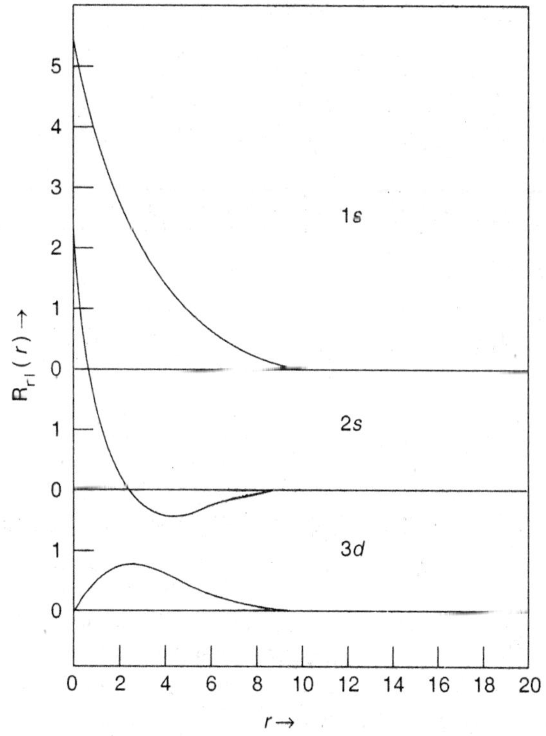

Fig 2.3 Radial wave functions for the $1s$, $2s$ and $3d$ states of the hydrogen atom.

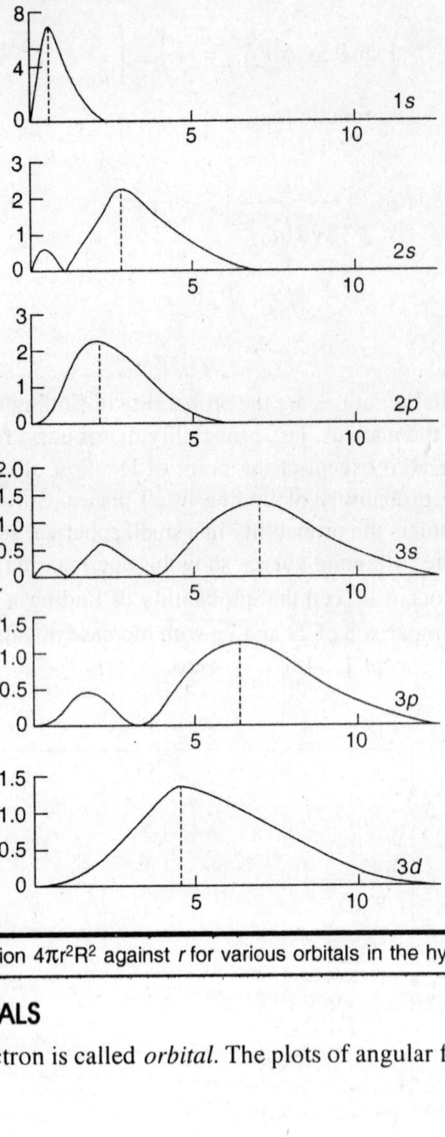

Fig 2.4 Radial distribution function $4\pi r^2 R^2$ against r for various orbitals in the hydrogen atom.

2.07 SHAPES OF ORBITALS

The wave function of an electron is called *orbital.* The plots of angular function θ and ϕ determine the geometry of the orbital.

s orbital

Ψ_{100}, Ψ_{200} are termed as $1s$ and $2s$ orbitals. These wave functions are independent of θ and ϕ but depends only on R. So s orbitals are spherically symmetrical and their projection in a plane gives circular figure and the charge distribution is symmetrical.

p orbital

All these orbitals have n as an integer and $l = 1$, $m = 0, \pm 1$. Thus for three possible values for m, there are three p-orbitals. For second quantum level $n = 2$, the wave functions are Ψ_{200}, Ψ_{210}, Ψ_{211}, $\Psi_{2,1,-1}$—all have equal radial part but different angular functions. Ψ_{210} (polar part $r\cos\theta$) Ψ_{211} ($r\sin\theta\cos\phi$) and $\Psi_{2,1,-1}$ ($r\sin\theta\sin\phi$) are termed as $2p_z$, along z-axis of correlation co-ordinate, $2p_x$ along x-axis and $2p_y$ along y-axis. p_z^2 will give the electron density along z-axis and it will have no probability along x- or y-axis. Hence, if we plot Ψp_z^2 against r for a complete rotation of ϕ, we shall get cosine curve and assume the shape of a dumb-bell with two lobes.

Similar is the case with p_y and p_z. In between the two lobes, at the nodal plane, probability of electron is vanishingly small. The two lobes are symmetrical about the nodal plane, one lobe is marked (+)ve and the other (–)ve symmetry of the angular functions, not the charge. Thus the *p*-electrons tend to concentrate along certain preferred directions, which are mutually perpendicular to each other. Maximum concentration can be obtained if sinϕ, sinθ, cosϕ, cosθ are made unity. If this is done, then relative electron density of *p*-orbital becomes √3 compared to s-orbital as one. Hence involvement of *p*-orbital forms more and more stable covalent bond.

Shapes of *d* orbitals

Here *n* is an integer, $l = 2$ and $m = 0, \pm1, \pm2$, i.e., *d*-orbitals are five degenerate states. All these orbitals have equal radial part, angular parts differ.

$$R_{3.2} = \frac{1}{81\sqrt{6\pi}\, a_0^{3/2}} \left(\frac{r}{a_0}\right)^2 e^{\frac{-r}{3a_0}} \qquad \Psi_{3.2.0} = 3d_{z^2} = R\left(3\cos^2\theta - i\right)$$

$$\Psi_{3,2,+2} = d_{x^2-y^2} = R \times \frac{1}{2} \times \sin^2\theta\, e^{-2i\phi} \qquad \Psi_{3,2,-2} = R \times \frac{1}{2} \times \sin^2\theta\, e^{2i\phi} = 3d_{xy}$$

$$\Psi_{3,2,1} = 3d_{xz} = R \times \sin\theta \cos\theta\, e^{i\phi} \qquad \Psi_{3,2,-1} = 3d_{zy} = R \sin\theta \cos\theta\; e^{i\phi}$$

In absence of field all these orbital wave functions when plotted against *r* have four equal lobes, but in presence of weak field, $d_{x^2-y^2}$, d_{xy}, d_{zx} and d_{yz} orbitals remain unaffected but d_{z^2} has two lobes aligned with the direction of applied field and the other two transverse to it, the later two are much affected. The aligned lobes are swollen up at the expense of the other two. The diagrammatic representations of these orbitals are as follows:

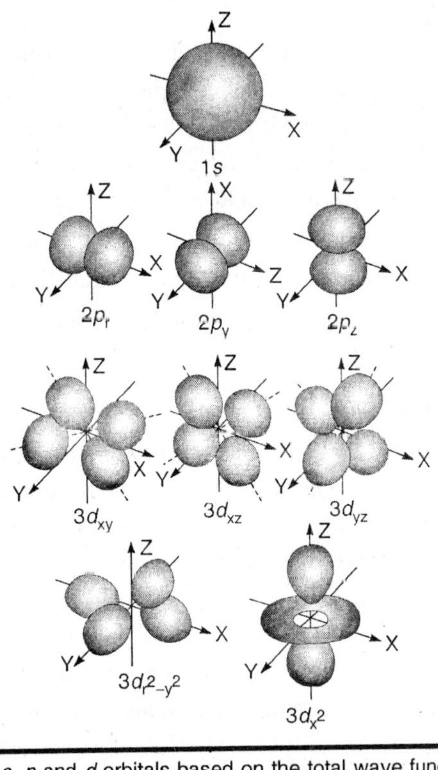

Fig 2.5 Conventional shapes for *s*, *p* and *d* orbitals based on the total wave function $\Psi_{n,\,l,\,m}$.

CHAPTER 3

NUCLEAR PROPERTY AND COMPOSITION

3.01 RADIOACTIVITY

What is radioactivity?

Radioactivity is a spontaneous, self-disruptive activity of natural elements of atomic weight greater than 206 with the emission of an invisible and complex type of powerful radiation as a result of continuous break-down of elements in an irreversible self-disintegration. This activity is a nuclear property and it is due to some intrinsic natural cause unaffected by any external agencies, like temperature, pressure, electric or magnetic field or any physical or chemical combination and it is not instantaneous but prolonged. The radiation consists of α- and β-particles and γ-rays.

Radioactivity is of two types: the natural and induced or artificial radioactivity. The former one is exhibited by natural elements of atomic weight greater than 206, and the latter one is exhibited by elements of lower atomic weights by bombardment with some projectile, like $\alpha, p, {}_0^1n, {}_1^2D,$ etc.

3.02 DISCOVERY OF RADIOACTIVITY

In 1896, Henri Becquerel studied the effect of X-rays on double salt of potassium uranyl sulphate and also the phosphorescence of various salts. He noticed that on exposure to light the above salt emitted radiation even in absence of light and exposed a photographic plate even when it was wrapped with black paper or thin layer of any opaque material, like aluminium. Becquerel continued his work with different salts of uranium in solid state or in solution and noticed the emission of same radiation indicating emission is independent of the physical or chemical state of the substance and also independent of temperature, pressure, etc. The radiations were first termed as Becquerel Rays and later changed to radioactive radiation to indicate all such rays emitted by several other substances discovered later.

In 1898, Marie and Pierre Curie observed the same phenomenon with thorium and found that the activity depended on the amount of the metal present in the sample. Again, they observed that Pitchblende, the natural source of uranium had activity many times more than the pure uranium salt. From Pitchblende, they separated a new element, Polonium (named after the native country Poland of Madame Curie) by co-precipitation with Bi_2S_3 and followed by sublimation in vacuum at 700°C. They again successfully separated another new element Radium from Pitchblende by co-precipitation with $BaSO_4$ and then fractional crystallisation of their halides. In 1898, Schmidt separated thorium. In 1900, Dobierne, following the work of Curie, discovered another element Actinium. Several scientists discovered different transuranic radioactive elements afterwards and at present more than 40 such radioactive elements are known in nature.

3.03 THE NATURE OF RADIOACTIVE RADIATIONS

Lord Rutherford in 1899–1900 investigated the penetrating power and ionisation power of radiation. He used thin tin foils as absorber and gold-leaf electroscope for ionisation power and showed two distinct types of radiation. One having intense ionisation and good absorbance power, called *α-rays* and the other having much more, about 100 times penetrating power, and much less ionising power, called *β-rays*, both of which affected the photographic plate. Villard later discovered another type of radiation of still higher penetrating power, called *γ-rays*.

Characterisation of these radiations by subsequent studies are tabulated below:

Properties	α-ray	β-ray	γ-ray
1. Penetrating power	less penetrating	more penetrating	most penetrating
2. Magnetic field	deflected less	deflected more in the opposite direction	no deflection
3. Nature	doubly charged helium nucleus $^{4}_{2}He^{+2}$	highly energetic electron	electromagnetic radiation of high frequency $\lambda = 10^{-8} - 10^{-11}$ cm
4. Charge	double positive charge, 3.2×10^{-19} coulomb	negative charge same as that of electron, 1.6×10^{-19} coulomb	uncharged
5. $\frac{e}{m}$ ratio	4.787×10^{7} coulomb/kg	same as that of electron 127×10^{11}coulomb/kg	nil
6. Mass	6.6×10^{-27} kg	$\sim10^{-31}$ kg	nil
7. Ionising power	very high	low	nil
8. Speed	2×10^{7} m/sec	2×10^{8} m/sec, but may vary	3×10^{7}m/sec

α-ray is a doubly charged helium nucleus He^{+2} shown by Rutherford and Royds in 1909. A thin-walled glass tube *A* is fitted in a wider tube *B* that tapers to a discharge tube *C* at one end as in figure 3.1. *A* contains some Radon gas which emits α-particles that pass through thin glass tube and neutralized to helium gas and collected in tube *B*. After some days, the discharge tube gives the spectrum of helium. This proves that α-particles are nothing but nuclei of helium atom.

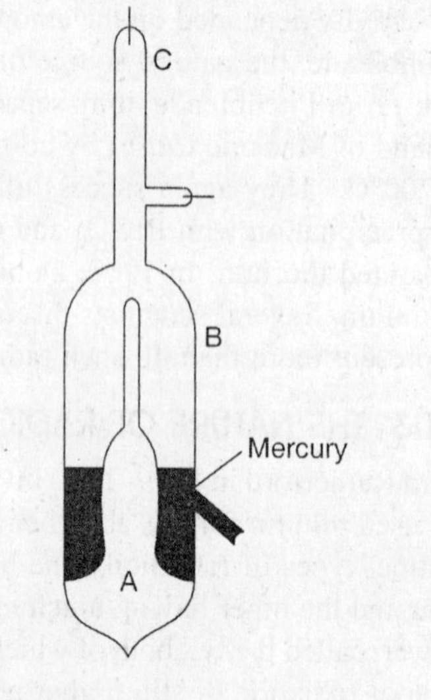

Fig 3.1 Design of the tube for studying the nature of a particle.

3.04 RADIOACTIVE DECAY CURVES

In 1900, William Crookes treated ordinary uranium salt with excess ammonium carbonate, $[(NH_4)_2CO_3]$. When the precipitate first formed, it dissolves mostly leaving a small amount of flocculent precipitate. The precipitate contains whole activity while the filtrate on evaporation, that contains the whole uranium, is inactive. On keeping for a long time, the precipitate gradually loses its activity while the inactive residue gradually gains the activity. The results can be graphically represented as follows in Fig 3.2(a):

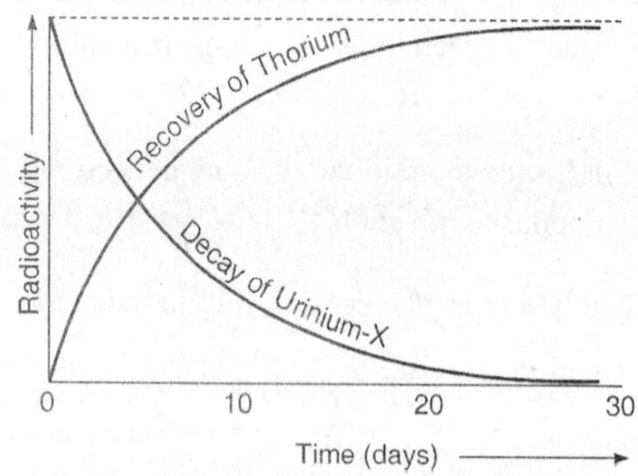

Fig 3.2(a) The nature of the decay-curve of U-X and growth-curve of U.

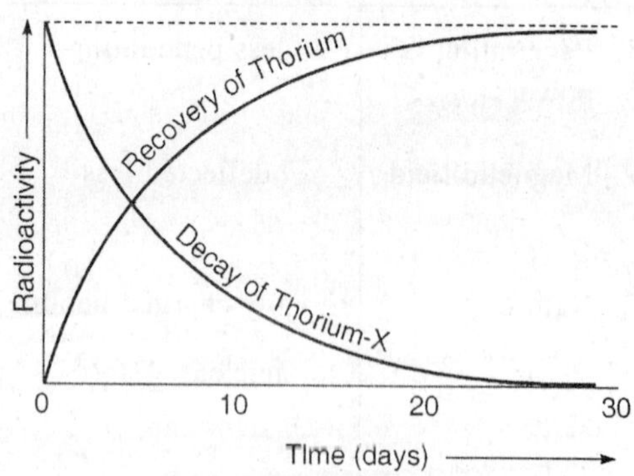

Fig 3.2(b) The nature of the decay-curve of Th-X and growth-curve of Th.

The precipitate is called U-X. Rutherford and Soddy in 1902 observed the similar phenomenon while treatment of thorium nitrate with NH_4OH, gives $Th(OH)_4$ precipitate which is inactive, but the filtrate on evaporation gives the residue, which is fully active, named as Th-X. The active Th-X loses its activity while the inactive $Th(OH)_4$ precipitate becomes active [Fig 3.2(b)]. The separation of active and inactive part can be repeated several times. The rate of loss or gain of activity is different for different radioactive elements (parent or daughter). The curves are hyperbolic or exponential in nature.

$$\text{Uranium salt} \xrightarrow{(NH_4)_2CO_3} \left[\begin{array}{l} ppt\,(active)\,after\,long\,time \rightarrow inactive \\ solution\,(inactive) \xrightarrow[\Delta]{} residue(inactive) \rightarrow active(after\,some\,time) \end{array} \right.$$

$$\text{Thorium salt} \xrightarrow{NH_4OH} \left[\begin{array}{l} ppt\,Th(OH)_4\,(inactive)\,after\,long\,time \rightarrow active \\ solution\,(active) \xrightarrow[\Delta]{} residue(active) \rightarrow inactive(after\,some\,time) \end{array} \right.$$

From these observations it can be concluded that activity of Uranium and Thorium are due to the presence of U-X and Th-X respectively in them. When these active species are separated, the original sample does not show any activity. But with time this U and Th regenerates U-X and Th-X, and these active species exhibit activity to the original salt. The active species again lose their activities and give rise to some new inactive elements or in turn may be active and will emit radiations. At the same time it was observed that thorium salts continuously emit an active gas, called *radioactive emanations*, which decay rapidly and induce activity to the surroundings. Radium or actinium was found to emit such emanations. The chemical and physical properties of these active and inactive masses were found to be different.

All these observations lead to the discovery of the radioactive disintegration theory that can be stated as:

Every radioactive element as a result of disintegration emits α-, β-particles producing a new element called *daughter element* which is chemically and physically different from that of the parent element. The daughter element may again emit in a similar manner to produce another new element and this process will go on until a final non-radioactive end product is reached. As the process goes on, the rate of formation of a daughter element becomes equal to the rate of its disintegration to form another new daughter element and as each has different rate of disintegration, the natural source of radioactive element shows a constant activity and we say, that radioactive equilibrium has reached. This can be presented as follows:

$$\text{Radio A} \xrightarrow{\alpha/\beta} \text{Radio B} \xrightarrow{\alpha/\beta} \text{Radio C} \xrightarrow{\alpha/\beta} \text{D (radiogenic inactive species)}$$

3.05 UNITS OF RADIOACTIVITY

1. Curie (C) (used since discovery) = 3.7×10^{10}dps.
2. Becquerel (Bq) = 1 dps. $\therefore$ $1C = 3.7\times10^{10}$ Bq.
3. Rutherford (Rd) = 10^6dps.
4. Roentgen (R) = It is the intensity of ionising radiation that produces 1 esu charge in 1cc dry air at S.T.P.

3.06 RADIOACTIVITY IS A NUCLEAR PHENOMENON

The radioactivity of an element is the intrinsic property of the element unaffected by any external cause, change of physical state or chemical combination. Moreover, we see that radioactive elements emit α-particle, which is ${}^4_2He^{+2}$, and outside nucleus nothing except electron is present in the accepted atom model, hence, emission of α-particle must have its origin in the nucleus of the radioactive element.

Radioactive elements emit β-particles, which are nothing but highly energetic electrons. For emission of extra nuclear electron requires very small amount of energy and the atom is ionised The ionised atom immediately takes up electron and reverts back to the original atom. Thus, it is reversible in nature. But β-ray emission by radioactive element is not of this type. β-ray emission is unaffected by external force and it is irreversible in nature producing other element of different physical and chemical properties. Thus, β-ray emission is a nuclear phenomenon.

γ-ray radiated by radioactive element has a very short wavelength electromagnetic radiation even shorter than X-rays, and radiated by extra nuclear electrons of deeply ionised atoms when revert to their normal state. Thus γ-ray emission is also a nuclear phenomenon.

3.07 COMPOSITION OF NUCLEUS

As radioactivity is a nuclear phenomenon, let us discuss the different theories regarding the composition of nucleus.

3.07(a) THE PROTON-ELECTRON THEORY

This theory was accepted before the discovery of neutron and gained support from the fact that the atom as a whole is electrically neutral and natural radioactive elements are β-emitter. According to this concept, the atom of mass *A* and atomic number *Z* would contain *Z* number of extra nuclear electron and *A* number of proton in the nucleus and as $Z \approx \frac{1}{2}A$, the excess positive charge is neutralized by the presence of (*A–Z*) number of electron in the nucleus. This led to the slight discrepancy of nuclear mass, which was attributed to mass defect.

This concept has the following defects:

(a) Electron can be considered to be a spherical charge of certain dimension. For heavier atoms, so large number of electrons cannot be fit into the comparatively small nucleus. Moreover, there will be every possibility to the combination of proton and electron to form a neutral particle, called *neutron*, as proposed by Rutherford in 1920.

(b) Second difficulty arises when wavelengths of electron, which is much larger than the nuclear radius, is considered.

(c) The third demerit of this concept arises when nuclear angular momentum or spin momentum is considered. To explain the very hyperfine structures in spectral lines, Pauli introduced the idea of nuclear spin. The total nuclear angular momentum $I = L + S$, where *L* and *S* are the orbital and spin angular momentum of the nucleus, like that of extra nuclear electrons. Calculation shows that for **odd** mass number nucleus, I = half-integral multiple of $\frac{h}{2\pi}$ and for **even** mass number I = integral multiples of $\frac{h}{2\pi}$. The nuclear angular momentum I of $^{14}_{7}N$, that contains $14\,^{1}_{1}p$ and 7 eletrons; total 21 particles; is not half-integral of $\frac{h}{2\pi}$.

(d) The fourth difficulty is observed when energy of the emitted β-particle is considered. If the electron is considered to be confined in a small volume of the nucleus, calculation shows its energy as 6×10^{-3} ergs, which is several times greater than the energy of emitted β-particle.

(e) Consideration of nuclear statistics precludes the existence of electron in the nucleus.

All these difficulties were solved when Chadwick in 1932 discovered neutron as a result of impact of α-ray bombardment with target nucleus. The proton-electron concept was discarded and new concept of proton-neutron was introduced.

3.07(b) THE PROTON-NEUTRON THEORY

This is the most accepted theory today. According to which nucleus contains *Z* number of protons and (*A–Z*) number of neutrons and *Z* number of extra nuclear electrons to balance the charge of the nucleus. The mass number *A* is the total mass of $^{1}_{1}p$ and $^{1}_{0}n$ which is an integer nearest to the actual mass of the nucleus and *Z* is the atomic number, which is the charge of the nucleus, i.e., number of $^{1}_{1}p$ in the nucleus. Here also slight discrepancy in the atomic mass is explained in terms of mass defect.

On the basis of this concept β emission and e^{+} emission by radioactive element is explained in this way:

The electron does not pre-exist in the nucleus but results from the conversion of $^{1}_{0}n$ to $^{1}_{1}p$ and likewise e^{+} emission by transformation of $^{1}_{1}p$ to $^{1}_{0}n$.

Thus, $_0^1n \underset{-e^+}{\overset{-\beta^-}{\rightleftharpoons}} {}_1^1p$.

Therefore, $_0^1n$ and $_1^1p$ are the manifestation of the same thing and together called *nucleons* and e^+ and β^- resonate between them. So, nucleus contains only one particle, nucleon, which is very close to Prout's (1815) old hypothesis that all elements are composed of the union of hydrogen and so atomic weight of all elements will be whole number (*called whole number rule*) taking hydrogen as unity.

The weakness of this concept is that how large number of $_0^1n$ and $_1^1p$ packed in small volume of the nucleus render stability in spite of strong $_1^1p - {}_1^1p$ repulsion and practically no $_0^1n - {}_0^1n$ attraction. Secondly, simply addition of e^+ or emission of β-particle, i.e., resonance of these particles between $_0^1n$ and $_1^1p$ cannot take place considering their relative masses. These weaknesses have been clearly explained when the nuclear forces are considered.

3.07(c) POSITRON-NEUTRON CONCEPT

Prof. Jean Perrin proposed this concept to solve the problem as to which of the two constituent particles, proton or neutron, would be considered as the primordial elementary particle. If proton is considered as the primordial particle, the neutron will be of complex nature containing proton and electron. As the mass of proton is less than that of neutron, the mass of stable nucleus will not fit the sum total mass of the constituent particles (as stable nucleus has less mass). If neutron is considered as the primordial particle, then proton becomes complex in nature being composed of neutron and positron. In that case, inertia of matter is not linked with electric charge and stable nucleus will have mass equal to the sum total mass of the neutron with some mass loss (equivalent to binding energy of the nucleus). Thus, neutron is the primordial particle. So, an atom of atomic weight A and atomic number Z, will have A number of $_0^1n$ and Z number of positron in the nucleus and Z number of extra nuclear electron to maintain neutrality. In the nucleus there will be a very close-range powerful attraction between neutron and positron, and it is quite natural as neutron contains a central positive core surrounded by a thin layer of negative charge.

But this concept will have same objections as like the proton-electron concept. Therefore, proton-neutron concept is accepted everywhere.

Thus, the nucleus of an atom contains only $_0^1n$ and $_1^1p$ in a small region of 10^{-36} cc (radius = 10^{-12} cm or $r = 10^{-15}$ m = 1 F, fermi).

$$\text{Density of Na nucleus} = \frac{M}{V} = \frac{\dfrac{23}{6.02\times10^{23}}}{\dfrac{4}{3}\times3.14\times\left(5\times10^{-12}\right)^3} \approx 1\times10^{14}\,\text{gm/cc}.$$

So, mass of the nucleus is too heavy that no container can withstand the mass if it is possible to isolate the nucleus.

3.08 MASS DEFECT AND PACKING FRACTION

In 1816, William Prout thought that hydrogen was prototype of all elements and all other elements were obtained as a union of hydrogen atoms. So taking hydrogen as unity, the mass of all elements will be an integer. This is known as Prout's *whole number rule*. This idea is not new because his eminent contemporary Humphry Davy stated that he proved, hydrogen might be actually present in many elements. Determination of accurate atomic weight by Aston shows that atomic weights of elements are not a whole number but slightly deviates from the whole number. Again, atomic weight of chlorine is fractional. All these observations temporarily discarded the whole number rule, but later it was found to be correct. The deviation of atomic mass from the nearest whole number was termed by Aston in 1927, as *mass defect.* Stability can be understood in terms of mass defect and binding energy. So, mass defect ΔM = actual mass of atom (M) – whole number (A).

$\therefore \Delta M = M - A$

ΔM may be positive or negative and is accordingly called *positive mass defect* and *negative mass defect*. Significant value being very small, is expressed in terms of packing fraction (f).

So, $f = \dfrac{M-A}{A} \times 10^4$.

The value is multiplied by 10^4 to make it easy to plot. It also expresses the average mass gain or mass loss per nucleon. In 1915, W. D. Harkins and E. D. Wilson referred this as packing effect, which is percentage deviation of atomic weight from the whole number. When packing fraction is plotted against mass number for stable nucleus, the curve obtained is shown below:

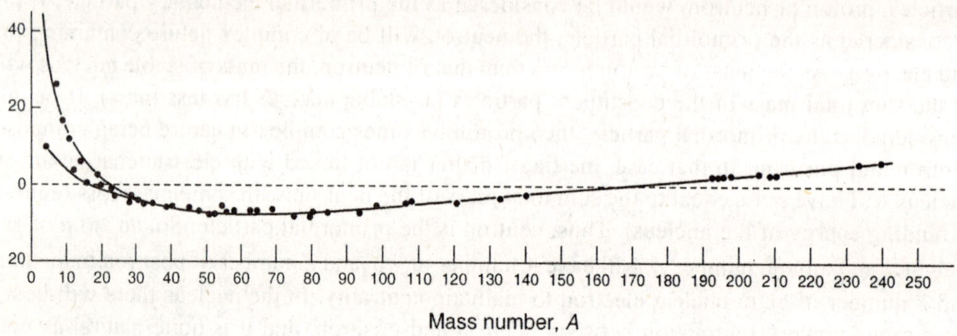

Fig 3.3 Packing fraction curve for most abundant isotopes.

The curve has four distinct portions:

(a) The portion for $A < 20$, here packing fraction is positive, means extra energy has been added to the nucleus to bind up the nucleons together and mass equivalent of this energy added to the nucleus increases their isotopic mass from their mass number. Hence, if this amount of energy is taken away, the nucleus will be unstable or rather these nuclei are to some extent unstable and susceptible to bombardment with a α-particle and others.

(b) The portion of the curve $20 < A < 180$, the packing fraction suddenly decreases, becomes negative and passing through flat minimum, again increases and becomes positive. Here the nucleons bind together at the expense of energy from inside and hence mass equivalent of this energy is lost, the isotopic mass becomes less than the mass number. The nuclei are exceptionally stable and not suitable for bombardment, hence the most abundant elements belong to this portion.

(c) The third part of the curve shows the positive packing fraction, i.e., above Hg (180). As above, extra energy is added to the nucleus to bind up the nucleons and mass equivalent of this energy increases the isotopic mass above mass number. These nuclei are unstable and show natural radioactivity. Even by bombarding these elements with suitable projectile, artificial elements with higher atomic number can be prepared.

(d) The fourth part is the bifurcation of the curve at lower atomic weight region where ${}^{4}_{2}He$, ${}^{12}_{6}C$ and ${}^{16}_{8}O$ show extra stability and form the branch. This is due to the fact that these nuclei have mass number multiple of 4 indicating their nuclear energy levels are closed. Nucleons multiple of 4 of kind ${}^{8}_{4}Be$ do not belong to this groups because the binding energy of this is only 6.4 MeV ($A = 9.01503$ amu, $M = 9$, $A - M = 0.01503$ amu, $f = 16.7$, $E = 6.4$ MeV) and hence it easily breaks up to helium atom.

If we plot mass defect with atomic weight, the curve will differ with the packing fraction curve. The cross point of curve with $\Delta M = 0$ are same for both curves but the minimum point here is Sn, not Fe, the former having large number of isotopes. There is no clear explanation for this.

The drawback of the mass defect curve is that the precise determination of atomic mass in the high atomic mass region is very uncertain and error is very high and so the curve will have no meaning.

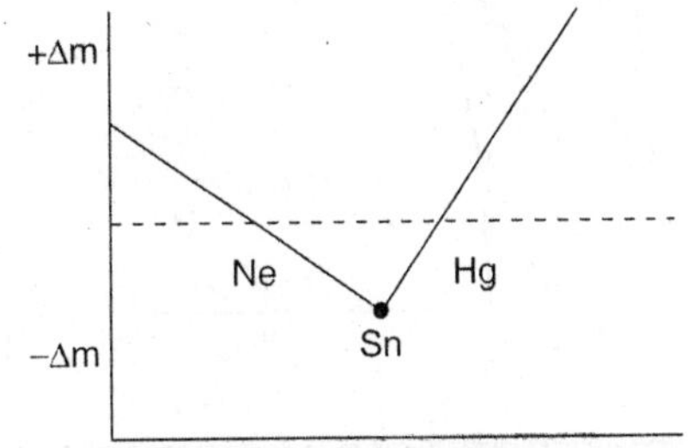

Fig 3.4 Plot of mass defect with atomic weight.

3.09 BINDING ENERGY

The stability of the nucleus with respect to mass defect is of historical interest. True mass defect will be the difference of total mass of neutrons, protons and electrons with the atomic mass.

Thus, true mass defect = $Zm_H + (A - Z)m_n - M$.

This difference in mass is converted to energy, which is required for the nucleus to bind up them to constitute the nucleus, or this amount of energy is to be supplied to the nucleus to break up the atom to the constituent particles. Thus, energy equivalent to the true mass defect is called the *binding energy of the nucleus*.

So, binding energy = (931 × true mass defect, MeV), as 1 amu = 931 MeV.

Calculation of B.E. of ${}^{20}_{10}Ne$ (19.992440 amu) and that of ${}^{209}_{83}Bi$ (208.9804 amu) and ${}^{52}_{24}Cr$ (51.5969 amu) can be illustrated as follows:

B.E. of Ne = 931.4(10×1.007825+10×1.008665−19.99244)
= 160 MeV

B.E. of Bi = 931.4(83×1.007825+126×1.008665−208.9804)
= 1640 MeV

B.E. of Cr = 931.4(24×1.007825+28×1.008665−51.9569)
= 469 MeV

But, if B.E. per nucleon is calculated, the constancy of the value indicates that for all stable elements the value is 8.5. So, if B.E. per nucleon is plotted against mass number, the curve is like that of packing fraction curve but in an inverted manner (Fig 3.5).

The curve shows that in the low atomic weight region $A < 40$ the value first increases, then reaches maximum to 8.5 over a large range of stable atomic nuclei $40 < A < 180$ and then again decreases for high atomic weight region at 7.5. Elements above atomic weight 180 are natural radioactive and are susceptible to bombardment. Here also ${}^{4}_{2}He$, ${}^{12}_{6}C$, ${}^{16}_{8}O$ form a separate branch, as they are more stable due to complete nuclear energy level and mass numbers are multiple of 4. Here again ${}^{9}_{4}Be$ does not belong to this branch, as B.E. per nucleon is $\frac{B.E}{A} = 6.158$ MeV.

Hence, the nucleus is unstable and easily breaks up to Helium atom.

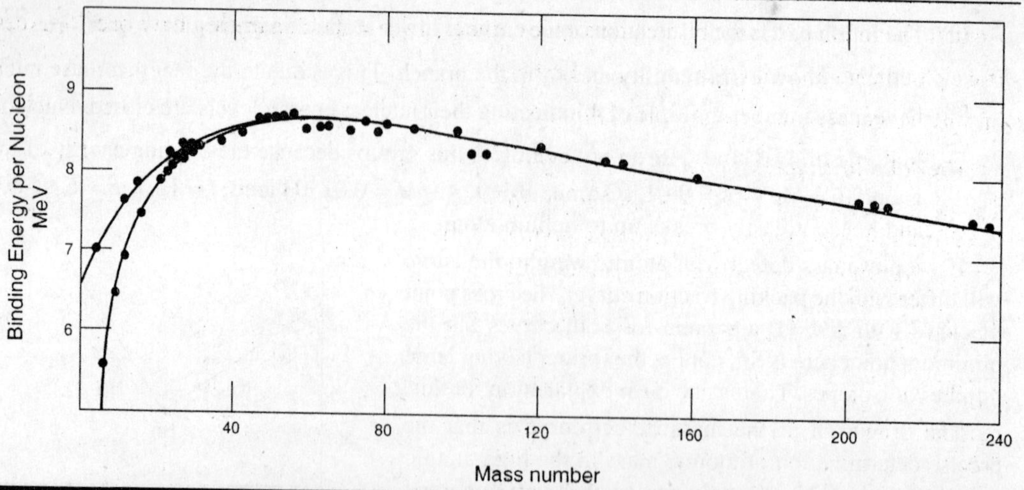

Fig 3.5 Nuclear binding energy curve.

Now it will be evident from B.E. why light elements are not α-emitter but heavy elements are α-emitter ^4_2He (4.0017).

B.E. = 931 [2 × 1.007825 + 2 × 1.008665 – 4.0017] = 0.0303 × 931 MeV = 28.1 MeV.

i.e., formation of α-particle from 2^1_1p and 2^1_0n requires 28.1 MeV. Emission of α-particle at an observable rate will require 5 MeV to penetrate or overcome the potential barrier. So to detect α-emission and detachment of α-particle require less than 28 – 5 = 23 MeV. For heavy elements above 210(*A*), say ^{209}Bi and ^{238}U, B.E. for Bi = 1640 MeV and for U=1802 MeV.

i.e., increase in B.E. is 1802–1640 = 162 MeV for additional 238–209 = 29 nucleons, i.e., energy per additional nucleon is $\frac{162}{29}$ = 5.6 MeV and for emission of α-particle 4 × 5·6 = 22.4 MeV is needed which is less than the required value of 23 MeV hence for heavy elements α-particle will be emitted. For lighter elements less than 210(*A*), similar calculation will show that for light element this energy will be greater than 23 MeV, e.g., $^{20}_{10}Ne$, B.E. = 160.6 MeV and for Cr = 469 MeV, i.e., increase is (469–160.4) = 308.6 MeV for additional (52 – 20) is 32 nucleons, i.e., 308.6/32 = 9.6 MeV per additional nucleon. So to emit α-particle for lighter element, it will require 9.6×4 = 38.4 MeV which is greater than the required 23 MeV, hence α-emission from lighter elements are not possible. But a few nuclei in the range of *A* = 150 are known which are α-emitter, their $\frac{n}{p}$ ratio is less than the optimum value and in these cases emission of α-particle will considerably decrease the B.E.

3.10 NUCLEAR INSTABILITY

The stability of nucleus can also be understood in consideration of the interaction between ^1_0n and ^1_1p. Large number of neutrons and protons are packed in the small nucleus, the nucleus must be unstable due to large repulsive force between protons. To get stability of the nucleus, attractive force between neutron and proton must exist and this attraction is via exchange of charge between them. This exchange of energy or charge will be maximum when number of ^1_0n and ^1_1p in the nucleus is equal which is actually found in low atomic weight region up to $Z = 20$, i.e., $\frac{n}{p} = 1$.

Actually within Z = 20, about 8 elements have number of ${}_1^1p$ = ${}_0^1n$ and the rest have one ${}_0^1n$ excess. If we plot the number of ${}_0^1n$ against the number of ${}_1^1p$, we get the curve as shown in Fig 3.6.

The solid line representing number of ${}_0^1n$ = ${}_1^1p$ is at an angle of 45° and elements up to Z = 20 follow this curve. The elements having Z > 20 the number ${}_0^1n$ exceeds the number of ${}_1^1p$ and $\frac{n}{p} > 1$ and in the heaviest stable nucleus ${}_{82}^{208}Pb$ and ${}_{83}^{209}Bi$, the $\frac{n}{p}$ = 1·5. This increasing number of neutron can be explained in this way: The electrostatic force of repulsion in a nucleus is proportional to $\frac{Z^2}{R}$, where Z is atomic number and R is the radius of the nucleus, and $R \propto A^{1/3}$ where A is the mass number. So repulsion energy is proportional to $\frac{Z^2}{A^{\frac{1}{3}}}$.

To overcome this $p \leftrightarrow p$ repulsion the nucleus should contain excess ${}_0^1n$ so that $n \rightarrow n$ and $n \rightarrow p$ attraction will compensate the $p \leftrightarrow p$ repulsion. The balance of this attraction and repulsion will make a nucleus stable.

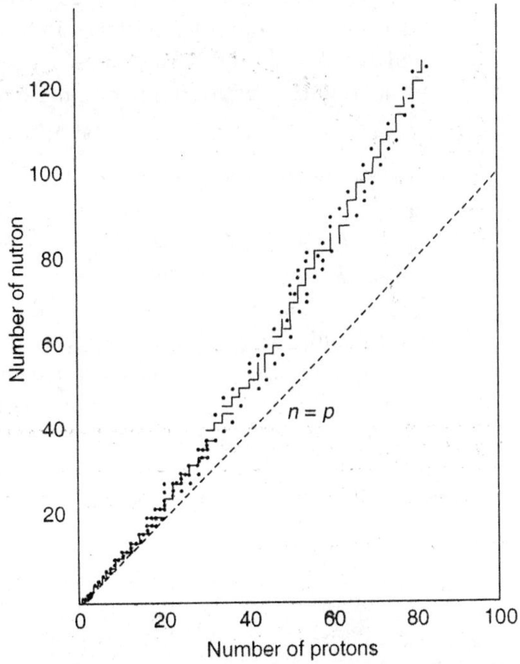

Fig 3.6 Relation between numbers of neutrons and protons in stable naturally occurring nuclei.

When Z > 83, the $\frac{n}{p}$ ratio is greater than 1.5, the repulsion exceeds the attraction and the nucleus is to some extent unstable and will have the tendency to adjust this unfavorable *n/p* ratio by emitting either or any charged particles, i.e., the nucleus will show radioactivity. Here, the nucleus can be looked upon as a liquid drop and susceptible to bombardment with projectiles.

The deviation of the nucleus from this ideal curve indicates instability and will adjust itself by emitting particles as given below:

(A) When the nucleus has high $\frac{n}{p}$ value it will adjust by:

(i) β *Emission*: This occurs due to conversion ${}_0^1n \rightarrow {}_1^1p + \beta^-$ and ν. This will increase the nuclear charge by unity without affecting mass number, so $\frac{n}{p}$ decreases, e.g., ${}_{11}^{24}Na \rightarrow {}_{12}^{24}Mg + \beta^- + \nu$. It is common to both natural and synthetic radioisotopes.

(ii) *Neutron emission*: It is uncommon and unimportant. It is observed in some fission reaction. This will decrease $\frac{n}{p}$ ratio without affecting the nuclear charge.

(B) *Nucleus having low n/p ratio*: It will adjust by decreasing nuclear charge by:

(i) *Positron emission*: It results from the conversion of a proton into a neutron ${}_1^1p \rightarrow {}_0^1n + e^+ + E$. Positron being emitted nuclear charge decreases by unity without affecting mass number. This type of conversion is observed among synthetic radioisotopes, e.g., ${}_{29}^{64}Cu \rightarrow {}_{28}^{64}Ni + e^+$.

(ii) *Proton emission*: This is highly unlikely, as the energy of the system does not permit.

(iii) *Orbital electron captures*: There are some instances where positron emission may occur due to conversion of a proton into a neutron and a positron. Another type of decay is observed with artificial nuclides. The nucleus captures one of the extra nuclear electrons that combines with proton forming neutron and some neutrino. Since the electron has some energy, the neutrino must have energy. The $\frac{n}{p}$ ratio increases and charge of the nucleus decreases by one unit without affecting mass number; $^{55}_{26}Fe + e^{-1} \rightarrow ^{55}_{25}Mn$. This process is called orbital electron capture. Since the electron is taken up from closest *K*-shell, it is called *K*-capture. Electron capture from *L*-shell may occur but it is uncommon. When *K*-electron is taken up, vacancy is left in the *K*-shell and electron from higher shell will immediately fill the vacancy and excess energy will be emitted by the corresponding X-radiation. Since, first capture takes place and then X-radiation occurs, the X-ray will be the characteristic of the product nucleus of atomic number one unit less than the parent nucleus. In most cases, the product nucleus after electron capture is found in the high-energy state and the excess energy is given off as γ-ray. Sometimes, this energy is internally converted, i.e., this energy is transferred to an orbital electron which is ejected giving line spectrum of electron associated with X-ray. But sometimes after electron capture the product nucleus is in the ground state and then it is called pure *K*-capture. Generally, radioactivity is found to be independent of physical and chemical condition. But one exception is the $^{7}_{4}Be$ isotope of which rate of orbital capture depends on its chemical combination, e.g., metal $^{7}_{4}Be$ has half-life for orbital capture 0.013% greater than that of BeO and 0.074% greater than that of BeF_2. This is because of small size of Be and 4 extra nuclear electron; the electron density close to nucleus is of the order Be > BeO > BeF_2.

3.11 HARKIN'S RULE OF ABUNDANCE OR ODD-EVEN RULE

It is supposed that elements and isotopes, which are abundant in nature, are very stable. Thus we are to look into the matter from two different sides:

(i) The relative abundance of elements with respect to one another and

(ii) The relative abundance of different isotopes within each element is the main section of stable isobar rule of Mattauch.

Chemical analysis of various igneous rocks and other geological materials and X-ray analysis by Noddack, Hevesy and other scientists led to the conclusion that, in general,

(a) the light elements are more stable and more abundant than the heavy elements and

(b) the elements of even atomic number are more stable and abundant in nature than the elements of odd atomic number.

In the atomic number region 1 – 83, about 274 stable isotopes have been recognized except the atomic number 5 and 8; 23 are simple and non-isotopic (Be, F, Na, Al, P, Si, etc). Careful examination of these stable isotopes of different elements, nuclei, with even number of protons and neutrons are much more common, stable and abundant than those with odd number of protons or neutrons, e.g., even $^{1}_{0}n$-even $^{1}_{1}p$ = 164, even $^{1}_{0}n$-odd $^{1}_{1}p$ = 50, even $^{1}_{1}p$-odd $^{1}_{0}n$ = 54 and odd $^{1}_{0}n$-odd $^{1}_{1}p$ = 4. Again, out of the most stable elements, the nuclei containing any type of nucleon multiple of 4, e.g., ^{12}C, ^{16}O, ^{24}Mg, ^{28}Si, ^{40}Ca, ^{48}Ti, ^{56}Fe constitute about 80% of earth's crust indicating α-particle play an important role in the stable constitution of nucleus as sub-unit.

The preferential stability of even-even nuclei may be understood in the light of Pauli exclusion principle, which states that for any set of quantum numbers or energy states of the nucleus there can be only $2^{1}_{0}n$ and $2^{1}_{1}p$, the spin being oppositely directed to each case. Thus, a pair of neutrons and a pair of protons form a kind of complete sub-shell and have greater stability than an incomplete sub-shell with odd number of ${}_0^1n$ or ${}_1^1p$. Hence, nuclei of even-even nature and $n/p = 1$ are most stable and abundant and also constitute a recognizable part of the binding energy curve. Although ${}_4^8$Be is stable but its binding energy being less than that of α-particle, so it easily breaks up to ${}_2^4$He. The only 4 odd-odd nuclei are ${}_1^2$D, ${}_3^6$Li, ${}_5^{10}$B and ${}_7^{14}$N and no nuclei of this kind above $A = 14$ are known. These are stable as $n/p = 1$ and extra unpaired ${}_0^1n$ and ${}_1^1p$ will have the same quantum number and parallel spin.

In general, whether a nucleus will attain stability by β^- emission, or through β^+ emission or ***K*-electron capture can be predicted by stable isobar rule of Mattauch**. These empirical rule states that stable isobar pair do not exist where nuclear charge differs by one unit or in other words, in a given set of isobars of adjacent atomic number, only one member is normally stable and abundant, and the next member will be unstable and radioactive. This can be explained on the basis of odd or even nature of mass number.

For odd value of mass number, only one nucleus is stable, no stable pair exists, e.g., ${}_{37}^{87}$Rb – ${}_{38}^{87}$Sr, ${}_{49}^{115}$In – ${}_{50}^{115}$Sn, ${}_{75}^{187}$Re – ${}_{76}^{187}$Os. Of these pairs, only the first is unstable and β^- active due to high nuclear spin and presence of odd number of ${}_1^1p$. The exception to this generalization is that ${}_{48}^{113}$Cd – ${}_{49}^{113}$In and ${}_{51}^{123}$Sb – ${}_{52}^{123}$Te both nuclei are stable and abundant because β^- transition is forbidden due to large difference in nuclear spin combined with small energy change, e.g., ${}_{48}^{113}$Cd has nuclear spin $+\frac{1}{2}$ and that of ${}_{49}^{113}$In is $+\frac{9}{2}$, ${}_{48}^{113}$Cd has unpaired ${}_0^1n$ and magnetic moment $\mu = -0.622$ and ${}_{49}^{113}$In has unpaired ${}_1^1p$ with moment $\mu = 5.52$. Actually, ${}_{48}^{113}$Cd is β^- active with very high $T_{½}$ and hence not observed. Similarly, for ${}_{51}^{123}$Sb – ${}_{52}^{123}$Te, Te is active with *K*-capture but $T_{½} = 1.2\times10^{13}$ years and hence not observed.

When mass numbers are even, there may be two stable isobars, their atomic numbers are even and differ by two units. The examples of even mass numbers and even atomic numbers are ${}_{40}^{96}$Zr – ${}_{42}^{96}$Mo – ${}_{44}^{96}$Ru, ${}_{50}^{124}$Sn – ${}_{52}^{124}$Te – ${}_{54}^{124}$Xe, all are stable as they have even number of ${}_0^1n$ and even number of ${}_1^1p$. But some naturally occurring triple isobars are known where successive nuclei differ by unity, e.g., ${}_{18}^{40}$A – ${}_{19}^{40}$K – ${}_{20}^{40}$Ca, ${}_{22}^{50}$Ti – ${}_{23}^{50}$V – ${}_{24}^{50}$Cr and ${}_{56}^{138}$Ba – ${}_{57}^{138}$La – ${}_{58}^{138}$Ce, the middle member of each set is an odd-odd species and shows slow β^- decay with long $T_{½}$ due to large difference in nuclear spin between parent and daughter nuclei, e.g., ${}_{23}^{50}$V has ground-state spin +6 whereas the decay products ${}_{22}^{50}$Ti and ${}_{24}^{50}$Cr have zero nuclear spin.

3.12 NUCLEAR FORCES

Both low and high atomic nuclei are stable although they contain large number of protons and neutrons. What hold these nucleons together to form stable nucleus? First, we see what kind of force is there in them.

1. Nuclear force cannot be electrical or gravitational in nature:

   Proton-proton interaction may be considered to be electrostatic in nature; and then what would hold the neutral particles—neutron? Moreover, electric or gravitational forces are many times weaker than the nuclear binding energy. Again, electrical or gravitational forces are long-range forces and their value will be roughly proportional to the square of the number of charges or masses. But in cases of nucleus, it has been found that nuclear forces or binding energy is proportional to the number of nucleons.

2. Nuclear forces are very short-range forces:
   Short-range forces are operative over a very short distance of 1–2 fermi (10^{-15} m). Nucleus has very short size (1 – 2 F) and very stable. The nuclear binding energy is proportional to the number of nucleons and binding energy per nucleon is constant, 8.5 MeV for all stable nuclei. This shows the almost saturation character of the forces between nucleons, that is to say that each nucleon attracts those in its immediate vicinity and is indifferent to other nucleons. So the nucleus can be considered as a liquid drop, each molecule in the drop attracts other molecules with which it is in contact but no interaction with other molecules at a higher distance.
3. The nuclear forces are charge-independent:
   In the nucleus, there are three types of interactions (*n-n*), (*n-p*) and (*p-p*) interactions. It can be proved that all these interactions are identical irrespective of their nature of charge, referred to as *charge symmetry principle*. Let us consider the nuclei:

Nuclide	*Composition*	*Interactions*	*Binding energy*
$^{2}_{1}H$	1-*n*, 1-*p*	*n-p*	2.22 MeV
			diff 6.26
$^{3}_{1}T$	2-*n*, 1-*p*	1 *n-n*, 2 *n-p*	8.48 MeV
			diff 0.76
$^{3}_{2}He$	1-*n*, 2*p*	1 *p-p*, 2 *n-p*	7.72 MeV

The addition of one neutron or one proton has increased binding energy to a large extent than that of $^{2}_{1}D$. Again, the interaction (*n-n*) or (*p-p*) is more or less identical, only difference is 0.76 MeV; the (*n-n*) force is slightly greater than (*p-p*) force.

From scattering experiment of $^{1}_{1}p$ and $^{1}_{0}n$ by target proton and $^{2}_{1}D$, in 1936 G. Breit, E. U. Condon and R. D. Present concluded that the nuclear forces between *n-n*, *n-p* and *p-p* are all equal, apart from electrostatic repulsion of *p-p*. This is referred to as the *charge-independence* of nuclear forces, which is more acceptable than the charge symmetry concept.

From the above discussion, it can be concluded that the nature of the nuclear force is very complex; *n-n*, *n-p* and *p-p* interactions are equal and charge-independent. This is in harmony with the constancy in binding energy/nucleon, constancy in nuclear densities and constancy in nuclear volume irrespective of number of nucleons present in them. If only attractive force is in the nucleus, the nucleus will collapse, and binding energy and nuclear volume will not be proportional to number of nucleons. So, there must be some repulsive force within the nucleus that is effective within a distance of 0.5 F.

ORIGIN OF NUCLEAR FORCE

In discussing the proton-neutron concept of nuclear structure, it was mentioned that proton and neutron are the manifestation of the same thing (called *nucleon*) and e^+ and β^- resonate between them (*Heisenberg*, 1932) $^{1}_{0}n \underset{+\beta^-}{\overset{+e^+}{\rightleftharpoons}} {}^{1}_{1}p$.

Thus, the force of attraction between $^{1}_{1}p$ and $^{1}_{0}n$ is due to resonance. But mass consideration shows that such light mass like e^+ or β^- cannot resonate between two heavy nucleons, since β-decay data showing this exchange force are very small to the factor 10^{14} to be significant in nuclear structure. In 1935, Japanese physicist Hideki Yukawa proposed the meson field theory of nuclear force, in analogy with the electromagnetic theory. The application of quantum mechanics to electromagnetic field surrounding a charged particle led to the conclusion that electrical force is exerted by the transfer of photon (called *field particle*), from one charged body to another. For nucleons the relative mass of the field particle can be obtained in this way:

The effective interaction range will be the dimension of the nucleus 2 F or 2×10^{-13} cm from wave mechanical relationship

$$r = 2\times10^{-13} = \frac{\lambda}{2\pi}, \quad \therefore \lambda = \frac{h}{mc}$$ [where λ equivalent wavelength of field particle travels with the velocity of light]

$$\text{So, } r = 2\times10^{-13} = \frac{h}{2\pi mc} = \frac{6.6\times10^{-27}}{2\times3.14\times m\times3\times10^{10}}$$

$$\text{or, } m = \frac{6.6\times10^{-27}}{2\times3.14\times3\times10^{10}\times2\times10^{-13}} = \frac{6.6\times10^{-27}}{37.68\times10^{-3}} = \frac{66\times10^{-28}}{37.68\times10^{-3}} \approx 2\times10^{-25}\text{ gm}$$

Mass of electron $= 9\times10^{-28}$ gm.

$\therefore$ Relative mass of field particle $= \dfrac{2\times10^{-25}}{9\times10^{-28}} = 2.2\times10^{2}$, i.e., about 220 times the mass of e or e^+.

Thus, e^+ or β^- cannot act as a resonating particle between nucleons. In 1936, muon having mass $206m_e$ was discovered in the cosmic ray and considered as the field particle in the nucleus. In 1947, E. Fermi and others showed that interaction of muon being slow and having longer life could not act as a field particle in strong nuclear force. In the same year, another new particle having mass $270m_e$ and longevity 10^{-8} sec called π-meson or pion was discovered in cosmic ray which has the requisite property of a field particle in the strong nuclear force. Pions are π^+, π^- or π^0 charged. According to the meson field theory, each nucleon is surrounded with meson field, where the following transformations / interactions may take place:

$${}^1_1p^+ + {}^1_0n \rightarrow \left[{}^1_0n \rightarrow \pi^+\right] + {}^1_0n \rightarrow {}^1_0n + {}^1_1p^+$$

$${}^1_1p^+ + {}^1_0n \rightarrow {}^1_1p + \left[\bar{\pi} \leftarrow {}^1_1p^+\right] \rightarrow {}^1_0n + {}^1_1p^+$$

$${}^1_1p + {}^1_1p \rightarrow \left[{}^1_1p \rightarrow \pi^0\right] + {}^1_1p \rightarrow {}^1_1p + \left[\pi^0 \rightarrow {}^1_1p\right] \rightarrow {}^1_1p + {}^1_1p$$

$${}^1_0n + {}^1_0n \rightarrow \left[{}^1_0n \rightarrow \pi^0\right] + {}^1_0n \rightarrow {}^1_0n + \left[\pi^0 + {}^1_0n\right] \rightarrow {}^1_0n + {}^1_0n$$

Thus, the transfer or exchange of π^+, π^- or π^0 takes place in an extremely short period that their free existence in the nucleus is undetectable and hence called *virtual meson*. When two nucleons surrounded by the cloud of virtual meson come very close (10^{-13} cm) to each other, transfer of π^+ or π^- or π^0 takes place and force of attraction results, hence these are short-range forces. The emission and absorption of π-meson take place so rapidly that the mass of ${}^1_0n$ and ${}^1_1p$ remain virtually constant. If mass of pion π_m transfers from one neutron to the other with conversion into energy by the relation $\Delta E = \pi_m c^2$ and pion travels the distance 1.7×10^{-13} cm (distance between two nucleons) with velocity c in the time $\Delta t = \dfrac{1.7\times10^{-13}}{c}$, then from uncertainly principle,

$$\Delta E \times \Delta t \simeq \frac{h}{2\pi} \quad \text{or,} \quad \pi_m c^2 \times \frac{1.7\times10^{-13}}{c} \simeq \frac{h}{2\pi}$$

$$\text{or,} \quad \pi_m = \frac{6.67\times10^{-27}}{2\times3.14\times3\times10^{10}\times1.7\times10^{-13}}$$

$$= \frac{6.67\times10^{-27}}{32\times10^{-3}} = \frac{66.7\times10^{-28}}{32\times10^{-3}} = 2.09\times10^{-25}$$

Mass of electron $\bar{e} = 9\times10^{-28}$ gm.

$$\therefore \quad \frac{\pi_m}{m_e} = \frac{2.09\times10^{-25}}{9\times10^{-28}} = \frac{20.9\times10^{-26}}{9\times10^{-28}} = 2.32\times10^2.$$

So the mass of π-meson is about 232 times the mass of an electron. The time of flight of the virtual meson without detection is 5×10^{-24} sec $\left(\Delta t = \frac{R}{c}\right)$ and the distance it travels is $3\times10^{10}\times5\times10^{-24} \approx 1.5\times10^{-13}$ which is within the dimension of the nucleus.

3.13 RATE OF DISINTEGRATION AND DECAY CONSTANT

One of the most helpful methods for determination and identification of radioactive element after its separation is to measure its rate of disintegration. From the study of the decay curve of Uranium and Thorium and their recovery, it is evident that the activity, i.e., the rate of decay of an active species diminishes exponentially. The number of atoms, which disintegrate in a unit interval of time, is proportional to the total number of atoms of that species present at that time. Since disintegration takes place continuously, the number of atoms present is changing and so also the decay rate.

Suppose, at a given instant there are N atoms of a particular radio element, dN, the number of atoms disintegrating in a small interval of time, dt, so the decay rate is dN/dt and this is proportional to the number of atoms N, present at that instant. As with time, rate of disintegration decreases, $-\frac{dN}{dt} \propto N$ or, $-\frac{dN}{dt} = \lambda N$, where λ is the proportionality constant, which is called *radioactive constant* or *disintegration constant* or *decay constant.* The decay constant is definite and specific property of the radioactive element. It depends on the nature of the species and is independent of physical or chemical condition. It is unaffected by external agencies.

$$\text{Now,} \quad -\frac{dN}{dt} = \lambda N \qquad \text{or,} \quad \frac{dN}{N} = -\lambda\, dt \qquad \text{or,} \quad [\ln N]_{N_0}^{N_t} = -\lambda [t]_0^t$$

$$\text{or,} \ \ln \frac{N_t}{N_0} = -\lambda t \qquad \text{or,} \quad \frac{N_t}{N_0} = e^{-\lambda t} \qquad \text{or,} \quad N_t = N_0 e^{-\lambda t} \quad \textbf{[exponential form]}$$

where N_0 is the number of atoms of radio element at any arbitrary initial time $t = 0$ and N_t the same remaining after time t. Expressing this in natural logarithm

$$\log \frac{N_t}{N_0} = -0.4343\lambda t \qquad \text{or, } \log N_t = \log N_0 - 0.4343\lambda t$$

Thus, if we plot log N_t against time, t with reference to some arbitrary zero, the plot will be a straight line and the slope of the plot will be equal to 0·4343λ from which λ can be calculated. Such determination is done on the assumption that each radio atom of a given species expel from its nucleus either one α- or one β-particle upon disintegration. So from the number of atoms disintegrating in a given interval of time, can be calculated by counting the number of particles emitted by the pure element.

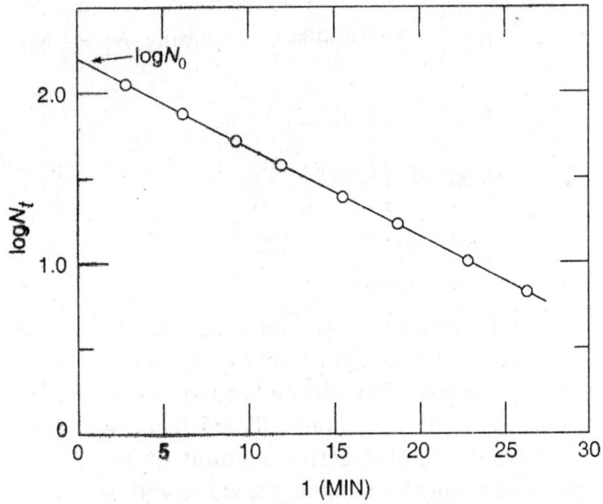

Fig 3.7 Graphical determination of radioactive (decay) constant.

In most cases, the radioactive elements are not pure and hence complicacy arises. The measurements are possible when $T_{½}$ varies from fraction of a second to few months. (a) If the radioactive element is impure and contains two or more with different $T_{½}$, the plot will be a combination of such lines with different slope characteristic of individual radioactive element merging into one another and forming a curve. By analyzing this curve at different point λ for each can be determined. (b) If the two elements have very close $T_{½}$, such analysis is not possible. (c) When the daughter element is also radioactive, then result is confusing. So, λ, the decay constant for the parent element, is to be determined before perceptible amount of the daughter element has accumulated. (d) When the radioactive element has long $T_{½}$, this procedure is not satisfactory as N_t changes very slowly with time and observation to be continued for several years to get sufficient data. In this case, absolute determination is made.

If ΔN is the number of atoms disintegrating, i.e., measured number of α- or β-particles emitted in a definite time t (say a day or week), then $\Delta N/\Delta t$ is the disintegration rate and $\lambda = \dfrac{\Delta N}{\Delta t \times N}$, where N is the Avogadro's number $\left[\dfrac{W}{A} \times N\right]$.

By extrapolation of the line to the ordinate log N, the intercept will be the value of N_0 (see the figure).

The physical signification of λ-equation:

(a) The exponential nature of the decay implies that there is a definite probability, determined by λ, that any particular atom will disintegrate at any given moment.

(b) The length of time, it can exist before it disintegrates, may have any value zero to infinity and it is impossible to say in advance when it will disintegrate.

(c) It is the fraction of the total number of atoms of the radioactive element disintegrating in unit time.

3.14 HALF-LIFE ($T_{½}$)

For the detection and study of radioactive elements, another constant, half-life or $T½$, was introduced by Rutherford. It is the time required for a certain quantity of radioactive element to decay to half its initial value. So the equation

$$\log \frac{N_t}{N_0} = -0.4343\lambda t \qquad (\text{putting } N_t = \tfrac{1}{2} N_0)$$

So, $\log \frac{1}{2} = -0.4343\lambda T_{\frac{1}{2}}$

or, $\log 2 = 0.4343\lambda T_{\frac{1}{2}}$

or, $T_{\frac{1}{2}} = \dfrac{\log 2}{0.4343\lambda} = \dfrac{0.693}{\lambda}$

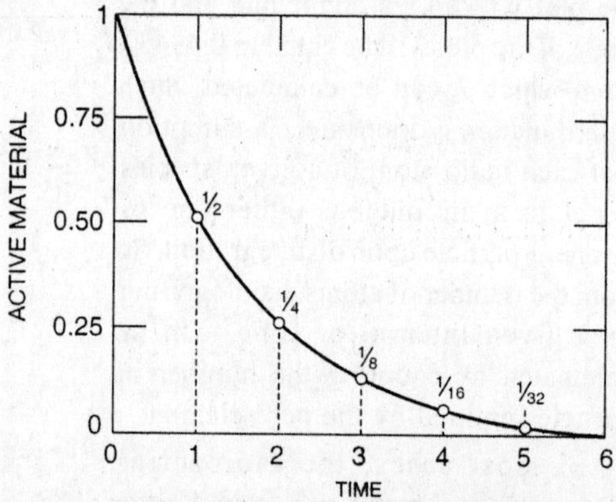

Fig 3.8 Radioactive decay, illustrating the principle of half-life.

For clarification of $T_{½}$, let a radioactive element of $T_{½}$= 1 hour and mass 1 gm was taken. So after 1 hour 0.5 gm will be left after decay and 0.25 gm after next 1 hour will be left and so on. Thus, if we plot active amount of the radioactive element with time, the curve will be like this as shown in Fig. 3.8. Thus after n such successive intervals, the remaining activity will be $(½)^n$. With higher value of n, the remaining active species will be very small, but never be zero theoretically. Thus after such 10 successive intervals, i.e., $nT_{½}$ hours, the remaining activity will be $(½)^{10} = 0.001$, i.e., 0.1%, which is very negligible.

By comparison with the help of radioactive equilibrium, $T_{½}$ can be determined from the following relation, $\dfrac{N_A}{N_B} = \dfrac{\lambda_B}{\lambda_A} = \dfrac{T_{\frac{1}{2}(A)}}{T_{\frac{1}{2}(B)}}$

If we can determine the equilibrium constants of two radioactive elements in a series with a long-lived parent and $T_{½}$ of one is known, then $T_{½}$ for other can be determined. Thus, say, uranium mineral, most of which are old enough for equilibrium to be established, contains one atom of Ra (A) to every 2.8×10^6 atoms of U (B) and $T_{½}$ for Ra has been observed from direct measurement to be 1620 years, $T_{½}(\text{U}) = 2.8 \times 10^6 \times 1620 = 4.5 \times 10^9$ years. Direct determination of $T_{½}$ is done for the determination of λ.

Thus, $T_{½}$ is independent of the initial amount of the radioactive element and as is the characteristic property, $T_{½}$ is also the characteristic property of the radioactive element.

3.15 AVERAGE LIFE OR MEAN LIFE

It is defined as the ratio of the total lifetime of all the radioactive atoms to the total number of such atoms present in it. So in an aggregate of different types of radioactive elements, average life or mean life is used.

We know, $-\dfrac{dN_t}{dt} = \lambda N_t$ or, $dN_t = -N_t \lambda dt$; the negative sign is neglected as number of atoms cannot be negative. All these atoms, say, lived for a time t, so total lifetime of all these atoms will be $t \times dN_t$.

Since t may have all possible values from 0 to ∞, total lifetime of all atoms will be $\int_{t=0}^{t=\infty} t \times dN_t$.

$$\text{So, average life, } \tau = \frac{\textit{total lifetime}}{\textit{total number of atoms}} = \frac{\int_0^{\infty} t \times dN_t}{N_0}$$

$$\text{or,}\quad \tau = \frac{\int_0^{\infty} t \times -\lambda N_t dt}{N_0} = \frac{-\int_0^{\infty} t \times \lambda \times N_0 e^{-\lambda t} dt}{N_0}$$

$$= -\lambda \int_0^{\infty} t\, e^{-\lambda t} dt \qquad \left[\text{Let } t = u \text{ or } dt = du \text{ and } v = -\frac{e^{-\lambda t}}{\lambda}\right]$$

So integrating by parts

$$\tau = -\lambda \left[t \times \frac{-e^{-\lambda t}}{\lambda} - \int \frac{-e^{-\lambda t}}{\lambda} dt \right]_0^{\infty}$$

$$\text{or,}\quad \tau = -\lambda \left[\frac{-te^{-\lambda t}}{\lambda} - \frac{e^{-\lambda t}}{\lambda^2} \right]_0^{\infty}$$

$$= -\lambda \left[\frac{-e^{-\lambda t}(\lambda t + 1)}{\lambda^2} \right]_0^{\infty} = -\lambda \times -\frac{1}{\lambda^2} = \frac{1}{\lambda}.$$

So the average life of a radioactive element is reciprocal of its decay constant. Average life τ, of natural radioactive element varies from 10^{-6} seconds to 10^{10} years.

Thus, relation between decay constant, half-life and average life of a radioactive element can be written as, $T_{\frac{1}{2}} = \frac{0.693}{\lambda} = 0.693\tau$.

From the study of decay constant Avogadro's number can be calculated thus: It has been found that 1 gm Ra-226 gives 11.6×10^{17} α-particles per year. So rate of disintegration, as one α-particle is given out by one radium atom and n is the number of atoms in 1 gm Ra-226. $T_{½}$ Ra-226 = 1590 years.

$$\therefore \lambda = \frac{0.693}{1590}. \qquad \text{So, we can write}$$

$$-\frac{\partial n}{dt} = \lambda n \qquad \text{or,}\quad 11.6 \times 10^{17} = \frac{0.693}{1590} \times n$$

$$\text{or,}\quad n = \frac{11.6 \times 10^{17} \times 1590}{0.693}$$

$$\text{So, } N = 226n = 11.6 \times 10^{17} \times 1590 \times 226 \times \frac{1}{0.693} = 6.0 \times 10^{23}.$$

3.16 RADIOACTIVE EQUILIBRIUM

In a natural radioactive decay process if

(i) the daughter element is not separated from the parent,

(ii) the daughter element itself is radioactive and

(iii) the parent and the daughter have widely different decay constant, a situation arises in this successive decay process where the rate of formation of daughter element is equal to the rate of its decay. Then, the system attains equilibrium and it is called *radioactive equilibrium* or *secular equilibrium*. Naturally, the rate of formation of the daughter element depends on the decay constant λ_p of the parent while the decay of the daughter depends on its own decay constant λ_d.

Let N_p, λ_p are the number of atoms and decay constant of parent at certain interval of time t and N_p^0 its initial value and $N_d \lambda_d$ and N_d^0 same for the daughter element, then we can write for the decay of the parent as,

i.e., $-\dfrac{dN_p}{dt} = \lambda_p N_p$, it is the rate of formation of daughter

and $\dfrac{dN_d}{dt} = \lambda_p N_p - \lambda_d N_d$, where $\lambda_p N_p$ is the decay of daughter.

So, the rate of net increase of daughter is

$$\frac{dN_d}{dt} = \lambda_p N_p - \lambda_d N_d$$

$$= \lambda_p N_p^0 e^{-\lambda_p t} - \lambda_d N_d$$

Multiplying by $e^{\lambda_d t} dt$,

$$e^{\lambda_d t} dN_d = \lambda_p N_p^0 e^{(\lambda_d - \lambda_p)t} dt - \lambda_d N_d e^{\lambda_d t} dt$$

or, $$e^{\lambda_d t} dN_d + \lambda_d N_d e^{\lambda_d t} dt = \lambda_p N_p^0 e^{(\lambda_d - \lambda_p)t} dt$$

On integration,

$$N_d e^{\lambda_d t} = \frac{\lambda_p N_p^0}{\lambda_d - \lambda_p} \times e^{(\lambda_d - \lambda_p)t} + C.$$

Putting the condition when $t = 0$, $N_d = 0$,

$C = -\dfrac{\lambda_p N_p^0}{\lambda_d - \lambda_p}$ and the equation takes the form

$$N_d e^{\lambda_d t} = \frac{\lambda_p N_p^0}{\lambda_d - \lambda_p} \left[e^{(\lambda_d - \lambda_p)t} - 1 \right]$$

$$N_d = \frac{\lambda_p N_p^0}{\lambda_d - \lambda_p} \left[e^{-\lambda_p t} - e^{-\lambda_d t} \right] \text{ dividing by } e^{\lambda_d t}.$$

Now,

(i) If $\lambda_d \approx \lambda_p$, i.e., average life of parent and daughter are of same order, then activity of the daughter increases gradually, reaches a maximum, then decreases at the decay rate of the longer-lived parent or daughter.

(ii) If $\lambda_p > \lambda_d$, i.e., rate of decay of parent, i.e., rate of growth of daughter is higher than the rate of decay of daughter, then activity of daughter will gradually increase until it becomes equal to that of parent. Then activity of daughter will increase and that of parent will decrease until the parent completely decays. So, no equilibrium will be established.

(iii) When $\lambda_p < \lambda_d$, i.e., the steady-state condition is attained among the genetically related radioactive elements. When ratio of N_p/N_d remains constant but activity decreases slowly, then it is called *transient equilibrium* and becomes zero and the equation becomes

$$N_d = \frac{\lambda_p N_p^0}{\lambda_d - \lambda_p} \times e^{-\lambda_p t} = N_p \times \frac{\lambda_p}{\lambda_d - \lambda_p} \quad \text{or,} \quad \frac{N_d}{N_p} = \frac{\lambda_p}{\lambda_d - \lambda_p}.$$

(iv) When $\lambda_p << \lambda_d$, then $\lambda_d - \lambda_p \approx \lambda_d$.

So, $\frac{N_d}{N_p} = \frac{\lambda_p}{\lambda_d}$. Then permanent or secular equilibrium is attained when decay constant of parent is negligibly small compared to that of daughter.

Examples of the transient equilibrium and permanent equilibrium are

$${}^{234}_{90}\text{Th} \xrightarrow[24.1\ day]{\beta^-} {}^{234}_{91}\text{Pa} \xrightarrow[1.18\ m]{\beta^-} {}^{234}_{92}\text{U}$$ (Transient equilibrium) series

$${}^{238}_{92}\text{U} \xrightarrow[4.5\times10^9\ yrs]{\alpha} {}^{234}_{90}\text{Th} \xrightarrow[24.1\ day]{\beta^-}$$ (Permanent or secular equilibrium) series

Radioactive decay curves show the equilibrium condition. Absolute or permanent equilibrium can never occur because $\frac{dN}{dt} \neq 0$ for the radioactive element.

Radioactive equilibrium is different from the chemical equilibrium. It is unidirectional or irreversible and independent of external conditions, takes longer time, energy change is higher, etc.

3.17 RADIOACTIVE DISINTEGRATION SERIES AND GROUP DISPLACEMENT LAW OF SODDY AND FAJANS

Towards the end of 1912, more than 40 naturally occurring radioactive elements are known with atomic numbers 81 to 92. If the periodic table was not to break-down in this high atomic region, several radioactive elements are to be placed in one position, at least in a few cases. The groups of elements, though have quite different radioactivity, are so identical in chemical properties that they cannot be separated by any chemical means. These groups of elements are to be placed in one position in the periodic table and termed **isotope** (**iso** = equal, **topes** = position) as the periodic table is based on the chemical analogy. Again among these 40 radioactive elements some relation was observed. Fajan and Soddy in 1913 stated some fundamental laws, called *displacement law*.

(i) In the radioactive transformation only one α- or β-particle will be emitted, never both or more than one of each kind is emitted at a time.

(ii) When α-particle is emitted by a radioactive element, the new element formed has the mass number less by 4 units and atomic number less by 2 units than the parent element.

(iii) When a β-particle is emitted, then also a new element is formed whose atomic mass remains the same but atomic number increases by one unit than the parent element.

With the help of this law, the natural radioisotopes can be classified into series, members of each series are genetically related where the mass numbers are divisible by *4 with remainders*. There are $4n$, $4n + 2$ and $4n + 3$ series, the series are named after the longest-lived parent element. Such as Thorium-, Uranium-, Actinium-series having $T_{1/2}$ 1.3×10^{10} years, 4.5×10^9 years, 7.1×10^8 years.

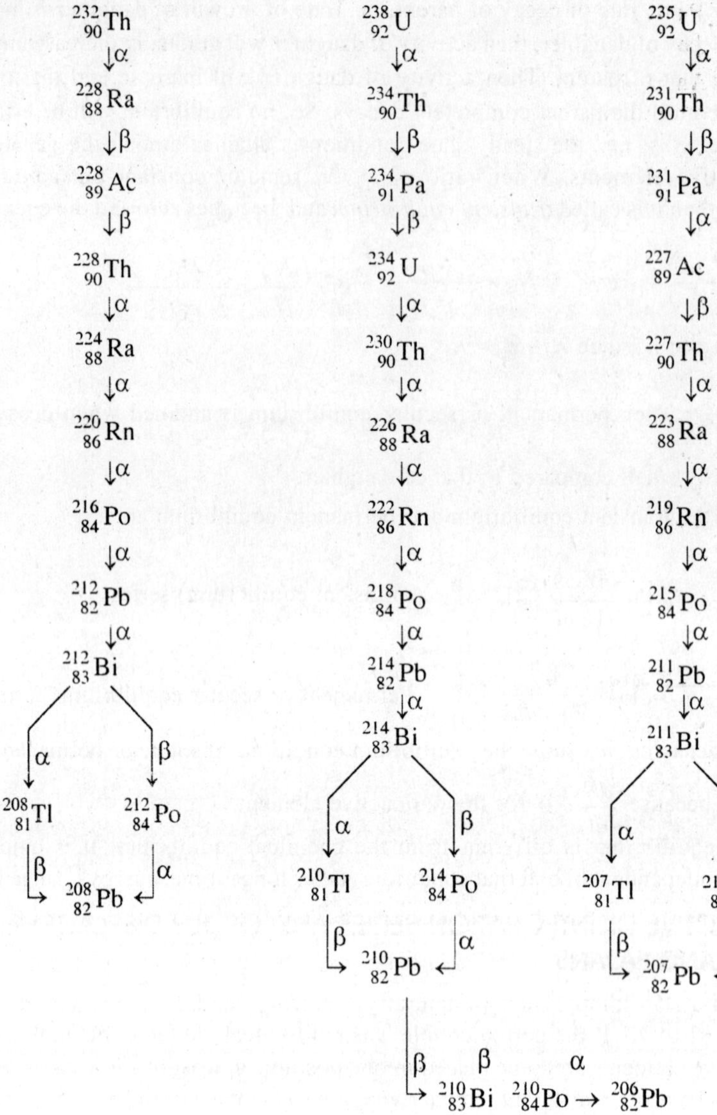

After the discovery of transuranic elements, the scientists are of the opinion that analogous to the naturally occurring radioactive series, the transuranic, i.e., synthetic radioactive elements will form a transformation series called *Neptunium series* ($T_{1/2} = 2.2 \times 10^6$ years) although the parent elements is Plutonium of $T_{1/2} = 13.2$ years. All members of this series are either unknown or very rare in nature. $^{237}_{93}Np$ was present in nature from the beginning of the earth and all its decay products are so infinitesimally small that they are beyond the possibility of detection.

All these heavy isotopes are placed in the periodic table in accordance with Soddy and Fajans' group displacement law: When the radioactive element emits an α-particle, its mass decreases by 4 units and atomic number by 2 units. Hence, the daughter element will be placed 2 positions left to the parent. When the radioactive element emits a β-particle, its mass number remains the same but atomic number increases by one unit and the product nucleus will be placed one place right to the parent element. Hence, if a radioactive element emits one α-particle and the successive 2β-particles, the end product will have same atomic number but different mass number, i.e., isotopes are formed.

$^{241}_{94}Pu$

$\downarrow \beta$

$^{241}_{95}Am$

$\downarrow \alpha$

$^{237}_{93}Np$

$\downarrow \alpha$

$^{233}_{91}Pa$

$\downarrow \beta$

$^{233}_{92}U$

$\downarrow \alpha$

$^{229}_{90}Th$

$\downarrow \alpha$

$^{225}_{88}Ra$

$\downarrow \beta$

$^{225}_{89}Ac$

$\downarrow \alpha$

$^{221}_{87}Fr$

$\downarrow \alpha$

$^{217}_{85}At$

$\downarrow \alpha$

$^{213}_{83}Bi$

$\swarrow \alpha \qquad \beta \searrow$

$^{209}_{81}Tl \qquad ^{213}_{84}Po$

$\beta \searrow \qquad \swarrow \alpha$

$^{209}_{82}Pb$

$\downarrow \beta$

$^{209}_{83}Bi$

CHAPTER

4

NUCLEAR REACTIONS

4.01 NUCLEAR REACTIONS

The Aristotelian theory, that all matters consist of some primordial substances of varying amounts of four qualities or principles, tempted the ancient alchemists to change the base metals into gold, although all attempts were in vain. With the development of the modern idea of fundamental particles of matter and their role in understanding the structure of atoms, the possibility of transmutation again gains momentum. Scientists became busy in changing and disintegrating various atoms but with little success. Only one case to be mentioned that some claimed to prepare gold by passing high-tension electric discharge through mercury vapour. After several discussions it was also ultimately disproved.

Rutherford carried out the first successful artificial transmutation of elements in 1919. He bombarded nitrogen gas by high velocity α-particles coming from $Ra-B\left({}^{214}_{82}Pb\right)$ and $Ra-C\left({}^{214}_{83}Bi\right)$, where long-range protons were obtained due to the change:

$$ {}^{14}_{7}N + {}^{4}_{2}He^{+2} \rightarrow {}^{17}_{8}O + {}^{1}_{1}H $$

Ordinary ${}^{1}_{1}H$ has the range up to 28 cm, but here ${}^{1}_{1}H$ emitted has the range up to 40 cm, although this was first predicted by Marsden in 1914. Later, it was found that long-range protons are also obtained as a result of α-ray bombardment from all elements up to ${}^{40}_{20}Ca$ with exception of Helium, Carbon and Oxygen as they are very stable with closed nuclear shells. The extent of this disintegration is very small as 1 in 3 00 000 α-particles expelled long-range ${}^{1}_{1}H$ from Nitrogen. The energy of ${}^{1}_{1}H$ emitted by this bombardment is sometimes greater than that of the α-particles indicating that they result from disintegration process and the excess energy they gain from the accompanying nuclear rearrangements. This type of transmutation can be carried out not only with α-particles but also with ${}^{1}_{1}H$, ${}^{2}_{1}D$, ${}^{1}_{0}n$ and even with γ-ray.

When charged particles are used as projectile, as they approach towards nucleus, which is also charged, they will experience a repulsive energy depending on the charge of both the projectile and the target nucleus, and the distance of separation between them $\left(E = \frac{Ze^2}{d}\right)$ [when the projectile is unit charged]. Thus, with the same projectile, higher and higher energy is required to bombard target of higher and higher nuclear charge, otherwise, it will be scattered. Thus to have an effective bombardment, the projectile will be sufficiently energized to cross that potential barrier. When $d > r$, in Fig 4.1 only coulombic repulsion acts on the projectile, but when within r, nuclear attractive force begins to operate and for successful bombardment, it should have energy to cross the point A, i.e., potential barrier, which is different for different target nucleus and different projectiles.

Regarding the mechanism of nuclear transmutation, two possibilities were considered: (1) the target nucleus merely loses a ${}^{1}_{1}H$ as a result of severe impact accompanying collision with the fast-moving α-particle and (2) α-particle enters the target nucleus; the resulting combined nucleus then eject a ${}^{1}_{1}H$ as a result of distribution of energy among the nucleons of the compound nucleus. From the photograph of the tracks of α-particles through N_2 in cloud chamber, it was observed that α-particles disappear in the disintegrating collision indicating the validity of the second possibility. Thus in nuclear transmutation, the particle of less than 50 MeV, first crosses the potential barrier, collides with the target nucleus where it is absorbed and forms the compound or combined nucleus which is the intermediate stage in most nuclear reactions.

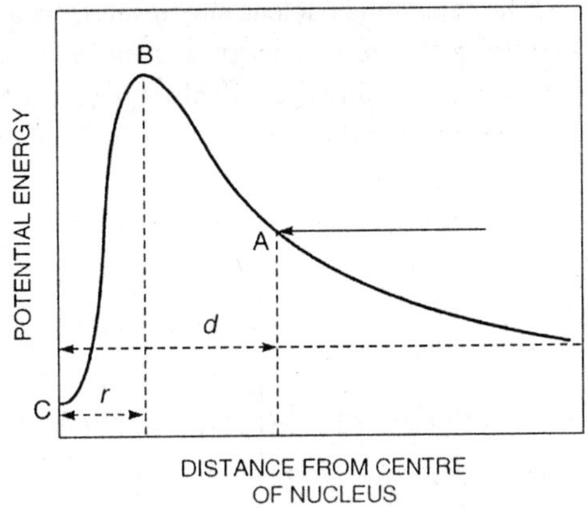

Fig 4.1 Potential energy curve for interaction between an atomic nucleus and an incoming charged particle.

The additional energy is very rapidly shared among the nucleons present in the compound nucleus and thus is energized. Generally, addition of a nucleon increases 8 MeV energy plus the kinetic energy to the compound nucleus.

For ${}^{1}_{1}H$ (8 MeV); ${}^{2}_{1}D\,(2\times 8-2)$ = 14 MeV. For ${}^{4}_{2}He^{+2}$ $[4\times 8 - 28(\text{B.E.})] = 4$ MeV plus K.E. is added. As a result of sharing of this energy one particular nucleon may acquire sufficient energy to be escaped leaving a more stable residual nucleus called *recoiled nucleus*. Average time taken for a projectile to be absorbed and emission of particle takes place within 10^{-14} sec. This is sufficiently long compared to the time taken 10^{-22} sec by a projectile to travel across the diameter of the nucleus 10^{-12} cm moving with the velocity of light, i.e., 10^{10} cm/sec.

Two important aspects regarding the emission of particle from compound nucleus are to be considered:

(a) The particles are emitted equally in all directions.

(b) The emitted particles do not all have the exactly same energy and may be compared to evaporation of liquid molecules at a specific temperature.

The nature of the particle emitted depends on

(i) the excitation energy of the particle,

(ii) height of the potential barrier, and

(iii) the stability of the recoiled nucleus.

When target nucleus is of high atomic number, the potential barrier will be large, and so emission of charged particle is low. Here $^{1}_{0}n$ with no charge will be a good projectile and the probability of its emission will be high.

When target nucleus is of low atomic number, the potential barrier will be relatively low and are more strongly bound to the compound nucleus. Hence charged particles may be used as projectile and their effectiveness is of the order $^{4}_{2}He^{+2} < ^{3}_{1}T < ^{2}_{1}D < ^{1}_{1}H$. Similarly, the ease of emission will be $^{1}_{1}H > ^{2}_{1}D > ^{3}_{1}T$ as relatively large energy is required for the formation of $^{2}_{1}D$ and $^{3}_{1}T$ particles, emission for low target nucleus is probable as the formation energy is only 4 MeV and potential barrier is low.

The compound nucleus may result from a number of processes, but once it is formed, its behavior depends only on the energy, not on the mode of formation. The compound nucleus formed will lose its individuality and will also break in a way dependent on the available energy (energy obtained from K.E. of the projectile and the gain of energy accompanying its combination with the target). The example is

$$^{63}_{29}Cu + ^{2}_{1}D \rightarrow \left[^{65}_{30}Zn\right] \rightarrow \begin{cases} \rightarrow ^{65}_{30}Zn + \gamma & (1) \\ \rightarrow ^{64}_{30}Zn + ^{1}_{0}n & (2) \\ \rightarrow ^{64}_{29}Cu + ^{1}_{1}H & (3) \\ \rightarrow ^{63}_{30}Zn + 2^{1}_{0}n & (4) \\ \rightarrow ^{61}_{28}Ni + ^{4}_{2}He & (5) \\ \rightarrow ^{62}_{29}Cu + ^{3}_{1}H & (6) \\ \rightarrow ^{63}_{29}Cu + ^{2}_{1}H & (7) \end{cases}$$

Of these reactions (4)-(7) are less probable, the other reactions, which will take place, depend on the total energy of the compound nucleus.

All artificial transformations or disintegrations are reversible in nature but natural transformations, i.e., natural radioactivity is irreversible in nature.

Nuclear reactions or transmutations can be carried out using different projectiles:

(a) α-particle, (b) $^{1}_{1}H$, (c) $^{2}_{1}D$, (d) $^{1}_{0}n$, and (e) γ-ray.

4.01(a) BOMBARDMENT WITH α-PARTICLE

α-particle being doubly charged will have high potential barrier, and hence suitable for low atomic weight target nucleus. With high atomic weight targets, they will be scattered. Moreover, addition of α-particle increases the energy of the compound nucleus by (4 × 8 – 28) = 4 MeV. So, sufficiently energized α-ray can be used as projectile for high atomic weight target nucleus, where more than one particle are emitted, e.g.,

$$^{14}_{7}N + ^{4}_{2}He \rightarrow ^{17}_{8}O + ^{1}_{1}H \qquad ^{9}_{4}Be + ^{4}_{2}He \rightarrow ^{12}_{6}C + ^{1}_{0}n$$

$$^{19}_{9}F + ^{4}_{2}He \rightarrow ^{22}_{10}Ne + ^{1}_{1}H \qquad ^{12}_{6}C + ^{4}_{2}He \rightarrow ^{15}_{8}O + ^{1}_{0}n$$

Also, (α, 2n), (α, np), (α, 3n) reactions are known.

In the above examples, after emitting particles, the recoiled nucleus becomes stable. In some cases, the recoiled nucleus is still unstable and emits particles exponentially even when the source of α-particle is removed. This is called the *induced radioactivity*.

$$^{25}_{12}Mg + ^{4}_{2}He \rightarrow ^{28}_{13}Al + ^{1}_{1}H \qquad ^{28}_{13}Al \rightarrow ^{28}_{14}Si + \beta^-$$

If the α-particle has energy more than 400 MeV, spallation reaction takes place, emitting large number of $^{1}_{1}p$ and $^{1}_{0}n$.

$$^{63}_{29}Cu + ^{4}_{2}He \rightarrow ^{35}_{17}Cl + 14^{1}_{1}H + 18^{1}_{0}n$$

4.01(b) BOMBARDMENT WITH $^{1}_{1}H$

Different types of reactions occur with proton as projectile. For lighter target, it has low potential barrier and depending upon the energy of $^{1}_{1}p$ the following reactions occur:

(i) Low energy $\quad ^{27}_{13}Al + ^{1}_{1}H \rightarrow ^{28}_{14}Si + \gamma \qquad ^{7}_{3}Li + ^{1}_{1}H \rightarrow ^{8}_{4}Be + \gamma$

(ii) For higher energy particle may cross the barrier.

$$^{23}_{11}Na + ^{1}_{1}H \rightarrow ^{23}_{12}Mg + ^{1}_{0}n$$

(iii) If energy is more than 20 MeV, more than one particle may be emitted.

$$^{63}_{29}Cu + ^{1}_{1}H \rightarrow ^{62}_{29}Cu + ^{1}_{1}H + ^{1}_{0}n$$

(iv) For low target and high-energy proton $^{4}_{2}He^{+2}$ or $^{2}_{1}D$ may be emitted although relatively uncommon.

$$^{14}_{7}N + ^{1}_{1}H \rightarrow ^{11}_{6}C + ^{4}_{2}He \qquad ^{9}_{4}Be + ^{1}_{1}H \rightarrow ^{8}_{4}Be + ^{2}_{1}D$$

(p, γ) and (p, $^{1}_{0}n$) type reactions give active isotopes.

4.01(c) TRANSMUTATION BY DEUTERONS

Deuteron may be considered as the loose combination of $^{1}_{1}H$ and $^{1}_{0}n$ with B.E. = 2 MeV. So entry of $^{2}_{1}D$ increases the energy of the compound nucleus by (2 × 8 – 2) = 14 MeV. Hence it is more effective projectile than either $^{1}_{1}H$ or $^{4}_{2}He$ and ejection of charged particles is possible. The most common reaction is (d, p) type. e.g.,

(i) $^{7}_{3}Li + ^{2}_{1}D \rightarrow ^{8}_{3}Li + ^{1}_{1}H \qquad ^{114}_{48}Cd + ^{2}_{1}D \rightarrow ^{115}_{48}Cd + ^{1}_{1}H$

Oppenheimer and Phillips suggested that being $^{2}_{1}D$ a loose combination of $^{1}_{0}n$ and $^{1}_{1}H$, as $^{2}_{1}D$ approaches the target nucleus, the electrostatic repulsive force of the positive target nucleus forces the proton away. But $^{1}_{0}n$ is not affected and enter the target nucleus, and thus recoiled nucleus is formed and the excess energy is emitted as γ-ray.

(ii) With still higher energy of $^{2}_{1}D$, sometimes $^{1}_{0}n$, $^{1}_{1}H$ both are emitted when target nucleus is of low atomic weight.

$^{2}_{1}D + ^{2}_{1}D \rightarrow ^{3}_{2}He + ^{1}_{0}n$ also $^{2}_{1}D + ^{2}_{1}D \rightarrow ^{3}_{1}T + ^{1}_{1}H$, this $^{3}_{1}T$ is β active and forms $^{3}_{2}He$, i.e., active isotopes are formed.

(iii) When ${}_{1}^{2}D$ of high energy is used as projectile, ${}_{2}^{4}He$ and ${}_{1}^{3}T$ also formed.

$${}_{20}^{40}Ca + {}_{1}^{2}D \rightarrow {}_{19}^{39}K + {}_{2}^{3}He \qquad {}_{10}^{20}Ne + {}_{1}^{2}D \rightarrow {}_{9}^{18}K + {}_{2}^{4}He$$

(iv) When ${}_{1}^{2}D$ is of 20 MeV, then (*d*, 2*n*), (*d*, 3*n*) type reactions occur and the reaction is complicated.

(v) ${}_{1}^{2}D$ with about 10 MeV with ${}_{15}^{31}P$, ${}_{4}^{9}Be$ (*d*, *T*) reaction occurs.

$${}_{4}^{9}Be + {}_{1}^{2}D \rightarrow {}_{4}^{8}Be + {}_{1}^{3}T$$

4.01(d) TRANSMUTATION BY NEUTRON ${}_{0}^{1}n$

As neutrons are neutral particles, it will not have any electrostatic potential barrier and hence are the most effective projectile for both low and high atomic weight target nuclei. Fast neutron and slow neutron can initiate neutron reaction. Slow ${}_{0}^{1}n$ is obtained by passing fast neutron through paraffin wax and this neutron is more effective as a projectile for useful purpose. When ${}_{0}^{1}n$ approaches the target, attractive force begins to operate, and ${}_{0}^{1}n$ is captured increasing the energy of the compound nucleus by 8 MeV plus kinetic energy. The nucleus is in excited state and then it emits either charged particle or γ-ray. The recoiled nuclei in most cases are radioactive.

Fast ${}_{0}^{1}n$:

1. Fast ${}_{0}^{1}n$ gives (*n*, α) type reaction with light target like N_2, O_2, F_2, Ne, etc.

$${}_{7}^{14}N + {}_{0}^{1}n \rightarrow {}_{2}^{4}He + {}_{5}^{11}B \qquad {}_{8}^{16}O + {}_{0}^{1}n \rightarrow {}_{2}^{4}He + {}_{6}^{13}C$$

2. When hydrogen-containing compounds are bombarded with fast ${}_{0}^{1}n$, γ-radiation is obtained and it is a radiative capture reaction. After passing through paraffin or water, the fast ${}_{0}^{1}n$ is slowing down which is then taken up by target with the liberation of γ-ray. This slow ${}_{0}^{1}n$ is also used in fission reaction to get nuclear energy. Simple radiative capture reaction takes place with hydrogen. ${}_{1}^{1}H + {}_{0}^{1}n \rightarrow {}_{1}^{2}D + \gamma$

   As in this radiative capture reaction $\frac{n}{p}$ ratio increases, the product nucleus is always active and β emitter.

$${}_{45}^{103}Rh + {}_{0}^{1}n \rightarrow {}_{45}^{104}Rh + \gamma \qquad {}_{45}^{104}Rh \rightarrow {}_{46}^{104}Pd + \beta^-$$

$${}_{92}^{238}U + {}_{0}^{1}n \rightarrow {}_{92}^{239}U + \gamma \qquad {}_{92}^{239}U \rightarrow {}_{93}^{239}Np + \beta^-$$

   ${}_{93}^{239}Np$ being unstable emits β^- and forms ${}_{94}^{239}Pu$. Thus, *trans*-uranic elements are prepared.

3. For the emission of charged particles, the compound nucleus will be sufficiently energetic to cross the potential barrier and so fast neutrons of varying energy dependent on the nature of the emitted particle are needed. For (*n*, *p*) type reaction, about 1 MeV ${}_{0}^{1}n$ is needed.

$${}_{2}^{3}He + {}_{0}^{1}n \rightarrow {}_{1}^{3}T + {}_{1}^{1}H \qquad {}_{7}^{14}N + {}_{0}^{1}n \rightarrow {}_{6}^{14}C + {}_{1}^{1}H$$

4. For (*n*, *d*) type reaction, ${}_{0}^{1}n$ of higher energy are required. For the expulsion of ${}_{1}^{2}D$ 14 MeV is required which is the B.E. of ${}_{1}^{2}D$. This reaction is unusual as before ${}_{1}^{2}D$ formation, ${}_{0}^{1}n$ and ${}_{1}^{1}p$ will be emitted. But when ${}_{0}^{1}n$ of 90 MeV is used ${}_{1}^{2}D$ is emitted. It

is suggested that here compound nucleus is not formed but it is a stripping or pick-up reaction, i.e., energetic ${}_0^1n$ approaching target, snatches or carries ${}_1^1p$ from the target and forms ${}_1^2$D.

$${}_{30}^{64}\text{Zn} + {}_0^1n \rightarrow {}_{29}^{63}\text{Cu} + {}_1^2\text{D}$$

5. ${}_0^1$n of 30 MeV can carry out $(n, 2n)$, $(n, 3n)$, $n, n2p)$ reactions.

$${}_{80}^{198}\text{Hg} + {}_0^1n \rightarrow {}_{80}^{197}\text{Hg} + 2{}_0^1n \qquad {}_{30}^{64}\text{Zn} + {}_0^1n \rightarrow {}_{30}^{63}\text{Zn} + 2{}_0^1n$$

6. When ${}_0^1n$ is of above 100 MeV, with moderate atomic weight targets, spallation reaction takes place.

4.01(e) TRANSMUTATION BY γ-RAY

γ-ray of suitable energy can bring about nuclear reactions emitting ${}_0^1n$, ${}_1^1p, \gamma'$, etc. This is known as *photodisintegration* or *photonuclear reactions* or *nuclear photo-effect*.

1. γ-ray of 2 MeV gives (γ, n) type reaction.

$${}_1^2\text{D} + \gamma \rightarrow {}_1^1\text{H} + {}_0^1n \qquad {}_4^9\text{Be} + \gamma \rightarrow {}_4^8\text{Be} + {}_0^1n$$

γ′-ray of 10 MeV can emit from any target.

2. When γ-ray cannot emit ${}_0^1n$, γ-ray is absorbed by the target, becomes excited and then emit γ′-ray of same or lower energy of the incident γ-ray.

$${}_{36}^{83}\text{Kr} + \gamma \rightarrow {}_{36}^{83}\text{Kr} + \gamma'$$

3. If γ-ray of very high energy is used for bombarding low targets, $2{}_0^1n, 3{}_0^1n$, etc., may be emitted.

$${}_8^{16}\text{O} + \gamma \rightarrow {}_6^{11}\text{C} + 3{}_0^1n + 2{}_1^1\text{H}$$

4.01(f) TRANSMUTATION BY PARTICLES OF HIGHER MASS AND CHARGE

${}_1^3$T or ${}_2^3$He can be used as projectile. With ${}_1^3$T the reaction (t, p), (t, d), (t, n), etc., is observed. The emission of ${}_1^1$H and ${}_1^2$D follows the Oppenheimer-Phillips mechanism, where ${}_1^1$H and ${}_1^2$D are repelled and the rest part enter the target.

$${}_{27}^{59}\text{Co} + {}_1^3\text{T} \rightarrow {}_{27}^{61}\text{Co} + {}_1^1\text{H} \qquad {}_{16}^{32}\text{S} + {}_1^3\text{T} \rightarrow {}_{17}^{34}\text{Cl} + {}_0^1n$$

4.02 STRIPPING AND PICK-UP REACTION

This is a type of nuclear reaction where the target nucleus captures one, two and sometimes three nucleons from the projectile due to interaction between them (the projectile and the target) leaving other part of the projectile and without formation of the compound nucleus as an intermediate stage. The remaining part of the projectile, i.e., the emitted particle is not isotropic, i.e., the spatial distribution of it is not uniform in all direction but towards the forward direction. Generally, energy of the emitted particle is higher. These are the main features of the stripping reaction, e.g., (d, p), (d, n), (t, p), (t, d) are known and mechanism of this is like the Oppenheimer-Phillips mechanism.

$${}_3^6\text{Li} + {}_1^2\text{H} \rightarrow {}_4^7\text{Be} + {}_0^1n \qquad {}_3^6\text{Li} + {}_1^2\text{H} \rightarrow {}_3^8\text{Li} + {}_1^1\text{H}$$

Pick-up reaction is just reverse of the stripping reaction. Here the projectile of high energy removes one, two or more nucleons from the target via the formation of compound nucleus, e.g., (p, d), (p, t), $(p, {}^{3}_{2}He)$, (d, t) are of this type.

$${}^{9}_{4}Be + {}^{1}_{1}H \rightarrow {}^{8}_{4}Be + {}^{2}_{1}H \qquad {}^{7}_{3}Li + {}^{2}_{1}H \rightarrow {}^{6}_{3}Li + {}^{3}_{1}H$$

When the projectile is of high energy, 50 MeV, then without forming compound nucleus, direct interaction will take place.

4.03 SPALLATION REACTION

The term spallation is derived from the verb "to spall", i.e., to break up by chipping off small fragments from the bigger one. When 200 MeV ${}^{2}_{1}H$ or 400 MeV ${}^{1}_{1}H$ bombards the target nucleus, unlike fission reaction, the target breaks up into smaller nuclei having mass numbers less by 10 to 20 and to a lesser extent 30 or more and emitting same number of nucleons—singly or in cluster, leaving a series of products. This is known as *spallation reaction.* With light target

$${}^{75}_{33}As + {}^{4}_{2}He \rightarrow {}^{38}_{17}Cl + 18{}^{1}_{1}H + 23{}^{1}_{0}n$$

With heavy target, fission and spallation reaction may occur at the same time. In fission reaction the mass of the product nucleus is much less than the target, usually half; but in spallation the mass of the product nucleus is less by 10 to 20 units from the target. Sometimes the product nucleus has slightly high mass number and atomic number.

Thus, $${}^{124}_{50}Sn + {}^{1}_{1}H(335\,MeV) \rightarrow {}^{126}_{53}I$$

Here ${}^{124}_{50}Sn$ emits ${}^{3}_{1}T$ as emitted particle in the first stage of spallation, then ${}^{3}_{1}H$ or ${}^{3}_{1}T$ unites with rest ${}^{124}_{50}Sn$ to form ${}^{126}_{53}I$ and ${}^{1}_{0}n$ in the secondary reaction.

Sometimes high-energy projectile with medium or high target gives product nucleus of mass numbers ranging from ${}^{18}_{9}F$ to ${}^{32}_{15}P$. The products are too lighter than the products of fission or spallation.

This is known as *nuclear fragmentation reaction*.

The mechanism of spallation reaction, proposed by R. Serber, is the cascade process. High-energy particle enters the target, strikes a single nucleon, known as *knock-on particle*; it may be ejected or may strike another which may in turn be ejected or share its energy with other nucleon. In this way several knock-on particle may be ejected. The whole process takes place within the nucleus of radius 10^{-12} cm within time 10^{-22} second. The product nucleus is in the highly excited state. It may then eject several individual nucleons or a group of nucleons after redistribution of energy within 10^{-14} second. leading to spallation reaction or after expelling a few nucleons, the excited nucleus may split into two parts, thus undergoing fission reaction.

4.04 FISSION REACTION

In the fission process, a nucleus when interacts with a projectile breaks up into two nuclei of roughly similar mass with liberation of tremendous energy and emit a few neutrons. The projectiles are ${}^{1}_{0}n$ (usually used in the fission of heavy nuclei like ${}^{235}U$, Ra, Th) charged particles like ${}^{1}_{1}H$, ${}^{2}_{1}D$, ${}^{4}_{2}He^{+2}$ of moderately high energy, 30 MeV usually used in the fission of light nuclei. Thus, ${}^{4}_{2}He^{+2}$ of 400 MeV and ${}^{2}_{1}D$ of 200 MeV can cause fission of Bi, Pb, Tl, Ta, etc. The fission products depend on the energy of the projectile and also on the nature of the target. For heavy targets, the fission products range from 70 to 160 (as in U), for light nuclei they range from 60 to 120 (as with Ta-181). When energy of the projectile is low, then fission becomes asymmetric, i.e., fission products have different masses. But fission of heavy targets with ${}_0^1n$ is more important and used in the preparation of atom bomb. The 6.3 MeV γ-ray cause fission, it is called *photo-fission*. Thus, photo-fission of ${}^{232}Th$, ${}^{235}U$ requires 5.2–5.4 MeV γ-ray that for Bi and Au requires 85 and 315 MeV γ-ray respectively.

In 1939, Otto Hahn and F. Strassmann observed that when ${}^{235}U$ is bombard with ${}_0^1n$, several daughter nuclei are obtained of which one can be precipitated by Barium carrier and thought that ${}_{88}Ra$ has been formed ($T_{½}$ -3.5 hour). This assumed ${}_{88}Ra$ then undergoes β-decay forming another element that can be precipitated with La as carrier and it was assumed that ${}_{89}Ac$ has been formed. But its fractional precipitation of oxalate from nitric acid solution Mme Joliot Curie and P. Savitch definitely proved that the daughter element is Lanthanum with $T_{1/2}$ 3.5 hours. Hahn and Stressmann also found the Barium in the fission fragments, and another supposedly trans-uranic element homologue of Rhenium, but later it was found to be missing element Technetium-43.

Thus, the fission fragments of heavy nuclei are not fixed. ${}_{92}^{235}U$ produces only (through fission) 6.4% nuclei with mass numbers 95 and 139. Only 0.01% of symmetrical products mass number 117, but whatever may be the products, they are not stable, undergo β-decay giving about 250 fission products.

$$
{}_{92}^{235}U + {}_0^1n \rightarrow \left[\begin{array}{ll} \rightarrow {}_{42}^{103}Mo + {}_{50}^{131}Sn + 2{}_0^1n & (1) \\ \rightarrow {}_{36}^{92}Kr + {}_{56}^{141}Ba + 3{}_0^1n & (2) \\ \rightarrow {}_{39}^{97}Y + {}_{53}^{135}I + 4{}_0^1n & (3) \\ \rightarrow {}_{42}^{95}Mo + {}_{57}^{139}La + 2{}_0^1n + 7e^{-1} & (4) \end{array}\right] + 200\text{ MeV}
$$

The fission yield curve of ${}_{92}^{235}U$ can be given as follows in Fig. 4.2.

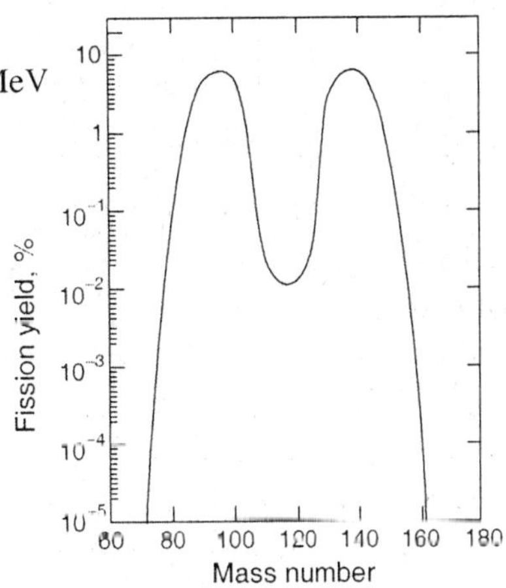

Fig 4.2 Yields in slow neutron fission of ${}_{92}^{235}U$.

In each case ${}_0^1n$ are emitted which in turn can initiate fission of other ${}_{92}^{235}U$ atom and in this way a self-propagating chain reaction occurs emitting tremendous energy and ultimately causing explosion. But if some ${}_0^1n$ absorber such as Cd is used, chain propagation can be controlled and a number of ${}_0^1n$ can be maintained at a constant level. The self-sustaining chain reaction can be presented in this way in Fig. 4.3.

The energy released in the fission reaction can be understood from total mass before fission and the total mass of stable product nucleus formed. In most cases the mass of the product nucleus ranges from 95 to 139, the energy released in reaction (4) in MeV is $E = 931[(235.0439 + 1.0087) - (94.9058 + 138.9061 + 2 \times 1.008)]$

$E = (236.0526 - 235.8279)\ 931$

$= 0.2247 \times 931 = 209.2$ MeV or approx 200 MeV

$E = 200$ MeV $= 200 \times 1.6 \times 10^{-13}$ J per nucleus [1 MeV $= 1.6 \times 10^{-6}$ erg]

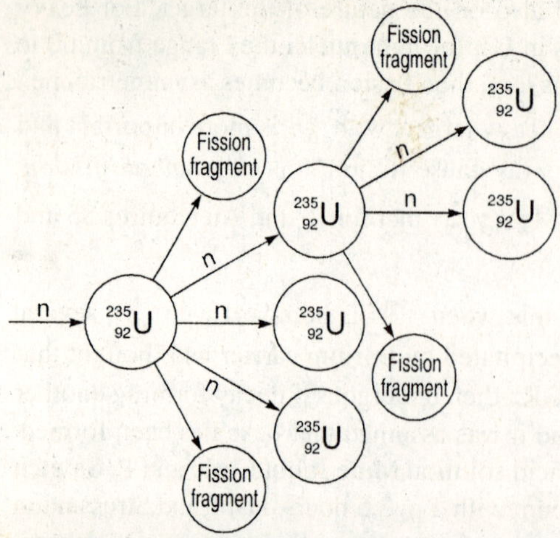

Fig 4.3 Schematic representation of chain process in nuclear fission.

So, fission of 1 kg ^{235}U gives $200 \times 1 \cdot 6 \times 10^{-13}$ $6.023 \times 10^{23} \times \frac{10^3}{235}$, i.e., 8.2×10^{13} J which is equivalent to the energy released in the explosion of 20 000 tons TNT. Again 1 J = 1 watt-sec. So 1 kg U-235 will yield 2.3×10^7 kilowatt-hour, i.e., 1000 megawatt per day; if only 30% of this energy is converted to electrical energy, 300 megawatt per day energy will be obtained. This amount of energy can be obtained by burning 2500 tons of coal per day.

It has been estimated that ^1_0n captured by ^{235}U requires 10^{-14} sec to emit secondary neutron. The average time to capture this secondary ^1_0n by ^{235}U takes place within 10^{-8} sec, so energy is emitted within 10^{-6} sec.

Now, if the piece of ^{235}U is large enough, the secondary ^1_0n produced will find targets and will be taken up and ultimately cause explosion. If the piece is small, the secondary ^1_0n produced will not find target and will escape and the chain will break and there will be no explosion. So, for causing explosion, there will be a limiting or critical size of the target for production of tremendous energy.

The liberation of energy due to fission of heavy targets can also be understood on consideration of binding energy values. The B.E. per nucleon for heavy target is 7.6 MeV per nucleon and that for stable fission fragments is 8.5 MeV per nucleon. So release of energy is

$$E = (8.5 - 7.6)\ 236\ \text{MeV} = 212.4\ \text{MeV}.$$

The mechanism of fission can be understood on the basis of liquid-drop model where surface tension led the liquid drop to be spherical. In the nucleus the attractive force between nucleons causes the nucleus to be spherical and the repulsive force causes the drop to elongate. In the stable nucleus these two forces balance each other. With increase in mass number, the repulsive force among the nucleons increases and it can be shown that no nucleus above $Z = 100$ is stable. High-target nucleus when captures ^1_0n to form the compound nucleus, the energy of the nucleus is now the binding energy of the added ^1_0n plus its K.E. The nucleus now either emits γ-ray or any particle if the energy is sufficient. But there is also the probability that the compound nucleus will undergo strong oscillation and the nucleus then passes from spherical shape *A* to the elongated stage *B*. If the gain in energy is insufficient to cause further deformation, then it will pass to *A* with the release of γ-ray or any particle. If the excitation energy is still higher, the stage *B* changes to dumb-bell shape stage *C*. From this stage restoration to *A* is impossible as the electrostatic repulsion at the two ends of *C* is high and the nuclear binding force operating in the constricted region cannot bring the two ends closer. Consequently, *C* changes to stage *D* with the release of high energy.

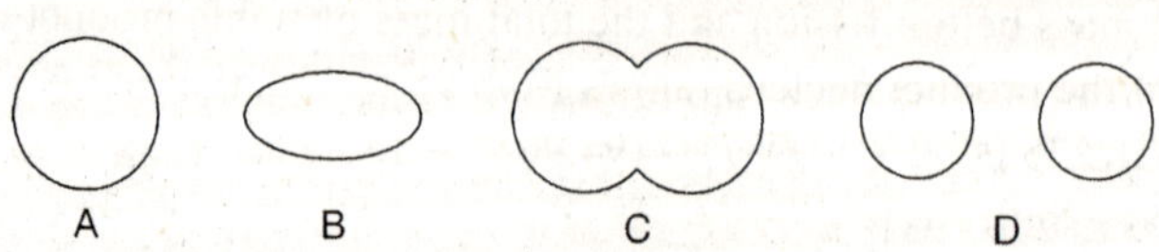

Fig 4.4 Interpretation of nuclear fission by liquid-drop model (*A*) Spherical compound nucleus (*B*) deformation into ellipsoidal (*C*) narrowing and critical state (*D*) separation.

Now every nucleus undergoing fission should have some critical energy or activation energy. This is the energy that must be supplied to the original nucleus in order to deform it to state C where the repulsive energy overcomes the surface energy. Now,

$$E_{\text{repulsion}} \propto \frac{Z^2}{A^{\frac{1}{3}}}; \qquad E_{\text{surface}} \propto A^{\frac{2}{3}}; \text{ and } \qquad \frac{E_{\text{repulsion}}}{E_{\text{surface}}} = \frac{Z^2}{A}.$$

Larger the value of $\frac{Z^2}{A}$, smaller will be the critical energy. When $\frac{Z^2}{A} = 50$, the nucleus is very much unstable and can be achieved by the element $Z = 115$ and $A = 265$. When $\frac{Z^2}{A} = 40$, it can undergo fission spontaneously without energy being supplied. The nucleus do not undergo fission if $\frac{Z^2}{A} < 30$. The values for ^{239}Pu = 37, ^{235}U = 36 and ^{233}U = 36.4. So these nuclei after capturing $_0^1n$, will undergo fission.

The value of critical energy can be calculated more accurately for ^{235}U and ^{238}U, and we can understand why ^{235}U undergoes fission but not ^{238}U.

$$E_{\text{crit}} = 0.89\,A^{\frac{2}{3}} - 0.02\frac{Z(Z-1)}{A^{\frac{1}{3}}}\text{ MeV} \qquad [A \text{ is the mass of compound nucleus}]$$

$$E_{\text{crit of U-235}} = 0.89(236)^{\frac{2}{3}} - 0.02\frac{92\times 91}{(236)^{\frac{1}{3}}} = 6.9 \text{ MeV}$$

$$E_{\text{crit of U-238}} = 0.89(239)^{\frac{2}{3}} - 0.02\frac{92\times 91}{(239)^{\frac{1}{3}}} = 7.3 \text{ MeV}$$

Now gain in binding energy is due to addition of $_0^1n$ and its kinetic energy. Additional B.E. can be calculated from the energy equivalent of mass difference.

For U-235, additional binding energy

$\Delta E = [236.0440 - (235.0439 + 1.0087)]$ 931 MeV = ·0086 × 931 MeV = 8MeV

This value is higher than the critical value and hence undergoes fission.

For U-238, isotope additional B.E.,

ΔE = [239.0530 − (238.0493 + 1.0087)] 931 MeV − [239.0530 − 239.0580] 931 MeV

= 0.005 × 931 MeV = 4.655 MeV

This is less than the critical value and will not undergo fission.

Now, the question arises why natural Uranium mineral does not undergo fission: (1) ^{235}U is not found pure in nature, but in the form of compound mixed with other compounds. (2) ^{235}U is less than 1% in nature and U-238 is 99%. So the critical mass for fission is not reached.

4.05 FUSION REACTION OR THERMONUCLEAR REACTION

The mass defect and packing fraction of different elements show that atoms of lower atomic weight and higher atomic weight region have positive mass defect, i.e., extra energy is needed to bind the nucleus together to get stability whereas elements of atomic weight of middle order $40 < A < 180$ have negative mass defect, i.e., energy is lost to bind the nucleons to get stable nucleus. So, also the binding energy curve indicates that the elements of low and high atomic weight region have less binding energy per nucleon and that of middle order elements have 8.5 MeV per nucleon. So fission of high order elements will liberate high energy and will be converted to stable elements of middle order. Similarly, there is possibility that union of two or more light nuclei to form an element of middle order will also liberate tremendous energy as a result of loss of some mass. This type of fusion reaction can hardly do actually accelerating the light nuclei because in that case the process will not be self-sustaining. Eddington in 1920 proposed that stellar energy is obtained from the fusion of four hydrogen atoms to form helium at a very high temperature, as large amount of hydrogen and helium is present in the sun. The reaction can be stated as $4\,{}^{1}_{1}H \rightarrow {}^{4}_{2}He + 2e^{-1} + 26.8$ MeV.

As this reaction occurs at a very high temperature in the stars, it is called *thermonuclear reaction*. In order to interact two light nuclei, say hydrogen, each one should have 0.1 MeV energy to cross the potential barrier. For higher atoms higher energy is required. The hydrogen atoms attain this 0.1 MeV energy if temperature is raised to 1000 million degrees Kelvin. In the stars the temperature is lower than this. But, as per Maxwellian distribution of energy most of the atoms are actually in the plasma state consisting of (+)ve charged ion and (–)ve electron. They have average energy of 0.002 MeV but a certain proportion will have energy exceeding the average energy and is at about 0.1 MeV and these can take part in thermonuclear reaction.

In 1938, H.A. Bethe proposed two sets of thermonuclear reaction to account for the stellar energy

(A) the carbon cycle process where carbon atoms act as a catalyst at above 20 million degrees Kelvin to unite four hydrogen atoms to form a helium atom as in hot stars and

(B) the proton-proton chain reaction that occurs at 15 million degrees Kelvin as is in cold or small stars. Thus for thermonuclear reaction to occurs three physical conditions are to be maintained:

1. Very high temperature must be attained.
2. Sufficient plasma density is required.
3. Plasma must be confined for the time to allow fusion reaction to occur.

THE TWO PROCESSES CAN BE GIVEN AS FOLLOWS

(A) *Carbon cycle process*: At 20 million degrees Kelvin as in hot stars carbon-12 acts as a sort of catalyst to unite protons to form helium atom with the liberation of fusion energy.

(i) ${}^{12}_{6}C + {}^{1}_{1}H = {}^{13}_{7}N + E_1$

(ii) ${}^{13}_{7}N = {}^{13}_{6}C + \beta^{+}\left(T_{\frac{1}{2}} = 10.1\,\text{sec}\right) + \nu$

(iii) ${}^{13}_{6}C + {}^{1}_{1}H = {}^{14}_{7}N + E_2$

(iv) ${}^{14}_{7}N + {}^{1}_{1}H = {}^{15}_{8}O + E_3$

(v) ${}^{15}_{8}O = {}^{15}_{7}N + \beta^{+}\left(T_{\frac{1}{2}} = 2.05\,\text{min}\right) + \nu$

(vi) ${}^{15}_{7}N + {}^{1}_{1}H = {}^{12}_{6}C + {}^{4}_{2}He$

So the overall reaction is

$4\,{}^{1}_{1}H = {}^{4}_{2}He + 2\beta^{+} + E\,(E = 26.7\text{ MeV})$ Mass of H = 1.00757, mass of He = 4.00386

(B) The proton-proton chain reaction occurs at 15 million degrees Kelvin as in small or cold stars where two ${}^{1}_{1}p$ fuse to form ${}^{2}_{1}D$ as follows:

(i) ${}^{1}_{1}H + {}^{1}_{1}H = {}^{2}_{1}D + \beta^{+} + E_1$

(ii) ${}^{2}_{1}D + {}^{1}_{1}H = {}^{3}_{2}He + E_2$

(iii) $2{}^{3}_{2}He = {}^{4}_{2}He + 2{}^{1}_{1}H + E_3$

Here, also total energy liberated is 26.7 MeV. In the sun, the inner core favours proton-proton chain reaction and the outer part carbon cycle reaction when temperature rises to proper degree.

The sun has 2×10^{30} kg H_2 in it and the rate at which energy is received from sun shows that 2×10^{19} kg of H_2 is used up annually. If it is taken that 5% of H_2 is consumed so far, i.e., 10^{29} kg has been consumed, so earth will receive solar energy still 5×10^{9} years. This fusion reaction is the main principle adopted in the preparation of hydrogen bomb.

In hydrogen bomb, prior to fusion, the reaction temperature is raised by fission of ^{235}U or ^{239}Pu by slow neutron from D_2O. Fusion reaction is initiated by the bombardment of Lithium deuteride with ${}^{1}_{0}n$.

$${}^{6}_{3}Li{}^{2}_{1}H + {}^{1}_{0}n \rightarrow {}^{3}_{1}H + {}^{4}_{2}He + {}^{2}_{1}H$$

A-20 megaton bomb contains about 140 kg ${}^{6}_{3}LiD$ and ^{235}U or ^{239}Pu of considerable amount. Energy liberated by 1 kg ${}^{2}_{1}D$ on fusion is equivalent to the energy liberated by 65 000 tons of TNT, which is much higher than the energy liberated by atom bomb.

The comparison of fission reaction and fusion reaction in regard to the energy liberated can be understood from the percent efficiency.

$$\% \text{ efficiency in fission reaction} = \frac{200 \times 100}{(235+1) \times 931} = 0.09\%$$

$$\% \text{ efficiency in fusion reaction} - \frac{26.7 \times 100}{4 \times 931} = 0.63\%$$

For the reaction $4{}^{1}_{1}H = {}^{4}_{2}He + 2\beta^{+} + E$

So the devastating power of hydrogen bomb is about 7 times than that of atom bomb.

COLD FUSION

When ${}^{2}_{1}D_2O$ is electrolyzed using Pd electrodes, much heat is produced and it was suggested that the energy is obtained due to fusion of ${}^{2}_{1}D$ into ${}^{3}_{1}T$. This is known as *cold fusion*, i.e., fusion at ordinary temperature.

$$2{}^{2}_{1}D = {}^{3}_{1}T + {}^{1}_{0}n + E$$

But this has been detected in a limited number of cases and cannot be established firmly.

4.06 INDUCED RADIOACTIVITY

When light elements like B, Mg and Al particularly are bombarded by α-particle from naturally occurring radioactive element Po, the reaction is (α, *p*) type as discovered by Rutherford.

$${}^{27}_{13}Al + {}^{4}_{2}He \rightarrow {}^{30}_{14}Si + {}^{1}_{1}H$$

But I. Joliot Curie and her husband F. Joliot in 1934 observed that along with $^{1}_{1}H$, $^{1}_{0}n$ β^+ is emitted. If source of α-particle is cut off, emission of $^{1}_{1}p$ and $^{1}_{0}n$ is stopped but β^+ emission still goes on for a short time and has the characteristic of a radioactive disintegration. It has been established that radioactive phosphorus $^{30}_{17}P$ has been prepared that undergo β^+ emission. The $^{30}_{15}P$ isotope does not occur in nature and hence called *man-made* or *artificial radioactive isotope*.

$$^{27}_{13}Al + ^{4}_{2}He \rightarrow ^{1}_{0}n + ^{30}_{15}P^*$$

$$^{30}_{15}P^* \xrightarrow{3.2\ min} ^{30}_{14}Si + \beta^+$$

The formation of $^{30}_{15}P^*$ was established in this way.

The irradiated Al-foil when dissolved in HCl, H_2 gas is evolved which carries with it the active species presumably $^{30}_{15}P^*H_3$ containing active $^{30}_{15}P^*$. Again, if the exposed Al is dissolved in a mixture of HCl + HNO_3 and small amount of Na_2HPO_4 is added as carrier followed by Zirconium nitrate solution, the whole activity is precipitated as Zirconium phosphate indicating the presence of radioactive phosphorus $^{30}_{15}P^*$. As radioactive isotopes of different types of the same element have been prepared now, it is customary to mention $T_{\frac{1}{2}}$ of the isotope with mass number, e.g., 3.2 min ^{30}P. Other reaction leading to the formation of radioactive isotope are:

$$^{10}_{5}B + ^{4}_{2}He \nearrow ^{13}_{6}C + ^{1}_{1}H$$

$$^{10}_{5}B + ^{4}_{2}He \searrow ^{13}_{7}N^* + ^{1}_{0}n \qquad ^{13}_{7}N^* \xrightarrow{11\ min} ^{13}_{6}C + \beta^+$$

$$^{24}_{12}Mg + ^{4}_{2}He \rightarrow ^{27}_{14}Si^* + ^{1}_{0}n \qquad ^{27}_{14}Si^* \xrightarrow{7\ min} ^{27}_{13}Al + \beta^+$$

Neutron with high energy $^{27}_{13}Al$ reacts as

$$^{27}_{13}Al + ^{1}_{0}n \nearrow ^{4}_{2}He + ^{24}_{11}Na^* \xrightarrow{15\ hr} ^{24}_{12}Mg + \beta^-$$

$$^{27}_{13}Al + ^{1}_{0}n \searrow ^{1}_{1}H + ^{27}_{12}Mg^* \xrightarrow{10\ min} ^{27}_{13}Al + \beta^-$$

With moderate neutron the following reaction occurs:

$$^{27}_{13}Al + ^{1}_{0}n \rightarrow ^{26}_{13}Al^* + 2^{1}_{0}n \qquad ^{26}_{13}Al^* \rightarrow ^{26}_{12}Mg + \beta^+$$

With slow $^{1}_{0}n$

$$^{27}_{13}Al + ^{1}_{0}n \rightarrow \gamma + ^{28}_{13}Al \xrightarrow{2.3\ min} ^{28}_{14}Si + \beta^-$$

The radioactive isotopes are very useful in biology, medicine, agriculture, metallurgy, etc.

The modes of decay of radioactive isotopes follow the same decay mode as natural radioactive elements and depend on the $\frac{n}{p}$ ratio. Sometimes they decay by (i) K-electron and (ii) internal conversion of γ-ray.

CHAPTER

5

ISOTOPES

5.01 ISOTOPES

As discussed in the radioactive disintegration series, any radioactive element emitting successively one α and two β particles produces the same element as the parent but with different atomic weights or mass numbers. Such elements are called isotopes (*iso* = same, *topes* = place), i.e., they occupy the same position in the periodic table. These have equal number of proton but different number of neutron in the nucleus. They are chemically identical and inseparable by any chemical methods, e.g., Radium-C′, Radium-A, Radium-F, Actinium-A, Actinium-C′, Thorium-A and Thorium-C′ all occupy the 84th position in Periodic Table. All are isotopic with $_{84}Po$.

ISOBARS

The nuclei, which have equal mass number but different number of proton and neutron are called *isobars*, e.g., $^{40}_{20}Ca$ and $^{40}_{19}K$. As these nuclei have different number of $^{1}_{1}p$ and $^{1}_{0}n$, their binding energies will be different.

As (a) atomic number Z increases, the number of excess $^{1}_{0}n$ in the nucleus will have less effect, (b) *p-p* repulsion will increase, (c) the application of odd-even rule will have the effect having odd number of $^{1}_{0}n$ and $^{1}_{1}p$. All these combined effects will be responsible for the difference in binding energies of the isobars. As the nucleus has different binding energies, the radioactive property will be different. The isobars of even-even nature will be more stable than the odd-odd nature.

ISOTONES AND ISODIAPHERES

The nuclei having the same number of neutron in the nucleus are called isotones. Isotones will have also different radioactivity. It is evident, therefore, that radioactivity is governed by the particular combination of neutron and proton in the nucleus, e.g., $^{226}_{88}Ra$ and $^{227}_{89}Ac$, $^{2}_{1}H$ and $^{3}_{2}He$.

Nuclei of different elements containing same excess of neutron over protons (A-2 *Z*) are called isodiapheres, e.g., ${}^{37}_{17}Cl$ and ${}^{39}_{18}Ar$.

HETERO-BARIC ISOTOPES

They are the general isotopes having same number of proton but different number of neutrons in the nucleus. So atomic number is same but atomic weight is different, e.g., ${}^{37}_{17}Cl$ and ${}^{35}_{17}Cl$.

5.02 ISOBARIC ISOTOPES OR NUCLEAR ISOMERS

Nuclear isomers undergo another type of radioactive decay. The isobaric isotopes having the same atomic number and same mass number have different radioactive property. These are called *nuclear isomers.* In radioactive disintegration series, the branched disintegration occurs due to the presence of nuclear isomers. Hahn in 1921, first established natural nuclear isomers were $U - Z$ and $U - X_2\left({}^{234}_{91}Pa\right)$. Artificial nuclear isomers were observed with Br_2 and established by cross-checking method.

$$\xrightarrow{\text{Slow } {}^{1}_{0}n} Br_2 \left[\begin{array}{l} \xrightarrow{\gamma} T_{\frac{1}{2}} = 17.6\,\text{min (Product – I)} \\ \xrightarrow{\gamma} T_{\frac{1}{2}} = 4.5\,\text{hr (Product – II)} \\ \xrightarrow{\gamma} T_{\frac{1}{2}} = 35.3\,\text{hr (Product – III)} \end{array} \right.$$

Bromine contains two isotopes 79 and 81. So, two different $T_{\frac{1}{2}}$ would be expected for 80 and 82 isotopes of Bromine.

$$\xrightarrow{{}^{1}_{0}D} Br_2 \left[\begin{array}{l} \xrightarrow{{}^{1}_{1}p} \text{Product – I, } T_{\frac{1}{2}} = 17.6\,\text{min} \\ \xrightarrow{{}^{1}_{1}p} \text{Product – II, } T_{\frac{1}{2}} = 4.5\,\text{hr} \\ \xrightarrow{{}^{1}_{1}p} \text{Product – III, } T_{\frac{1}{2}} = 35.3\,\text{hr} \end{array} \right.$$

Here, also target nuclei are 79 and 81 isotopes and product nuclei are 80 and 82 isotopes. Hence, it is clear that one of 80 and 82 isotopes decays in two different ways. Again,

$$\xrightarrow{\gamma\text{-rays}} Br_2 \left[\begin{array}{l} \xrightarrow{{}^{1}_{0}n} \text{Product – I, } T_{\frac{1}{2}} = 6.4\,\text{min} \\ \xrightarrow{{}^{1}_{0}n} \text{Product – II, } T_{\frac{1}{2}} = 17.6\,\text{min} \\ \xrightarrow{{}^{1}_{0}n} \text{Product – III, } T_{\frac{1}{2}} = 4.5\,\text{hr} \end{array} \right.$$

The product nuclei have mass numbers 78 and 80. It is evident that ${}^{80}_{35}Br - 80$ isotope decays in two different ways of $T_{\frac{1}{2}} = 17.6$ minutes and 4.5 hours and exhibit nuclear isomer. The difference between the isomers is due to the difference in nuclear energy state, the lower energy state is called the *ground state* and the higher one is called the *metastable state*, which often undergoes transition to the ground state emitting γ-rays within very short time 10^{-13} sec. If this transition is forbidden to some extent, the excited state will have appreciable stability.

About 150 cases of nuclear isomerism have been detected in medium and moderately high atomic weight region as a result of various types of nuclear reactions.

e.g., ${}^{82}_{36}Kr\,(d, p)\,{}^{83}_{36}Kr^{m}$, ${}^{80}_{34}Se\,(\alpha, n)\,{}^{83}_{36}Kr^{m}$

Nuclear isomers are of three different types:

(A) Isomers with independent decay: In this class of isomers, the transition from metastable state of the parent to the ground state is highly forbidden and transition takes place to a very small extent. So, the metastable and the ground state decay independently with their own $T_{\frac{1}{2}}$ with β-decay and led to the same ground state decay product. Sometimes the parent nucleus yields excited decay product which then emit γ-ray,

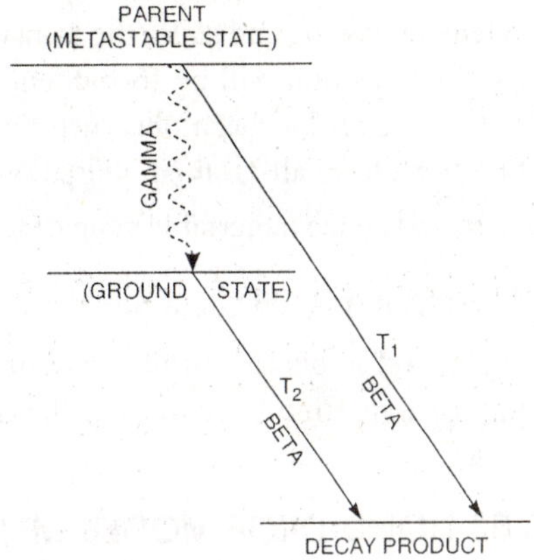

Fig 5.1 Nuclear isiomerism with independent decay.

e.g., $^{52}Mn\left(\beta^+, 5.7\text{days};\ \beta^+, 21\text{min}\right)$

$^{71}Zn\left(\beta^-,\ 4.1\text{hr};\ \beta^-, 2.4\text{min}\right)$

$^{91}Mo\left(\beta^-, 65\text{sec};\ \beta^-,\ 15.5\text{min}\right)$

(B) These are the transitions of nuclear isomers where the metastable parent element decays to the ground state with definite $T_{\frac{1}{2}}$, say, T_1, with the emission of γ-ray. This is known as isomeric transition, IT. In most cases, the γ-ray is internally converted to give the line spectra of electron together with X-ray characteristic of the parent. The ground state then decays to the product nucleus with different $T_{\frac{1}{2}}$, say T_2, emitting β-ray. Examples are as follows:

$^{69}Zn\left(\text{IT}, 13.8\text{hr}; \overline{\beta}\ 55\text{min}\right)$

$^{80}Br\left(\text{IT}, 4.5\ \text{hr}; \overline{\beta}\ 17.6\text{min}\right)$

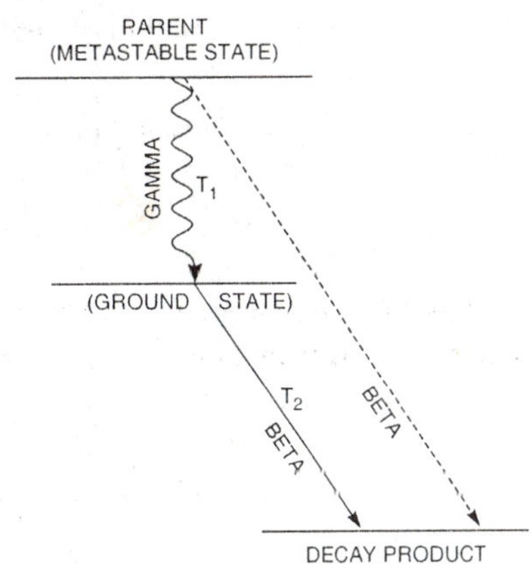

Fig 5.2 Genetically related isomers: isomeric transition.

Isomeric transition may sometimes cause the breaking of chemical bond that makes it possible to separate the isomeric nuclei.

(C) The third type of isomerism is observed where the metastable parent element decays to the stable ground state by isomeric transition, accompanied by γ-radiation. If γ-radiation is internally converted then, as before, line spectrum of electron and X-radiation is observed, e.g., ^{83}Kr, ^{107}Ag, ^{197}Au. Investigation shows that the isomeric pairs have more or less the same maximum energy, e.g.,

^{80}Br – 17.6 m, 2 MeV and 4.5hr, 2.05 MeV

^{104}Rh – 45 sec, 2.754 MeV and 4.3 m, 2.76 MeV

All these isomeric pairs can be separated. This indicates two things: (1) the isomers have same atomic number and atomic mass, same chemical and physical properties but can be considered as different atoms and (2) generally the short-lived isomers grow from the long-lived isomers as indicated in many cases. This indicates that the nuclear isomers are genetically related.

REASONS FOR NUCLEAR ISOMERISM

The transition from the excited state to the ground state of the nucleus depends upon the energy difference between the states and the multipolar character of radiation. We shall consider only the spin character of the nucleus.

In general, if the spin quantum number of the parent nucleus is the same as that of the daughter nucleus, in the ground state, the transition is allowed and no γ-radiation is obtained. If they differ, then the transition will be forbidden, there will be γ-radiation and the daughter nucleus is in the excited state. It has been observed that nuclei having even number of neutrons and even number of protons, have all of them paired, as a result, resultant spin in the ground state and excited state is zero and hence metastable state of long $T_{1/2}$ is not observed. Nuclei with odd number of ${}_0^1n$ or ${}_1^1p$, the resultant spin (odd multiple of ½) is very large, e.g., $\frac{11}{2}$ in excited state, it will have moderately long $T_{1/2}$ and so nuclear isomers are observed. Now the spin of the nucleus is determined from the study of hyperfine structures in the atomic spectra and study of alternating intensities in band spectra.

5.03 LIQUID-DROP MODEL OF NUCLEUS

Inadequacy in explaining quantitatively the forces between nucleons and hence its stability, Meitner and Frisch and later modified by Bohr and Wheeler (1936) introduced the liquid-drop model of the nucleus having the following similarities with a liquid drop:

1. Like a liquid drop, nucleus has close packed particles, constant density, and interaction between nucleons are short range, i.e., they interact with particles in its close vicinity but no effect on others apart. Hence, sometimes called *independent particle model.*
2. Evaporation of liquid can be imagined analogues to the radioactive emanations to release the excess energy of the nucleus.
3. Like liquid drop, nucleus is homogeneous and incompressible. Like surface tension of liquid, the interaction of nucleons at the surface of the nucleus renders it to have spherical shape.
4. With large liquid drop, the surface tension force is less than the dissipative force and hence the spherical drop becomes unstable. Similarly, for a large nucleus the *p-p* repulsive force (electrostatic) is the dissipative force that exceeds the *n-n* and *n-p* force analogous to surface tension force and hence the nucleus is unstable and shows the natural radioactivity.
5. When two small liquid drops come in contact, it may fuse to form a bigger drop. Similarly, for a small nucleus where the surface tension force prevails, the small nuclei may fuse and undergo fusion reaction. When the *p-p* electrostatic repulsive forces have upper hand (as in case of bigger nucleus) the nuclei undergo fission reaction even simply by bombardment with a projectile.

   Bohr and Wheeler calculated the surface tension force and electrostatic force of different nuclei and concluded that the surface tension force predominates up to mass number 110, hence very stable nucleus and electrostatic repulsive force prevails for nuclei beyond 110 and hence potentially unstable.
6. If the nucleus gets external energy, the excess energy on a nucleon is readily shared around the neighbours and so the energy level may be assumed as a quantum states of the nucleus.

Thus, liquid-drop model can better explain the different nuclear reaction and also helps to calculate the average binding energy per nucleon.

5.04 DEMERITS OF THE LIQUID-DROP MODEL AND INTRODUCTION OF SHELL MODEL, MAGIC NUMBER AND ISLAND OF ISOMERISM

After the discovery of neutron in 1932, T.H. Bartlett indicated the possibility of some quantum group, or shells for nucleons just as do the extra nuclear electrons.

(a) In 1948, M.G. Mayer and others came to the conclusion that nuclei containing magic number of nucleons of any kind 2, 8, 20, 50, 82 and 126 are exceptionally stable and present the closed shell in the nucleus.

(b) It has been observed that nucleus containing 20 and 50 protons and 20, 50 and 82 neutrons have the largest number of isotopes, e.g., ${}_{50}Sn$ has 10 stable isotopes of which ${}^{115}Sn$, ${}^{117}Sn$ and ${}^{119}Sn$ having odd number of neutrons are stable. Most abundant elements of the universe being ${}^{16}_{8}O$, ${}^{40}_{20}Ca$, ${}^{88}_{38}Sr$, ${}^{89}_{39}Y$, ${}^{90}_{40}Zr$, ${}^{118}_{50}Sn$, ${}^{138}_{56}Ba$, ${}^{208}_{82}Pb$ have magic number of either ${}^{1}_{1}p$ or ${}^{1}_{0}n$.

(c) If the elements have magic number of either ${}^{1}_{0}n$ or ${}^{1}_{1}p$, in heavy nucleus generally, the energy change during α-particle emission is exceptionally high.

(d) Spins and magnetic moments of nucleus having even number of ${}^{1}_{1}p$ and ${}^{1}_{0}n$ is almost zero, but with odd number of either ${}^{1}_{0}n$ or ${}^{1}_{1}p$ are all ½ integral. Again, isotopes of same odd number of ${}^{1}_{1}p$ but different even number of ${}^{1}_{0}n$ have same magnetic moment; and isotopes with same even number of ${}^{1}_{1}p$ and odd ${}^{1}_{0}n$ or even ${}^{1}_{0}n$ and odd ${}^{1}_{1}p$ have the same magnetic moment. This is clear that, as regard the moment, ${}^{1}_{1}p$ and ${}^{1}_{0}n$ behave independently and last odd ${}^{1}_{0}n$ or ${}^{1}_{1}p$ move in the potential energy field of either ${}^{1}_{0}n$ or ${}^{1}_{1}p$. Inside or outside of this potential energy field no force acts at the surface, only a nucleon experiences a strong attraction. Thus the nuclear energy states are similar to the extra nuclear energy shell of electron, the only difference is that electron moves independently in an electrostatic field of the nucleus, whereas nucleons move independently in the potential energy field of the nucleus, which is of short range. This potential energy field determines the quantum states of nucleons and hence its number in closed shells.

Like electronic shells, the nucleons are moving in orbits and also on their own axis. For any orbital quantum number l, it splits up into two states having total angular momentum quantum number J and $J = l + ½$ and $J = l - ½$ where ½ are the spin quantum number, each of these states can hold total $(2J + 1)$ number of either ${}^{1}_{1}p$ or ${}^{1}_{0}n$. The $l + ½$ is the lowest energy state and filled first, while with higher l value, separation energy of $l + ½$ and $l - ½$ is so large that $l - ½$ state goes to different shells. The different energy states and their filling with nucleons are as follows:

Table 5.1 : The different energy states in the nucleus and the number of nucleons that can be accommodated

Radial quantum number	Orbital quantum number	States	$2J + 1$ Total number of nucleon	Total number up to and including levels	J Total angular momentum
1	0	1*s*	2	2	$\frac{1}{2}$
1	1	1*p*	4	2 + 6 = 8	$\frac{3}{2}$ $\frac{1}{2}$
			2		
1	2	1*d*	6	8 + 6 + 2 + 4 = 20	5/2
2	0	2*s*	2		1/2

(Contd.)

1	2	1*d*	4		3/2
1 2 1 2 1	3 1 3 1 4	1*f* 2*p* 1*f* 2*p* 1*g*	8 4 6 2 10	20 + 8 + 4 + 6 + 2 + 10 = 50	7/2 3/2 5/2 1/2 9/2
1 2 3 1	4 2 0 5	1*g* 2*d* 3*s* 1*h*	8 6 4 2 12	50 + 8 + 6 + 4 + 2 + 12 = 82	7/2 5/2 3/2 1/2 11/2
1 2 3 1	5 3 1 6	1*h* 2*f* 3*p* 1*i*	10 8 6 4 2 14	82 + 10 + 8 + 6 + 4 + 2 + 14 = 126	9/2 7/2 5/2 3/2 1/2 13/2

Thus, closed shell configuration of the nucleus justifies the magic number.

In this concept, nucleus is assumed to be sphere. In fact, due to nucleon-nucleon interaction, it is deformed and the nucleus is considered as a drop with deformation. This is termed as *collective model*.

ISLAND OF ISOMERISM

Magic number of either ${}^{1}_{1}p$ or ${}^{1}_{0}n$ has closed nuclear energy shell and are very stable. It has been observed that nuclear isomers of odd mass number, i.e., having an odd number of either ${}^{1}_{1}p$ or ${}^{1}_{0}n$ can be classified into three groups: number of protons or neutrons in these groups are—(i) 39–49, (ii) 69–81 and (iii) 111–125, i.e., each group ends just before the magic number of nucleon, i.e., closed nuclear energy shell. Thus the isomers occur in clustures just before the closed shell. These are called *island of isomerism*.

5.05 DETECTION AND STUDY OF ISOTOPES

POSITIVE RAY ANALYSIS

Positive rays are the gaseous ions of substances obtained by passing high voltage electricity through the gas in an evacuated discharge tube using perforated cathode when some luminous beam is obtained that travels away from the cathode. When these positive rays are subjected to electric and magnetic field simultaneously, they are deflected and from the extent of deflection, $\frac{e}{m}$ of the ray and hence, mass m of the positive particle can be evaluated. Thomson in 1912 first developed the method and established the existence of ${}^{20}_{10}\text{Ne}$ and ${}^{22}_{10}\text{Ne}$ isotopes.

A narrow beam of positive ray travels in straight line in absence of any field and strikes the photographic plate at a point O (in Fig 5.3), called the undeflected spot, while moving downward the plane of the paper. For convenience, let us take a single positive particle of $\frac{e}{m}$ and velocity v. It is subjected to an electric field of strength X in the direction perpendicular to the ionic beam. The beam will be deflected in the direction of the field, it will not strike at the point O, but, say at M, at a certain distance x from O.

So, $OM = x$.

Therefore, $x = \frac{1}{2}ft^2$, considering initial velocity of the particle to be negligible compared to the accelerated velocity after application of the field.

$\therefore\ x = \frac{1}{2}\frac{Xet^2}{m} = \frac{1}{2}\frac{Xe}{m}\left(\frac{l}{v}\right)^2$ [l = distance of screen from the field and t = time of flight]

$= \frac{l^2}{2}\frac{Xe}{mv^2} = K_1\frac{Xe}{mv^2}$ [K_1 is constant depending on the dimension of the apparatus]

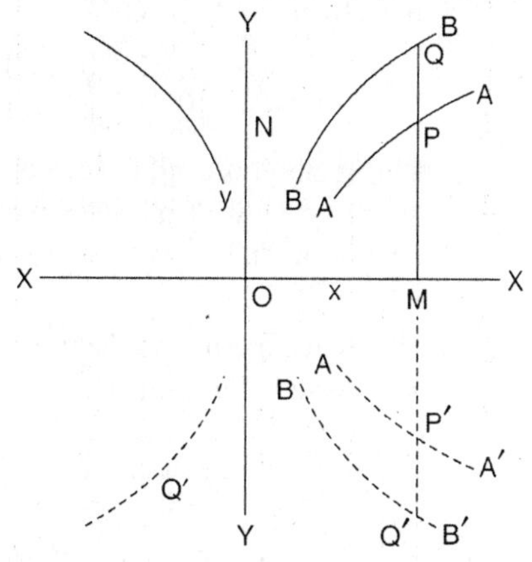

Fig 5.3 Thomson's positive ray analysis.

Now, the particle is subjected to magnetic field H in such a way that the deflection due to magnetic field is perpendicular to the direction of the beam itself and also perpendicular to the direction of deflection due to electric field. Let $y = ON$ is the deflection along Y-axis and Hev is the magnetic force.

$\therefore\ y = \frac{1}{2}ft^2 = \frac{1}{2}\frac{Hev}{m}\frac{l^2}{v^2} = \frac{l^2}{2}\frac{He}{mv} = K_2\frac{He}{mv}$ [K_2 is constant depending on the dimension of the apparatus]

If the electric and magnetic field is applied simultaneously, the positive particle will strike at P whose co-ordinates are (x, y). By eliminating v from these two equations

$$v^2 = \frac{K_1Xe}{mx} = \frac{H^2e^2K_2{}^2}{m^2y^2} \quad \text{or,} \quad \frac{y^2}{x} = \frac{K_2{}^2}{K_1}\times\frac{H^2}{X}\times\frac{e}{m} = K\times\frac{H^2}{x}\cdot\frac{e}{m}.$$

Here K, H, X all are constant. So, $y^2 = px$.

Now,

(a) Instead of one positive particle, a beam having same $\frac{e}{m}$ but different velocity, will have different deflections and will fall at different points constituting the parabola AA; Particles of low velocity will be deflected more and of high velocity will fall at close point from the no deflection point O.

(b) The particles having same $\frac{e}{m}$ and v will fall on the same point.

(c) Particles having same $\frac{e}{m}$ but different from the above considered will be deflected to the different extent and will form another parabola BB. If the $\frac{e}{m}$ is greater, the displacement $y > x$ and so the parabola BB will be at higher position than AA. If charge or mass is different, different parabola can be obtained.

(d) This method can be used for comparing the masses of the particles. We know,

$$x = K_1\times\frac{Xe}{mv^2} = K_1\times\frac{Xe}{m}\times\frac{y^2m^2}{H^2e^2K_2{}^2} = K\times\frac{X}{H^2}\times\frac{m}{e}\times y^2$$

So, for the particle of smaller e/m, the displacement of y will be greater than x. So, we may write,

$$(PM/QM)^2 = \left(\frac{Y_A}{Y_B}\right)^2 = \frac{(e/m_A)}{(e/m_B)} = m_B/m_A$$

This method is used for detecting the presence of or even identifying and determining the atomic weight of particles whose masses differ from each other. In most cases, H_2 was present and taken as standard and by comparing the parabola, the masses of other particles are determined.

(e) In actual experiment, the reference line OX is not obtained on the plate. The magnetic field is reversed after same exposure when the mirror image curves $A'A'$ and $B'B'$ are obtained.

From this it is evident that $PM = \dfrac{PP'}{2}$ and $QM = \dfrac{QQ'}{2}$.

So, we may write, $\dfrac{m_B}{m_A} = \left(\dfrac{PP'}{QQ'}\right)^2$.

Here the charges of particles are considered to be same.

(f) If the isotope forms a doubly charged particle, then the deflection will be proportional to $\dfrac{2e}{m}$, i.e., $\dfrac{e}{\frac{m}{2}}$, i.e., the deflection will be like that of singly charged isotope with $\dfrac{m}{2}$ mass and the parabola will be a weaker one.

(g) In the experiment of preparing positive ray, when the gaseous ion passes through the cathode, it may collide with it and capture electron producing negative ion. The deflection will be in opposite direction and so another parabola in the other quadrant will be obtained. This type of curve is obtained with electronegative elements like C, O, Cl, etc.

(h) With the help of this method, some valuable chemical information can be obtained. If CH_4 is introduced in the discharge tube, parabolas corresponding to the mass of CH, CH_2 and CH_3 are obtained indicating that such species exist in the discharge tube for a considerable time.

LIMITATIONS OF THE METHOD

Positive ray analysis method has some drawbacks:

1. In the discharge tube although the pressure is very low, it is not negligible. The positive particles may collide with one another, gain or lose charge passing through the deflecting fields and may give secondary rays that produce fog on the photographic plate or may produce parabola that may mislead the result.
2. During passage through canal, the positive rays are lost by collision, the intensity becomes very low and hence, accurate measurements are not possible.

In spite of these difficulties, the method is most suitable for the indication of the presence of isotopes.

With the help of positive ray analysis method, several instruments have been developed for the detection and study of isotopes of which some are described here.

5.06 DEMPSTER'S MASS SPECTROGRAPH

In this mass spectrograph, the ion current of the element to be studied is prepared by vaporizing the substance electrically and then bombarding with electron. The narrow beam is obtained by

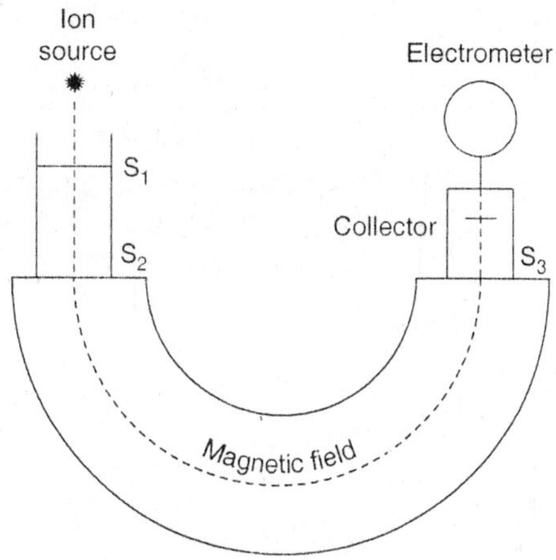

Fig 5.4 Sketch of Dempster's mass spectrograph.

passing it through the slit and then accelerated by the electric plates S_1, S_2, with high voltage 800–2000 volts. Thus the ion beams emerging from S_2 will have the same energy.

If V is the electrical potential and e the charge of the ion, then electrical energy is Ve that accelerates the beam, i.e., kinetic in nature. So we can write,

$$Ve = \frac{1}{2}mv^2 \text{ or, } v^2 = 2\frac{V_e}{m}.$$

The particles of same energy emerging through S_2 are then subjected to strong magnetic field acting perpendicular to the plane of the paper. According to left-hand rule, the beam will be deflected and describes a semi-circular path. The particle of suitable mass after traversing the path passes through the slits S_3 and reaches the plate of electrometer and produces ion current. The intensity of the current is the measure of positive ions reaching the plate, P. If H is the magnetic field and v the velocity, then

$$Hev = \frac{mv^2}{r} \quad \text{or, } r = \frac{mv}{He} \quad \text{or, } r^2 = \frac{m^2v^2}{H^2e^2}$$

$$\text{or, } \quad r^2 = \frac{m^2}{H^2e^2}\frac{2Ve}{m} = \frac{2V}{H^2}\frac{m}{e} \quad \text{or, } \frac{2V}{H^2r^2} = \frac{e}{m}$$

i.e., the particles of $\frac{e}{m}$ moving with definite radius can reach the electrometer. Hence this process is direction focusing (a) When H and V are constants, the particles of particular $\frac{e}{m}$ can pass through slit S_3 and reach the electrometer plate. Then H remaining constant V is varied, then positive particle of another mass will reach the plate. (b) As the intensity of current is proportional to the number of positive ion reaching plate, then plot of current against mass number gives the curve like this. Peak of the curve gives the relative abundance of isotopes, (Fig. 5.5). With the help of this method, Li, Mg, Ca, Zn isotopes are separated having the mass number of nearly integral value.

5.07 BAINBRIDGE MASS SPECTROGRAPH

This is the velocity focusing mass spectrograph and the arrangements are made in such a way that positive particle of different velocities describes semicircular path of different radius and reaches the photographic plate.

The positive beam after passing through two slits S_1 and S_2 becomes narrow, then through simultaneous electric field by P_1P_2 and the magnetic field H perpendicular to the plane of the paper. The two fields are arranged such that deflections produced by them balance each other. Only these undeviated ions will pass through the slit S_3.

Thus, electrical field and magnetic field balance each other. So, $Xe = Hev$ or, $v = \frac{X}{H}$. All others will be prevented by S_3. The ions of same velocity emerging from S_3 are then subjected to magnetic field perpendicular to the plane of the paper such that it traverses a semicircular path.

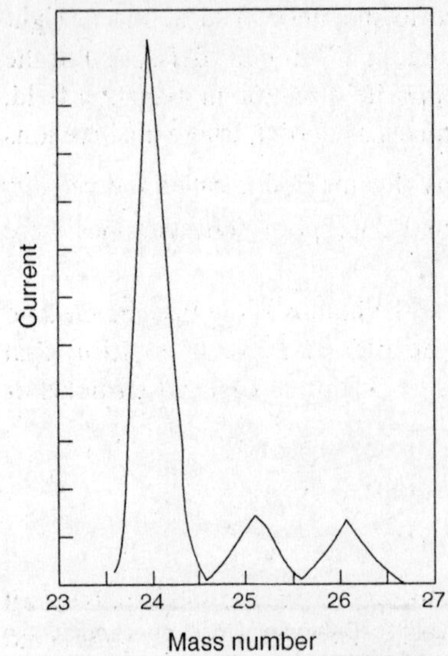

Fig 5.5 Curve obtained for magnesium with Dempster's mass spectrograph.

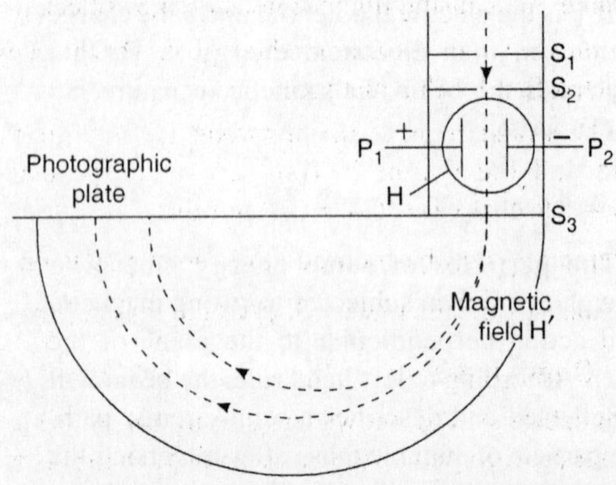

Fig 5.6 Plan of Bainbridge's mass spectrograph. S_1, S_2, S_3, slit; P_1P_2 electric field and H, H′ magnetic fields.

So, $H'ev = \frac{mv^2}{r}$ and $\frac{e}{m} = \frac{v}{H'r} = \frac{X}{HH'r}$ [v and $H'$are constant]

So, $\frac{e}{m} \propto \frac{1}{r}$, i.e., isotopes of different mass reach the plate at different points, heavier isotope describes path of greater radius.

5.08 ASTON'S MASS SPECTROGRAPH

In other process, the ions of same $\frac{e}{m}$ but different velocities are brought to focus on different points constituting the parabola and hence the intensities are very low. In Aston's apparatus, arrangements are made in such a way that all isotopes of same $\frac{e}{m}$ but different velocities are brought to focus on one point and hence has higher intensities. Aston applied electric field and magnetic field successively at right angles to each other such that the deflection due to electric field is reversed by the magnetic field. The electric field produces a slight dispersion of the positive ions through a certain angle depending on the velocity of the ions. Ions of low velocity spend greater time in the field and will be deflected to a greater extent than the ions of higher velocity and the beam will spread out into an electric spectrum.

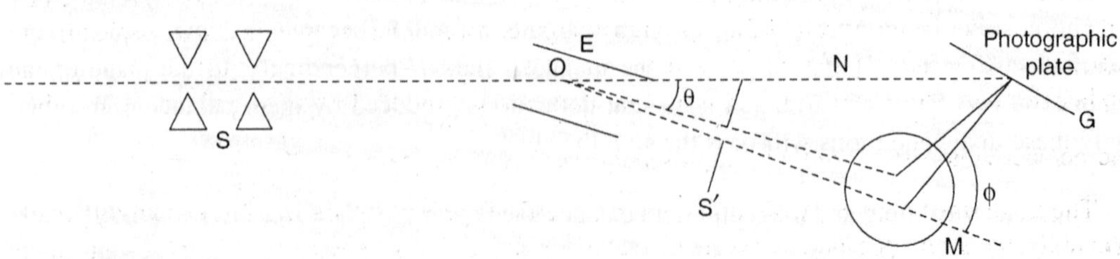

Fig 5.7 Schematic arrangement of Aston's mass spectrograph.

The divergent beams passing through slit S' are subjected to magnetic field acting at right angles to the plane of paper and deflect the beam in opposite direction. More the deflection of the beam in electric field, more is the deflection of the beam in opposite direction in magnetic field. The result is that the divergent beam becomes convergent and meets at a point, where positive ions of same $\frac{e}{m}$, causing higher intensity, are collected. That is why this method is called the *velocity focusing method.* As this method gives sharp focusing of ions at one point, rare isotopes were detected first by this method.

The arrangement of the apparatus is shown in the Fig 5.7. Calculations show that the electric field and the magnetic field are to be adjusted and to be placed at such position that $\phi = 4\theta$. In that case, $ON = GN$ and the plate should be placed at G. In that case, all particles of different $\frac{e}{m}$ will be focused at different points on the plate.

Analysis of this plate will give the mass of the unknown isotopes.

(a) Direct method

Here substances of known masses are mixed with unknown isotopic mass and then spots of all masses are recorded on the same plate. Then the distances of the spots of known masses are measured from a point on the plate and distances are plotted against mass numbers. This is called the *calibration curve*. From measuring the distance of unknown masses and by extrapolation of the curve, their masses can be determined.

(b) The co-incidence method

Here the spots of some known mass M is obtained on the plate with certain electric field X and magnetic field H. Then unknown mass M' is introduced and its spot is obtained on the same position for the known mass by varying electric field X' and magnetic field H'. Under that condition we can write,

$$\frac{X}{X'} \times \frac{H'^2}{H^2} = \frac{M'}{M}.$$

In practice, magnetic field is kept constant but electric field is varied. So, $\frac{X}{X'} = \frac{M'}{M}$, from which M' can be calculated.

5.09 SEPARATION OF ISOTOPES

Early attempts to separate isotopes met with little success. Later improved methods are helpful in the separation of isotopes in high state of purity. Very close similarity in physical and chemical properties of isotopes render their separation very difficult. As mass difference impart greater difference in physical properties than in chemical properties, the separation methods are physical in character. Chemical procedures are dependent upon the rates of particular reaction rather than upon different chemical behaviour.

The extent to which separation of isotopes is achieved in a particular process is represented by separation factor. It is the ratio of abundance of a given isotope in the enriched state to that in the initial state. If before separation, the mixture contains n_1 atoms of lighter isotope and n_2 atoms of the heavier one per gram of the mixture and, after application of the process, n_1' and n_2' of lighter and heavier isotopes respectively, then separation factor f is given by $f = \frac{n_1'/n_2'}{n_1/n_2} = \frac{f_1}{f_2}$.

The larger the value of f from unity, the more efficient is the separation. If f is slightly greater than unity, the extent of separation is small, and many successive stages are necessary for appreciable concentration. If the numbers of stages are x, then overall separation F is given by $F = f^x$. The different methods are as follows:

5.09(a) GASEOUS DIFFUSION METHOD

According to Graham's law of diffusion, the rate at which a gas diffuses through an aperture, which is small in comparison with the mean free path of the molecules, is inversely proportional to the square root of the density or molecular weight, i.e., $f = \sqrt{\frac{M_2}{M_1}} = \sqrt{\frac{D_2}{D_1}}$.

In a mixture of two isotopes, the lighter isotope will diffuse more readily than the heavier one, as at the same temperature, the lighter will move with higher speed than the heavier. As the diffusion proceeds, the concentration of the lighter isotope increases and probability of back-diffusion becomes greater, separation factor decreases and after certain time concentration of the isotopes on both sides of the barrier becomes equal. In actual practice, half of the mixture is allowed to diffuse and taken out to another vessel for diffusion.

From the above equation it can be seen that larger the ratio of the atomic weight of the two isotopes, greater will be their separation. So for ^1_1H and ^2_1D the separation will be $\sqrt{2}$, i.e.,1.414 and for ^{235}U and ^{238}U it is only 1.0014 for single diffusion. If n such successive operations are done, then $f = (1.0014)^n$. In case of volatile UF_6, all 4000 such operations give 99% enriched ^{235}U isotope. Here UF_6 is taken because the compound is volatile and fluorine is non-isotopic, hence separation will be due to the two isotopes of uranium.

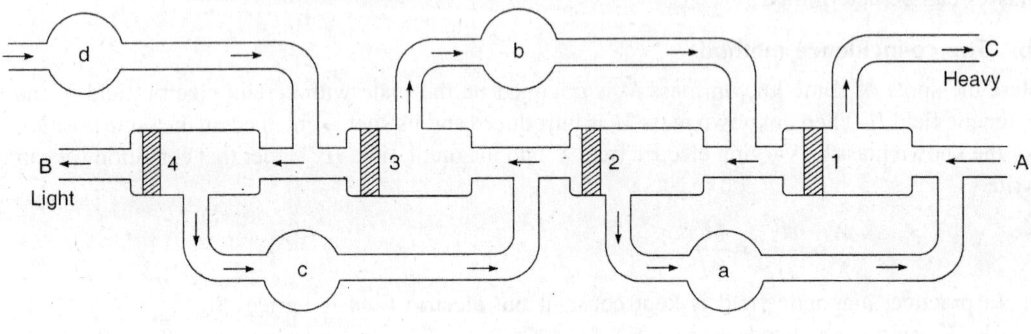

Fig 5.8 Separation of isotope by gaseous diffusion method.

Regarding the diffusion barrier it can be mentioned that pores of the barrier will be 1/10th of the mean free path, i.e., in the order of millionth part of a centimetre in diameter, at ordinary pressure. At decreased pressure, the mean free path increases, and so also the diameter of the pores of the barrier. If a suitable barrier is made by etching a very thin sheet of Ag-Zn alloy with HCl, then a large number of Zinc atoms are dissolved out leaving sub-microscopic holes in the sheet.

The gaseous isotopic mixture is fed into any intermediate unit. After diffusion the diffusate enriched in lighter one is fed to next unit in right side and the residue enriched with heavier isotope is pumped into the left unit. In this way the extreme right unit will contain lighter one and the extreme left will contain the heavier one. By this method O_2, C, N_2, Ne, Ar, etc., isotopes are separated effectively.

5.09(b) ELECTROMAGNETIC METHOD

The principle of this method is like that of mass spectrograph in which each isotopic species present in the ion beam is bent through different path by suitable arrangement of electric and magnetic field. Then by placing small receivers at suitable positions, separate isotopes can be collected, although amounts are very small.

Following are to be noted here:

(a) It is capable of giving large separation factor but quantities collected are very small.

(b) When strength of the positive ion beam is increased, efficiency of separation decreases.

(c) As the instrument operates in high vacuum, strong beam cannot be obtained and hence quantity of isotopes collected cannot be increased.

(d) The isotopes of high mass number region with high state of purity and large separation factor are achieved by this method.

(e) Greater efficiency of the method is achieved if the feed material, e.g., isotope, is partially enriched by other method, e.g., thermal diffusion.

Apparatus used for the purpose is like that of Dempster's mass spectrograph. A suitable compound of the element whose isotopes are to be separated is first vaporized. An electron beam from a hot filament is passed through it when beam of positive ion is produced at source. Then the beam passes through plates S_1, S_2 accelerated by the potential V. The ion beam then passes through the magnetic field H acting perpendicular to the plane of the paper. When the beam bent through different path depending upon the isotopic mass, the lighter one is bent more and collected at the closer distance. The equation is

$$\frac{e}{m} = \frac{2V}{H^2 r^2}.$$

In practice the collecting pockets and source are at the same positions, and by varying magnetic field, isotopes of different masses are collected in the same position replacing the collecting pockets.

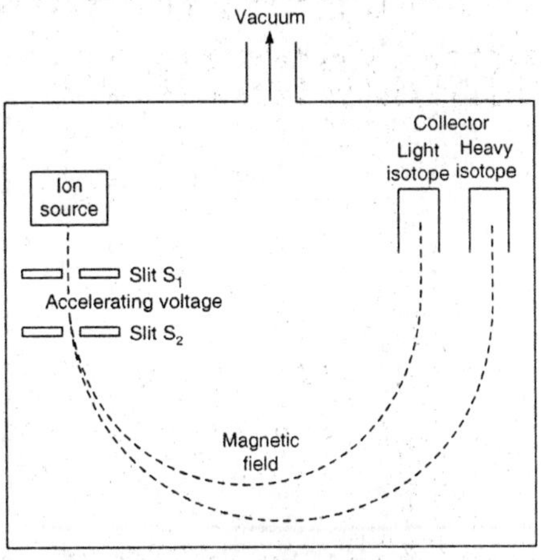

Fig 5.9 Separation of isotopes by the electromagnetic method.

5.09(c) CENTRIFUGAL METHOD

The gravitational force acting on a particle is proportional to its mass and so molecules or atoms of higher mass tend to be collected or stay in the lower level and the lighter mass tend to stay in the upper level, thus partial separation is possible. If centrifuge is used, this separation is more effective as this provides force analogous to gravitational force but much more powerful. Thus it is a type of pressure diffusion. If a gas or vapour flows through a rapidly rotating cylinder, the heavier molecule, due to the action of the force, produces an increased concentration at the lower level of the wall and the lighter one at the central axis in the upper level, if the cylinder is vertical. For horizontal one, the heavier is at the periphery and the lighter at the central axis. The theoretical separation factor $f = e^{v^2(M_2 - M_1)/RT}$, where v is the peripheral velocity, M_2 and M_1 are the masses of isotopes, R and T have usual meanings. Thus at constant temperature and velocity, the extent of separation is proportional to the mass difference and not on the ratio like other processes. Thus this process is better effective for heavy isotopes and also applicable to lighter ones. In this way isotopes of chlorine in CCl_4 and bromine in C_2H_5Br have been separated effectively.

If liquid substances are taken, then evaporative centrifuge technique is applied where liquid is slowly evaporated, centrifuge spins at a very high speed. The light fraction is drawn off from the centre and heavier condenses at the periphery.

5.09(d) THERMAL DIFFUSION

The basic idea of this method is that if a gaseous mixture of isotopes is placed in a vertical cylinder, part of which is hotter than the other, the lighter isotopes are collected at the hotter region and the heavier at the colder region. The separation factor f is given by $\ln f = K \frac{m_1 - m_2}{m_1 + m_2} \ln \frac{T_2}{T_1}$, where $m_1 > m_2$ and $T_2 > T_1$. This diffusion is purely independent of kinetic property of gases but finally balances the thermal diffusion.

The apparatus is a long vertical tube with a central wire that can be heated to 500°C electrically and sets up a temperature gradient between the hot wire and the cold wall (Fig 5.10). The isotopic mixture placed in the tube undergoes thermal diffusion, the lighter one collects near the hot wire and rises up and the heavier one at cold wall, then moves downward due to thermal convection current. Lighter isotope is drawn off from upper layer and the heavier one from the lower layer. In this way effective separation of isotopes of Cl_2, Ne, C, N_2 are done. ^{235}U and ^{238}U partially enriched by this process, are used as feed material in electromagnetic method.

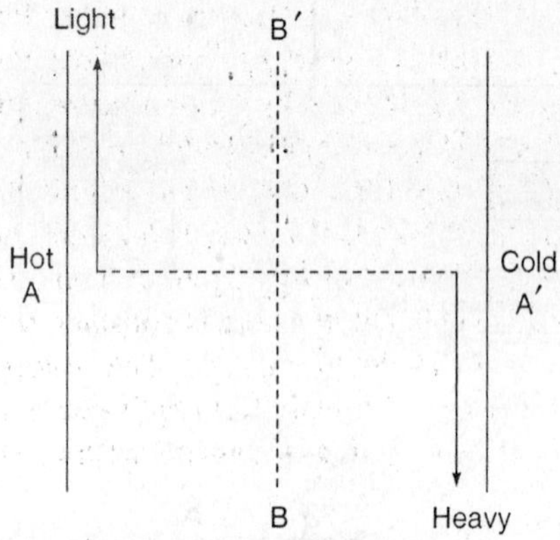

Fig 5.10 Separation of isotope by thermal diffusion method.

The main drawback of the method is high consumption of power and hence large-scale production is not done.

5.09(e) DISTILLATION METHOD

The principle of the method is that the rate of escape of atoms or molecules from a liquid surface are inversely proportional to the square root of their masses, and hence isotopic concentration can be effected by distillation. Using elaborate fractionating column operating at low pressure, neon isotopes are separated. The boiling point of D_2O is about 1.4°C higher than that of H_2O. Although the difference is small, separation of D_2O from H_2O can be done using efficient fractionating column, but the amount of D_2O separated is very small.

5.09(f) ELECTROLYTIC METHOD

Electrolysis of water results in preferential evolution of lighter isotope of hydrogen and hence deuterium concentration gradually increases in the residual water. The electrolysis is carried out using nickel electrode, and KOH is used as electrolyte. Separation factor is 6 and higher concentration of D_2 from initial value of 0.015% to 99.9% can be obtained. Since current consumption is 3–4 k amp-hour per gram of heavy water in this method, it is applicable where electricity is cheap.

Continuous electrolysis is disadvantageous as KOH concentration becomes more and more, and hydrogen gas liberated at the end contains much D_2, which cannot be neglected. So electrolysis is carried out in stages. When the volume of water is reduced to $\frac{1}{10}$th the initial volume, the residual water is treated with CO_2 to neutralize KOH, enriched water is distilled. The distillate then goes to next stage of electrolysis and in this way repeated H_2 gas evolved at the later stage containing higher percentage of D_2 is burnt in O_2 to form H_2O and D_2O and again electrolyzed. In this way D_2O is obtained in high state of purity.

5.09(g) CHEMICAL EXCHANGE METHOD

The zero-point energy of isotopic molecular species is different and hence small difference in their physico-chemical properties can be utilized to separate different isotopic species. In the reversible reaction between $^{16}O_2$ and $H_2{}^{18}O$, the equilibrium constant at 25°C is 1.012.

$$^{16}O_2 + H_2{}^{18}O \rightleftarrows {}^{18}O_2 + H_2{}^{16}O \qquad K = 1.012$$

In order to achieve greater exchange, the gas and the liquid are treated by counter-current process in a long column. Similarly, other reactions are:

$$\underset{(g)}{CO_2^{16}} + \underset{(l)}{H_2^{18}O} \rightleftarrows \underset{(g)}{C^{18}O_2} + \underset{(l)}{H_2^{16}O} \qquad K = 1.014 \text{ at } 0°C$$

$$\underset{(g)}{H\overset{12}{C}N} + \underset{(soln)}{\overset{13}{C}N^-} \rightleftarrows \underset{(g)}{H\overset{13}{C}N} + \underset{(soln)}{\overset{12}{C}N}$$

5.10 USES OF ISOTOPES

Isotopes are of two types radioisotopes and non-radioactive isotopes, and hence their fields of application in chemical industry and also in physiological and medicinal field are noticed. Isotopes differ in their zero-point energy and so the activation energy in a chemical reaction is higher for a heavier isotope and hence the heavier isotope reacts slowly. The rate of exchange of non-radioactive heavier isotope with lighter one and also the course of reaction can be detected by mass spectroscope studies; for radioisotopes, as they are β-emitter, they are detected by G-M counter. $T_{1/2}$ for a radioisotope must be long enough so that without decay it can persist for a long time to follow the course of process. Non-radioactive isotopes generally used are ^{17}O, ^{18}O, ^{15}N, etc., as they do not have radioisotopes of suitable $T_{1/2}$ for tracer studies. Besides, isotopes of hydrogen, carbon and sulphur are also used. Generally, the isotopic composition of an element in nature is constant, but this can be changed by artificial means and then the fate of the element is followed. The isotope so used is called an *isotopic tracer* and the element is thus labelled. The analytical applications of isotopes are classified as (a) Isotopic dilution, (b) Activation analysis and (c) Radioactive titration.

5.10(a) ISOTOPIC DILUTION

When quantitative separation of a mixture is difficult or impossible, this method is applied. The greatest advantage of the method lies in the fact that it is only necessary to obtain the product in pure form, for any loss in the purification procedure is automatically considered. The method implies

1. a known weight of one of the compounds, labelled with isotope, is added to the mixture of materials.
2. this compound is purified from remaining constituents by some procedure. Its weight and isotope constant are determined.

DIRECT METHOD

Let A^* gm of same substance, as the unknown A, containing the active isotope is added to the unknown and then thoroughly mixed. A sample A' gm pure is isolated from the mixture.

The activity of unknown substance $A = 0$, that of labeled A^* is Q^*. Activity of the portion taken out is Q'. After mixing the total amount are $A + A^*$. As the amount A^* is negligible compared to A, we can neglect A^*.

Specific activity of added substance, $S_{A^*} = \dfrac{Q^*}{A^*}$

Specific activity of mixture $= \dfrac{Q^*}{A + A^*}$

Specific activity of pure substance taken out, $S_{A'} = \dfrac{Q'}{A'}$

We can write, $\dfrac{A'}{A + A^*} = \dfrac{Q'}{Q^*} = \dfrac{S_{A'}A'}{S_{A^*}A^*}$

$$\text{or,}\quad \frac{1}{A+A^*}=\frac{S_{A'}}{A^*S_{A^*}}$$

$$\text{or,}\quad A=\frac{A^*S_{A^*}-S_{A'}A^*}{S_{A'}}$$

$$\text{or,}\quad A=\frac{A^*\left(S_{A^*}-S_{A'}\right)}{S_{A'}}=A^*\left(\frac{S_{A^*}}{S_{A'}}-1\right)$$

When $S_{A^*} >>> S_{A'}$, then $A = A^*\dfrac{S_{A^*}}{S_{A'}}$.

As long as the added isotopic material is identical with the compound to be analyzed and does not undergo any exchange reaction with other material, an atom balance can be set up where U stands for unknown R for amount recovered.

$$\frac{N_{R^*}}{N_R}=\frac{N_{U^*}+N_{A^*}}{N_U+N_A}.\ \text{Here}\quad N_{U^*}=0\ \text{and}\ N_A <<< N_U.$$

So we can write, $N_U = N_R\dfrac{N_{A^*}}{N_{R^*}}$

If the added isotopic compound is large compared to small unknown compound, then the determination is uncertain.

Isotopic dilution analysis is used in the determination of percentage of glycine in the hydrolysis product of protein (using synthetic glycine labelled with ^{14}C.

Total volume of blood in an animal or human body is determined by isotopic dilution method injecting readily available iron compound labelled with ^{59}Fe.

5.10(b) ACTIVATION ANALYSIS

When high energetic projectile like ^1_1p, ^2_1D, α or ^1_0n is bombarded on the target element, a large number of target elements becomes radioactive; measurement of their activity provides data for their quantitative estimation. As an example, when thermal neutron collides with a target nucleus, it captures the neutron to give a larger nucleus with same nuclear charge. The compound nucleus thus formed is radioactive and emits either a particle or γ-ray with widely varied $T_{1/2}$ and measurement of this decay constants provides data for their quantitative estimation. The rate of production of active nucleus N_A^* will be proportional to the neutron flux ϕ, nuclear cross-section σ of the target element and number of atoms N_t present in the target. So rate of production $R = N_t\phi\sigma$. If λ be the decay constant of active nucleus N^*, their decay will be λN^* and observed decay constant

$$\frac{dN^*}{dt}=R-\lambda N^*\qquad\text{or,}\quad \int\frac{dN^*}{R-\lambda N^*}=\int dt$$

$$\text{or,}\quad -\frac{1}{\lambda}\ln\left(R-\lambda N^*\right)=t+K$$

Imposing boundary condition $N^*= 0$ for $t = 0$, we have, $K = \dfrac{1}{\lambda}\ln\dfrac{1}{R}$ or, $K = -\dfrac{1}{\lambda}\ln R$

Therefore, we can write,

$$\ln\left[\frac{R}{R-\lambda N^*}\right] = \lambda t \qquad \text{or, } N^* = \frac{R}{\lambda}e^{-\lambda t}\left(e^{\lambda t}-1\right)$$

$$\text{or,}\quad N^* = \frac{R}{\lambda}\left(1-e^{-\lambda t}\right) = \frac{RT_{\frac{1}{2}}}{0.693}\left(1-e^{\frac{-0.693t}{t_{\frac{1}{2}}}}\right) = \frac{Rt_{\frac{1}{2}}}{0.693}\left(1-2^{\frac{-t}{t_{1/2}}}\right)$$

When time of irradiation $t = t_{1/2}$, only 50% saturation is reached, for $t = 2t_{1/2}$, only 75% saturation and for $t = 5t_{1/2}$, 97% saturation is reached.

When $t_{1/2} >>> t$, then $2^{-t/t_{1/2}}$ converted to a power series and

$$N^* = \frac{Rt_{\frac{1}{2}}}{0.693}\left[1-(1-0.693n+......)\right] = \frac{Rt_{\frac{1}{2}}}{0.693}\times 0.693\frac{t}{t_{\frac{1}{2}}} = Rt$$

So, activity produced varies linearly with time of irradiation.

$\therefore\ N^* = \phi\sigma N_t \times t,\qquad$ when $t_{\frac{1}{2}} >>> t$

and $N^* = \phi\sigma N_t \times \dfrac{t_{\frac{1}{2}}}{0.693}\qquad$ when $t >>> t_{\frac{1}{2}}$.

5.10(c) RADIOACTIVE TITRATION

Radioisotope can be used in volumetric determination of some metal ions. Radio phosphorus ^{32}P was converted to soluble phosphate and mixed with Na_2HPO_4 standard solution. This solution is used to titrate Be^{+2}, Pb^{+2}, Th^{+3}, Mg^{+2} and UO_2^{+2}. After each addition of phosphate, a sample of clear filtrate was taken in G-M counter and activity of the filtrate determined which remains essentially constant until whole metal ion is precipitated. When titration is complete, the activity will suddenly increase after each addition of titrant, activity plotted against titrant. Intersection of the two linear curves gives the equivalent point.

5.10(d) STUDY OF STRUCTURE AND REACTION MECHANISM

(i) In the photosynthesis of plants it was believed that O_2 comes from CO_2. But labelled O_2, ^{18}O in water and $C\overset{18}{O}_2$ when used it was found that ^{18}O is retained in the plant body.

$$x\mathrm{C}\overset{18}{\mathrm{O}}_2 + x\mathrm{H_2}\overset{16}{\mathrm{O}} \xrightarrow{h\nu} \mathrm{C_6H_{12}}\overset{18}{\mathrm{O}}_6 + \overset{16}{\mathrm{O}}_2$$

$$x\mathrm{C}\overset{16}{\mathrm{O}}_2 + x\mathrm{H_2}\overset{18}{\mathrm{O}} \xrightarrow{h\nu} \mathrm{C_6H_{12}}\overset{16}{\mathrm{O}}_6 + \overset{18}{\mathrm{O}}_2$$

(ii) Hydrolysis of ester with labelled $H_2\overset{18}{O}$ shows that O^{18} is retained in acid. Hence the reaction takes place as

$$\mathrm{RC}\begin{smallmatrix}\nearrow \mathrm{O} \\ \searrow \mathrm{OEt}\end{smallmatrix} + \mathrm{H_2}\overset{18}{\mathrm{O}} \rightarrow \mathrm{RC}\begin{smallmatrix}\nearrow \mathrm{O} \\ \searrow \overset{18}{\mathrm{O}}\mathrm{H}\end{smallmatrix} + \mathrm{EtOH}$$

(iii) When SO_3^{-2} is boiled with S*, then S^*SO_3 is formed and when decomposed with acid the activity goes to sulphur deposited. Then S_2O_3 can be written as,

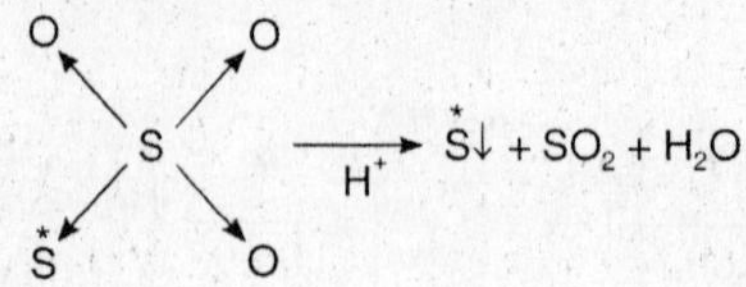

5.10(e) MEDICINAL APPLICATIONS

Radioisotopes are used in both diagnosis and treatment:

(i) Dysfunction of blood circulatory system, location and extent of blood clot can be deleted by injecting $^{24}NaCl$ solution. It is injected to the forearm of a patient and a γ-ray detector is attached to one of the feet. If blood circulation is normal, γ-ray will be detected to the feet within less than an hour. If there is any obstruction, γ-ray activity will increase very slowly. By moving the counter over the whole body, position of the obstruction can be located.

(ii) The same procedure is applied in observing the pumping action of the heart where an automatic pen recorder is placed over the heart. From rise and fall of the recorder and the curve thus drawn, abnormalities in the heart can be detected.

(iii) Thyroid gland easily takes up all iodine. By using NaI^{131}, the function of thyroid gland can be studied. NaI^{131} is also used for the treatment of malignancy and cancer of the thyroid.

(iv) Iodo-fluorescein labelled with $\overset{131}{I}$, is easily taken up by unusually grown brain cell, thus brain tumor can be detected.

5.10(f) INDUSTRIAL APPLICATION

Radioisotopes are used:

(i) to determine the effectiveness of washing of precipitates and rinsing laboratory apparatus.

(ii) to determine the uniformity and thickness of coating on metal surface.

(iii) to measure the thickness of sheet of paper, steel, rubber, etc., by placing radioactive source behind the sample and measuring the amount of radioactivity absorbed by the sheet. Thickness is also measured by determining the back scattering where the radioactive source and the shielded detector are placed in the same side of the sample.

(iv) to determine the surface area of solid.

(v) to determine the age of rocks and thus the age of the earth.

Age of certain rocks or minerals can be determined by knowing the amount of end product of the radioactive decay. Here age means the time that has been elapsed since the mineral has been solidified. It is known that 1 gm Uranium after decay produces (a) 1.35×10^{-10} gm radiogenic ^{206}Pb or (b) 1×10^{-7} cc Helium as gas per year (this gives the minimum age as some gaseous He may escape from the mineral due to diffusion). Thus, knowing the ratio uranium-lead, $^{206}Pb : ^{235}U$, age of the mineral and hence age of the earth can be determined. Non-radiogenic ^{204}Pb, i.e., normal lead is subtracted from the total amount and thorium in the mineral due to decay can be obtained from the relation $N = N_0 e^{-\lambda t}$. If the mineral is of great age, then decrease in the quantity of Uranium content is Nu-I and the radiogenic Lead is N_U-II, then original Uranium content will be N_U-I + N_U-II as one atom of U gives one atom of Pb. So, the above equation takes the form,

$$N_U - I = (N_U - I + N_U - II)\, e^{-\lambda t} \quad \text{or,} \quad t = \frac{1}{\lambda} \times 2.303 \log_{10}\left(\frac{N_U - II}{N_U - I} + 1\right)$$

Similarly, knowing the ratio $\frac{N_{Th} - D}{N_{Th}}$ or, $\frac{N_{Ac} - D}{N_{U^{235}}}$ and λ for Th or ^{235}U, the age can be determined, which are 2.52×10^8, 2.6×10^8 and 2.8×10^8 years respectively. Similarly, the He content in the mineral gives the value as 1.29×10^9 years.

5.11 RADIOCARBON ^{14}C DATING

This is another method of determining age of antique objects having carbon derived from plant and animal bodies and it is used extensively in archeology.

A.V. Groose in 1934 stated that radioactive isotopes are formed in nature by the action of cosmic rays. In 1946, Libby made specific suggestion that atmospheric nitrogen ^{14}N (99.6% in atmosphere) produces ^{14}C by the $_0^1n$ coming from cosmic ray $^{14}_{7}N + {}^1_0n \rightarrow {}^{14}_{6}C + {}^1_1p$. This ^{14}C is converted to $^{14}CO_2$ by atmospheric O_2 and is taken up by the plants during photosynthesis to form carbohydrate that is taken up by herbivorous animals and then by carnivorous animals. A part of this $^{14}CO_2$ is returned to atmosphere during respiration and also in excreta. Thus, in this plant-animal carbon cycle an equilibrium is established in course of time and carbon in all living matter—plant and animal—will contain a constant equilibrium concentration of ^{14}C. Some of this ^{14}C is naturally converted to ^{14}N. Thus the amount of ^{14}C in living being is determined by the rate of formation of ^{14}C from cosmic ray $_0^1n$ and the natural decay of ^{14}C to ^{14}N.

$$^{14}_{6}C \rightarrow {}^{14}_{7}N + \beta^-$$

In the atmosphere the ratio of $^{12}CO_2 : {}^{14}CO_2 = 1 : 10^{-12}$.

An examination of specimens of fresh wood collected from various parts of the world indicates the constant radioactivity equivalent to 15.3 disintegration per minute per gram of carbon. This value was obtained prior to 1952, after which different thermonuclear explosions increased the $_0^1n$ content in atmosphere and also the ^{14}C in the atmosphere and in living matters. Since 1962 after termination of thermonuclear explosions by USA, UK and USSR, this ^{14}C content has been decreasing as a result of dissolution of $^{14}CO_2$ in ocean, but several years more will be required to attain the pre-1952 stage of ^{14}C as $T_{1/2}$ of ^{14}C = 5700 years.

When plants and animals die, it is separated or isolated from carbon cycle. It will not take up ^{14}C any more, but its ^{14}C content decreases due to decomposition ($T_{1/2}$ = 5700 years). The old specimen of age ranging from 1000 to 50000 years, i.e., maximum 8 times of $T_{1/2}$, ^{14}C disintegration gives us the idea of its age, but of very old origin such as petroleum or coal, which has been removed from carbon cycle hundreds and million years before, ^{14}C content is extremely small and cannot have detectable count.

It is assumed that the new carbon specimen will have the count equivalent to 15.3, which is taken as N_0, and the present count as N_t after t years has elapsed since the article was made.

So, $N_t = N_0 e^{-\lambda t}$ or, $t = \frac{T_{\frac{1}{2}}}{0.301} \log \frac{N_0}{N_t}$.

Thus for Egyptian tombs, i.e., pyramids, $N_0 = 15.3$, $N_t = 8.8$, $T_{\frac{1}{2}} = 5700$ years, $t = 4549$ years.

This method is successfully applicable to determine the ages of textiles, seeds, etc., from pyramids (5000 years) examining woods, pits and mud for latest glacial age, also from scrolls discovered in caves near Dead Sea, which is 1900 years old.

Determination of age of different materials in the period between 3000–2000 B.C. gives erroneous result. This is because earth's magnetic field changed to some extent during that period. This causes the variation of intensity of cosmic ray near earth's surface, so also the ^{14}C content in atmosphere.

CHAPTER

6

PERIODIC TABLE AND PERIODICITY IN PROPERTIES

6.01 PERIODIC CLASSIFICATION OF ELEMENTS

Systematic study of elements and a large number of their compounds led scientists in different times to develop some scientific methods to classify the elements according to their properties or correlate the properties with atomic weights, although atomic weights of elements were not known with accuracy at that time. The elements then were broadly classified as metals and nonmetals having characteristic properties. Later discovery of many other elements showed that some have no such characteristic properties and some have both the properties of metal and nonmetal. So this attempt failed. Afterwards scientists classified elements according to their chemical properties with regard to their atomic weights.

In 1815, Prout stated that different elements are made up of the union of integral number of hydrogen atoms and so the atomic weights will be a whole number (whole number rule), but accurate determination of atomic weights of some elements were found to be a fraction and hence this idea was discarded at that time, although it gained momentum afterwards after the discovery of isotopes. Prout's hypothesis enlightened scientists that there must be some relation between the properties of elements and their atomic weights. So scientists tried to classify the elements according to their chemical properties with regard to their atomic weights and this method becomes more important when Cannizzaro established the relation between the atomic weight and the equivalent weight of the elements in his book "Sketch of a course of chemical philosophy".

The periodic classification of elements was made on the basis of the observations on the properties of elements and their compounds (prior to the development of modern atomic theory) and later supported by their electronic configurations. This has placed them upon sound theoretical foundations. In 1817, Dobereiner first noted that within a group of three chemically similar elements the atomic weight of the middle element is approximately the arithmetic mean of the atomic weights of the other two. This is known as "The Law of Triad", e.g., atomic weight of bromine (80) is the mean of the atomic weights of chlorine (35) and iodine (127). He also pointed out that chemically similar elements might have approximately equal atomic weights, e.g., Fe (56), Co (59) and Ni (58.6). In 1850, Pattenkofer modified Dobereiner's concept and suggested that among chemically similar elements, the successive differences in atomic weights are either constant or some multiple of the constant, i.e., among oxygen (16), sulphur (32), selenium (80) and tellurium (128), the difference in oxygen and sulphur is 16 and for others 16×3 = 48. After the development of accurate atomic weight determination of elements by Cannizzaro in 1858, attempts were made to correlate the physical and chemical properties with atomic weights of elements. In 1862, B. de Chancourtois arranged elements in order of increasing atomic weights in the form of a spiral around a cylinder and accidentally observed that elements of analogous properties occupy the identical position in the spiral. This is the telluric screw.

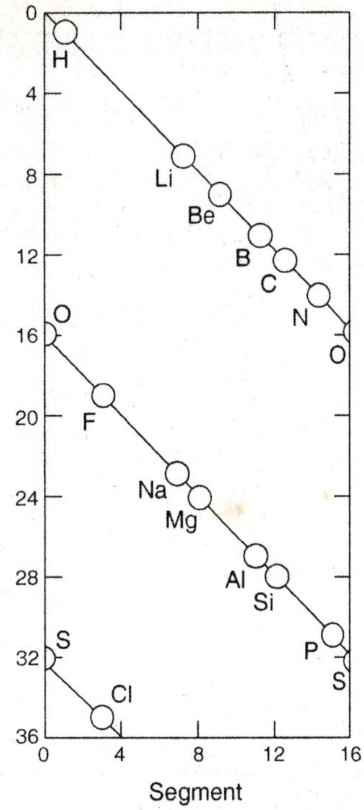

Fig. 6.1 The Telluric Screw of de Chancourtois.

In 1865, J.A.R. Newland quite independently noted that when elements were placed in order of increasing atomic weights, the elements with physical and chemical similarities reappear after interval of eight elements, like a series of octaves in music. This is the " law of octave". Newland's arrangements contain seven horizontal series each containing eight members. A salient part of Newland's arrangement is as follows (Table 6.1).

This arrangement has some inconsistencies due to the incorrect atomic weight data during his time and his failure to realize the existence of some new elements hitherto unknown. The careful study of chemical facts and scientific insight, led Dmitry Mendeleeff in 1869 to put forward the periodic classification of elements as the function of their atomic weights. Again, contemporarily another scientist Lother Meyer independently classified elements based on their physical properties.

Table 6.1 : Newland's table

H	F	Cl	Co, Ni	Br	I	Pt, Ir
Li	Na	K	Cu	Rb	Cs	Os
Be	Mg	Ca	Zn	Sr	Ba, V	Hg
B	Al	Cr	Y	Ce, La	Ta	Tl
C	Si	Ti	In	Zr	W	Pb
N	P	Mn	As	Dy, Mo	Nb	Bi
O	S	Fe	Se	Rh, Ru	Au	Th

However, these two classifications were more or less alike. As regards the fundamental understanding of periodicity, Mendeleeff's approach was far reaching useful. Again, Lothar Meyer's atomic volume curve is better fruitful for development of chemical thought.

LOTHER MEYER'S CURVE AND MENDELEEFF'S ORIGINAL TABLE

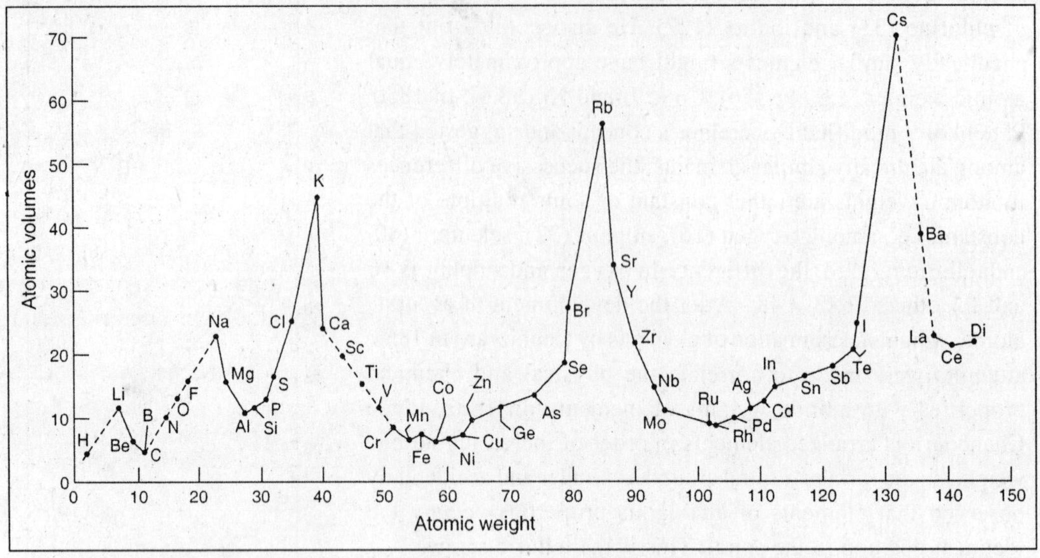

Fig. 6.2 Atomic volume curve of Lother Meyer.

Table 6.2 : Mendeleeff's original table of the elements

I	II	III	IV	V	VI
			Ti = 50	Zr = 90	? = 180
			V = 51	Nb = 94	Ta = 182
			Cr = 52	Mo = 96	W = 186
			Mn = 55	Rh = 104.4	Pt = 197.4
			Fe = 56	Ru = 104.4	Ir = 198
			Ni = Co = 59	Pd = 106.6	Os = 199
H = 1			Cu = 53.4	Ag = 108	Hg = 200
	Be = 9.4	Mg = 24	Zn = 65.2	Cd = 112	
	B = 11	Al = 27.4	? = 68	Ur = 116	Au = 197?
	C = 12	Si = 28	? = 70	Sn = 118	
	N = 14	P = 31	As = 75	Sb = 122	Bi = 210
	O = 16	S = 32	Se = 79.4	Te = 128	
	F = 19	Cl = 35.5	Br = 80	I = 127	
Li = 7	Na = 23	K = 39	Rb = 85.4	Cs = 133	Tl = 204
		Ca = 40	Sr = 87.6	Ba = 137	Pb = 207
		? = 45	Ce = 92		
		?Er = 56	La = 94		
		?Yt = 60	Di = 95		
		?In = 75.6	Th = 118?		

Newland's arrangement of elements was changed to vertical form with the introduction of group 0 for the noble gases in the old periodic table. The pertinent points are to be noted:

1. Periodicity of properties.
2. Correspondence to the valence of elements.

3. Arrangements to follow the atomic numbers.
4. Rectification of doubtful atomic weights from the suitable position in the periodic table.
5. Chemically similar elements have either nearly same atomic weights (e.g., Os, Ir., Pt) or have atomic weights increasing regularly (e.g., Li, Na, K).
6. Undiscovered elements are predicted accurately.
7. It is noteworthy that such periodic arrangement of the elements is supported by the later developed electronic configuration to know the whereabouts of atoms.

The major discrepancies of Mendeleeff's classification were: (a) Chemically different elements placed in the same group (Na with Cu, Ag and Au). (b) Anomalous discrimination of sub-groups within a group. (c) Unsuitable position of metals and non-metals. (d) Unique position of the 'rogue element' hydrogen requiring clarification. (e) To account for the presence of inverted position of four pairs of elements [Ar, K; Co, Ni; Te, I; Th, Pa]. (f) Suitable positions of transition elements are wanting. (g) Non-accommodation of lanthanides and actinides.

It was established that the atomic weight (which is the average of isotopic composition in nature) is not the fundamental characteristic of an element but atomic number, i.e., number of external electron or number of proton in the nucleus. Taking into the consideration of atomic number, Z, discovered by Moseley's work on X-rays, most of the earlier inconsistencies can be removed. Hence, the periodic law was modified in terms of atomic number that properties of elements are the periodic function of their atomic number. Bohr arranged the elements according to electronic configuration, which is honoured as Bohr modern periodic table.

Table 6.3 : Modern Long-form of the Periodic Table based on Thomson model (1895) with the inclusion of super-heavy elements predicted by Seaborg (1968)

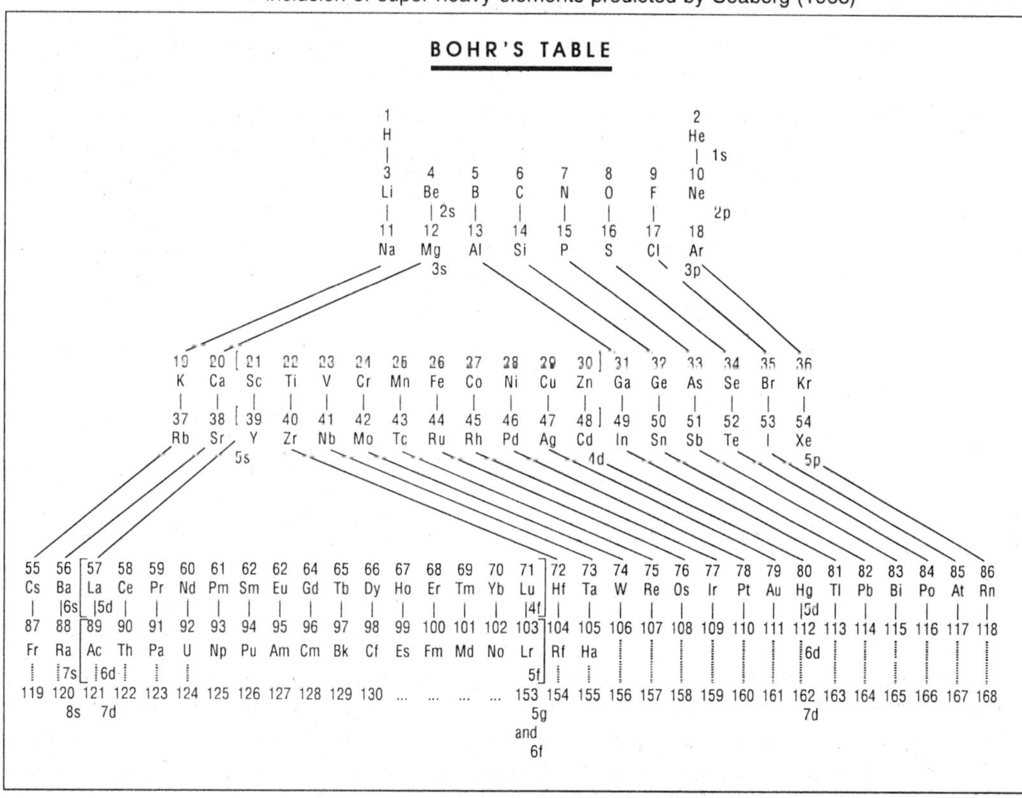

Table of atoms, i.e., the extended form of the periodic table placed in the chapter "classification of elements on the basis of electronic configuration" to see Chapter-1, article-1.26.

6.02 PERIODICITY IN PROPERTIES

In the old concept, atoms are considered as rigid spheres in contact with each other in solids and hence will have volume, radius, etc., and this concept will lead to the development of crystal chemistry and to study and correlate many properties both physical and chemical in nature. According to modern concept, atoms are electron clouds of continuously varying intensity and so meaningless in the conception of volume, radius, etc., and all atoms will have same size. The radial distribution curves indicate that the probability of electron cloud density is maximum at the location of principal quantum numbers. If this position of quantum numbers (n) were considered as the sphere of influence, we can readily develop the idea of atomic or ionic size and also develop the crystal model of the substances.

The electron clouds have fluctuating densities and are also influenced by external conditions. So the atomic or ionic dimensions depend upon both the method of measurement and also on the environment. The factors on which the atomic or ionic dimensions depend are (i) multiplicity of the bond, (ii) the oxidation number of the ion, (iii) relative amount of covalent and ionic character, (iv) the number of neighbours on which the particle under consideration is associated and (v) the crystal structure.

The physical and chemical properties of elements vary periodically with size of the element, which in turn depends on the atomic number, i.e., electronic configuration. Such properties are (i) melting and boiling points, (ii) density, (iii) atomic, volume, atomic radius, ionic radius, (iv) ionisation potential, (v) electron affinity, and (vi) electronegativity.

6.02(a) MELTING POINT AND BOILING POINT

Melting or boiling point of a substance is that temperature where solid state starts changing into liquid or liquid into gaseous state, i.e., force of holding the constituent unit in respective position is overcome by thermal energy at normal atmospheric pressure and this temperature depends on the packing of units, lattice energy and degree of association through hydrogen bond or forces like that. The periodic variation of melting point with atomic number is represented by the curve, Fig. 6.3.

This curve has the following characteristics:

(A) Inert gases have very low melting points. They have complete electron shell and bond order according to MO conception is zero, i.e., no tendency to unite among them.

(B) The melting points of Group I alkali metals are low and gradually increase up to group IV (maximum), then gradually decrease to Group VIII (noble gases). The alkali metals have one free valence electron that takes part in the formation of weak metallic bond and the strength of this metallic bond increases with increase in number of valence electron participation which is 2 for group II metals.

Hence melting point of alkali metals is low compared to alkaline earth metals. Be > Li, Mg > Na and so on.

(C) Boron, carbon and silicon have exceptionally high melting points. These elements form giant three-dimensional network stiucture and tremendous energy is required for complete rupture of the bonds and hence very high melting points.

(D) Nitrogen, oxygen and fluorine form diatomic molecules with triple bond, double bond and single bond respectively. The inter-atomic distance between the atoms of respective molecules increases, hence stability decreases; the repulsion between non-bonding electrons

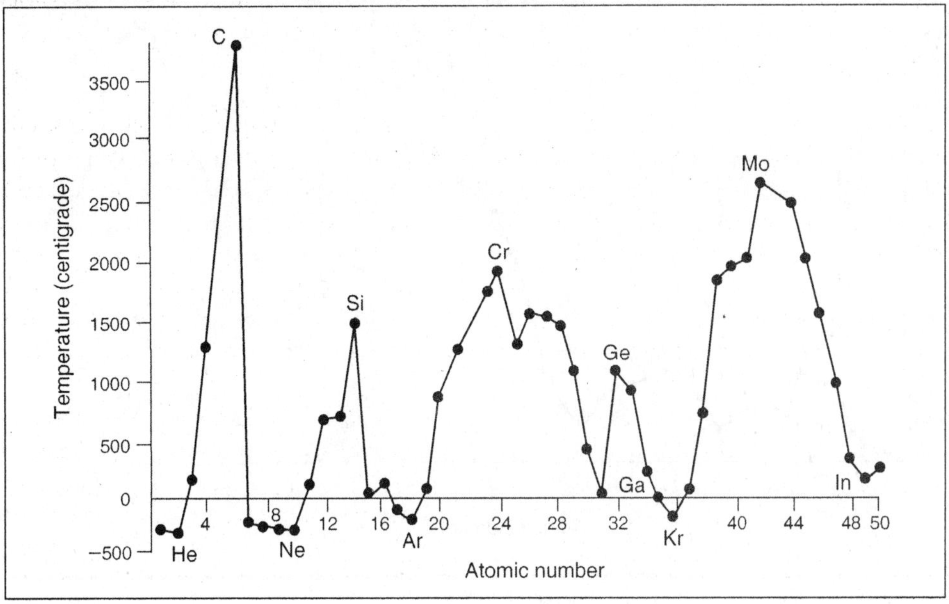

Fig. 6.3 Periodic variation of melting points of elements.

are high and do not form three-dimensional giant molecule; hence they have very low melting points.

(E) The melting point decreases down the group due to loosely packed metal lattices because of their large atomic sizes, viz., Li > Na > K > Rb > Cs. The exceptions are P > N, S > O, I > Br > Cl > F. Phosphorus and sulphur form polyatomic molecules through single covalent bond (due to larger size cannot form multiple bonds), hence higher melting points. For halogens larger size is associated with increase in van der Waals' interaction. So the melting point increases in the order.

(F) The melting points of the transition metals are very irregular due to varying extent of *d*-orbital participation in the metallic bond formation. It is highest for d^5 configuration where the number of unpaired electrons is maximum. After completion of *d*-orbital there is sharp decrease in melting point, e.g., Cu 1083°C, Zn 419°C, i.e., complete *d*-orbital does not take part in metallic bond formation.

6.02(b) DENSITY

The periodic variation of densities of elements with atomic number is noted. The density depends on the nature of packing of the atoms, stronger and compact bindings increases the densities that depend on the number of valence electrons. Hence the alkali metals have lower densities then increase with the increase in group number up to group IV and then decrease up to group VIII. Within the same group, the density increases downward due to the fact that increased in atomic mass is not compensated by the increase in atomic volume.

6.02(c) ATOMIC VOLUME AND MOLECULAR VOLUME

It is the ratio of gram-atomic weight or gram-molecular weight to density; i.e., the volume occupied by Avogadro's number of atoms or molecules. The unit is cubic centimetre or millimetre. It is a rough indication of the size of an atom or a molecule because the value depends upon several factors, e.g., physical state, allotropic form, nature of packing, etc. The curve has been given in the Fig 6.2,

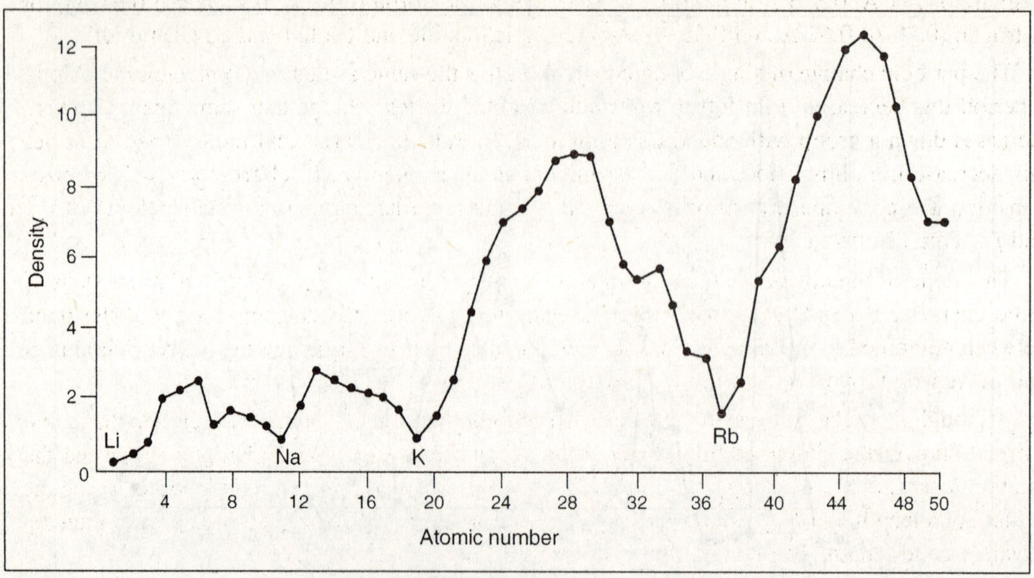

Fig. 6.4 Periodic variation of density with atomic number.

in Page 130. The molecular volume of a series of isomorphous compounds is helpful in evaluating the radii of component materials. For series of covalent compounds, molecular volume data can be correlated with the melting point, boiling point and other physical properties.

6.02(d) ATOMIC AND COVALENT RADII

Atomic radius may be considered as the distance of valence electrons from the nucleus. As isolation of single atom and evaluation of its radius is impossible, this evaluation is to be done when it is in its natural environment or the element is in association with other like atoms, free atom, or in covalently bonded substance. It has been found that size of free atom and that obtained for the same atom in the purely covalent compound remain unchanged, the atomic radius and the covalent radius are the same.

X-ray, electron or neutron diffraction study, band spectroscopy or nuclear magnetic resonance can measure the inter-nuclear distance between the atoms of a free element or of a covalent compound. Assuming atoms are touching each other such inter-nuclear distance is then apportioned between atoms, for free element half the distance will represent atomic radius and for covalent molecule substitution procedure is employed. It is supposed that the atomic radius remains constant in a purely covalent compound, provided other factors remain unchanged. For example, C–C single bond distance in diamond is 1.542 Å, so the covalent radius of carbon is 0.771 Å. In various hydrocarbons C–C distance varies from 1.52 Å to 1.55 Å, i.e., more or less constant. Pauling showed that C–C = 1.542 Å and Si–Si in free silicon is 2.34 Å, so C–Si distance in a compound should be 1.94 Å and experimentally is 1.93 Å in $Si(CH_3)_4$.

Huggins assumed that the covalent radius of sulphur to be 1.05 Å and used this value to determine the radius of other elements in combination with S having Wurtzite (ZnS) structure.

The covalent radii of an element depend on various factors: (i) multiplicity of the bond, (ii) extent of ionic character of the bond and (iii) involvement of different orbitals undergoing hybridization that determines the geometry of the molecule.

Pauling and Huggins pointed out that if inter-nuclear distance decreases, covalent radius decreases from single bond to double bond to triple bond. Thus for single, double and triple bond Carbon radii are 0. 771 Å, 0.665 Å and 0.602 Å respectively, for Boron (0.88 Å, 0.71 Å and 0.64 Å), for Nitrogen (0.70 Å, 0.62 Å and 0.55 Å) are the single, double and triple bond covalent radii.

The periodic change of single bond covalent radii is the same as that in atomic volume. Along a period this decreases from left to right due to added nuclear charge and same quantum level; increases down a group with added quantum level. In transition series and inner transition series this decrease in radius is less and less prominent as outer quantum level are constant. For post-transition and post-inner transition elements this is not remarkable due to poorer shielding of d^{10} and f^{14} core electrons.

The value of the covalent radius of an atom varies with chemical environment, i.e., extent of ionic character depending upon the electronegativity of the elements. Ionic nature of the bond between combined atoms depends on the difference of their electronegativity value. Schomaker and Stevenson proposed the empirical relation $r_{AB} = r_A + r_B - 0.09\ (x_A - x_B)$.

Although this relation was an empirical one and not theoretically sound, it indicates that some correlation over the simple additivity of covalent radii are required. M. L. Higgins suggested the relation as $r^*_{AB} = r_{AB} + 1/2\ \log_{10}\ E_{AB}$ where r^*_{AB} is the covalent bond energy distance, r_{AB} is the actual bond length and E_{AB} is bond energy. The constant energy distance is 0.8 Å, greater than the covalent bond length. The bond lengths of the following halides are given:

CF_4 (1.41 Å, expt. 1.48 Å); CCl_4 (1.76 Å, expt. 1.76 Å); CBr_4 (1.91 Å, expt. 1.91 Å); CI_4 (2.10 Å, expt. 2.11 Å); SiF_4 (1.81 Å, expt. 1.56 Å); $SiCl_4$ (2.16 Å, expt. 2.01 Å).

Covalent radii of elements depend on the geometry of the molecules also. Molecules may have tetrahedral, square planar or octahedral arrangements and consequently the radii are termed as tetrahedral, planar or octahedral radii. In general for metal complexes, $r_{tet} > r_{oct} > r_{sq}$, e.g., for Au ($r_{sq}$ = 1.31 Å, r_{oct} = 1.49 Å r_{tet} = 1.50 Å); for Ag (r_{oct} = 1.49 Å, r_{tet} = 1.53 Å) and for Sn (r = 1.40 Å for all). For non-metals such radii do not vary significantly.

6.02(e) IONIC RADIUS

According to the wave nature of the atom, atom or the ion will have no definite size. If we consider that ions are touching each other in crystals and are spherical in shape, then the term ionic radius becomes meaningful. In ionic compounds, the cations have smaller and anions have larger radius than that of the neutral atom. With higher charge, the radius will change to a greater extent depending on the force of attraction of the nucleus. As the ions of spherical size are touching each other in solid crystals, X-ray diffraction study of the ionic crystal shows the inter-ionic distance (i.e., distance between the nucleus of cation and anion). This distance is then apportioned between the ions such that their sum will be the measured inter-ionic distance and this necessitates some predetermined standard radius measured independently and the ionic radius of the other to be determined by substitutional procedure provided the ions are in identical chemical environment.

In 1920, Bragg measured the distance between two O^{2-} in silicate and half this value 1.35 Å was taken as the ionic radius for O^{2-}. The radius for the cations is then measured by substitutional procedure for different oxides. But this procedure is limited and does not give correct value.

In 1920, Lande took LiI and measured the I^-– I^- distance and half of this is the ionic radius of I^- assuming Li^+ being very small in size, I^- are touching each other and Li^+ is placed between the gaps. In this way he measured the radius of halide ions.

In 1922, Wasastjerna gave a different approach. He stated that the molar refractivity of an ion is more or less proportional to its volume and so apportioned the measured inter-ionic distance for alkyl halides and alkaline earth oxides in the ratio of their molar refractivity and thus obtained the radius F^- (1.33 Å) and O^{2-} (1.32 Å).

In 1926, Goldschmidt used these values of F^- and O^{2-}, and calculated radii for different ions following Bragg's method. These values although differ from present-day values, it helped crystallographic study to interpret properties of inorganic compounds.

In 1927, Pauling approached theoretically stating that ionic radii will be such that which after suitable corrections, when added together will give the measured inter-ionic distance. He thought that the inter-ionic distance depends on the following factors:

1. Types of electron distribution in the ions
2. Types of crystal structure
3. Radius ratio, $\rho = \frac{r^+}{r^-}$

Pauling has chosen standard crystals NaF, KCl, RbBr and CsI, because

(a) crystals have same ionic character;

(b) same crystal arrangement like NaCl (6 : 6) faces centered cubic lattice except CsI;

(c) they have same radius ratio $\rho = 0.75$;

(d) the cation and anion in the crystals have same type of electron arrangement, i.e., isoelectronic but different nuclear charges.

The inter-ionic distances of these crystals are 2.31 Å, 3.14 Å, 3.43 Å and 3.85 Å respectively. The size of these ions are inversely proportional to their effective nuclear charge Z_{eff} which is in turn actual nuclear charge Z less by screening constant S obtained from refraction data and X-ray terms but empirically calculated using Slater's rule.

Thus, $r_{ion} = \frac{C_n}{Z_{eff. of ion}} = \frac{C_n}{Z - S}$ where C_n is the proportionality constant depending upon principal quantum number n of valence electrons.

Thus, $\rho = \frac{r_c^+}{r_a^-} = \frac{Z_{eff. for anion}}{Z_{eff. for cation}} = \frac{Z_a^* - S}{Z_c^* - S}$, where S is same for both the ions. Knowing inter-ionic distance $r^+ + r^-$ and from radius ratio value, thus, theoretically calculated radii are Na^+ (0.95 Å), K^+ (1.33 Å), Rb^+ (1.69 Å), F^- (1.36 Å), Cl^- (1.84 Å), Br^- (1.95 Å) and I^- (2.16 Å). Using these values, radius for other ions can be calculated from measured inter-ionic distances. Thus Li^+ (0.60 Å), the radius for O^{-2} obtained 1.4 Å. The radii obtained in this way for multivalent ions are relative to those for alkali metal ions and halide ions, and their sum does not always give the measured inter-ionic distance. These multivalent ions behave in such a way that their coulombic forces are like that in univalent ions. Hence these radii calculated relative to alkali metal ions are called univalent radii R_1. Pauling gave a relation between univalent radii and crystal radii R_x (i.e., radius or inter-ionic distance in crystal observed). $R_x = R_1 Z^{\frac{-2}{n'-1}}$. Z is the ionic charge, n' is the Born exponent or repulsion exponent dependent on the repulsive force arising from interpenetration of the two ions. Its value differs for different valence shell configurations, for He ($n' = 5$), for Ne ($n' = 7$), for Ar and Cu^+ ($n' = 9$), for Kr or Ag^+ ($n' = 10$). These empirical values calculated from crystal consideration are very close to each other, e.g., Be^{2+} (exp, 0.34 and calc 0.31), Na^+ (exp 0.98 and calc 0.95), Mg^{2+} (exp 0.78 and calc 0.65).

The ionic radii thus obtained are not constant but vary with: (a) coordination number in crystal, (b) oxidation states or valences of the neighbouring ion and (c) radius ratio of the ions concerned.

The compounds or the crystals are stable when the attractive force between the oppositely charged ions increases, that increase is with increasing coordination number of the ion, and the repulsive force between the like ions when squeezed together. The optimum conditions are when the ions in the crystals just touch each other and it depend on the relative size of the ions. Let us consider the case of CsCl, a body-centered cubic lattice with coordination number 8 : 8 (where O are Cs^+ and Cl^- are ●).

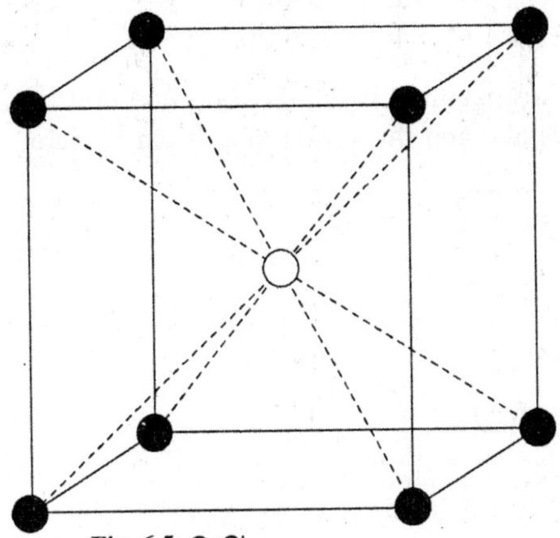

Fig. 6.5: CsCl

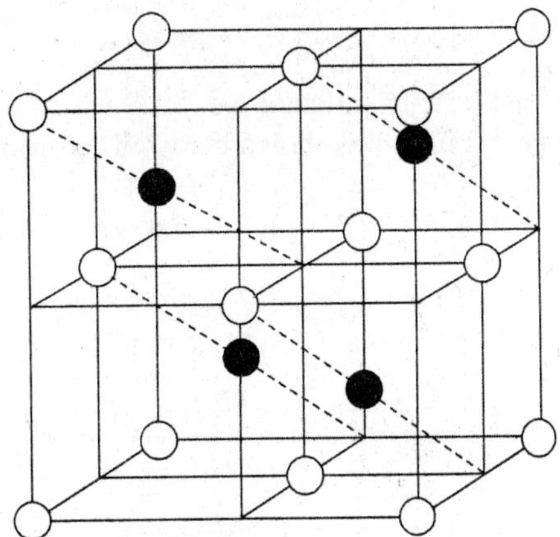

Fig. 6.6: ZnS

○ = Cs^+

● = Cl^-

● = Zn^{2+}

○ = S^{2-}

Fig. 6.5 and 6.6 The cesium chloride and zinc blende structures.

The two Cl^- is touching along the edge (a) and Cs^+ and Cl^- along the diagonal, *d*, of the cube. Diagonal,

$$d = \sqrt{3a^2} = \sqrt{3}a$$

$$\text{So,} \quad r^+ + r^- = \frac{d}{2} = \frac{1}{2} \times \sqrt{3}a = \frac{\sqrt{3}a}{2} = \frac{\sqrt{3}}{2} \times 2r^- = \sqrt{3}r^-$$

$$\text{or,} \quad r^+ = r^- (\sqrt{3} - 1) \qquad \text{or,} \quad r^+ / r^- = 0.732.$$

Knowing inter-molecular distance or inter-ionic distance r^+ and r^- can be determined in the crystal, i.e., crystal radii in 8 : 8 system.

For face-centered cubic system as in NaCl, here O are the Cl^- and ● are the Na^+.

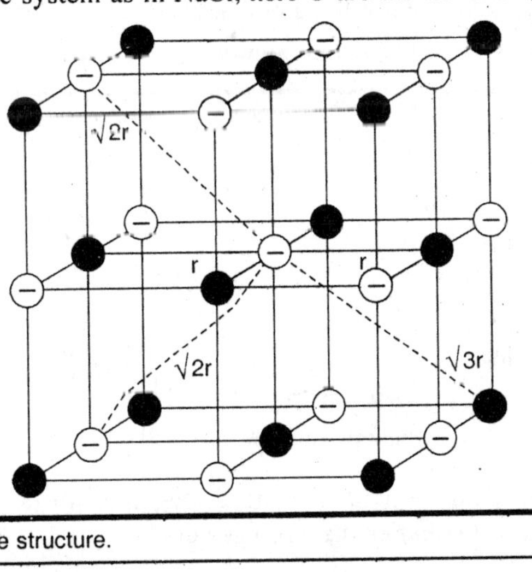

Fig. 6.7 The sodium chloride structure.

Here, $d^2 = 2a^2$ or, $d = \sqrt{2}a$, $\therefore\ r^+ + r^- = \frac{d}{2} = \frac{\sqrt{2}a}{2} = \frac{a}{\sqrt{2}} = \frac{2r^-}{\sqrt{2}} = \sqrt{2}r^-$

i.e., $r^+ + r^- = \sqrt{2}r^-$ or, $r^+ = \sqrt{2}r^- - r^- = (\sqrt{2} - 1)\ r^-$ or, $\frac{r^+}{r^-} = 0.414$

For Wurtzite structure, i.e., ZnS where the crystal arrangement is tetrahedral, the radius ratio can be calculated as shown below. The open circles are S^{-2} and the solid circle is Zn^{+2}. Here,

$$r^+ + r^- = \frac{d}{2} = \frac{\sqrt{3}a}{2} \quad \text{and } 2r^- = \sqrt{2}a \quad \text{or, } a = \frac{\sqrt{2}r^-}{\sqrt{2}} = \sqrt{2}r^-$$

$$r^+ + r^- = \frac{\sqrt{3}a}{2} = \frac{\sqrt{3}.\sqrt{2}r^-}{2} = \frac{\sqrt{6}r^-}{2}$$

$$\text{or, } r^+ \frac{\sqrt{6}r^-}{2} - r^- = r^-\left(\frac{\sqrt{6}}{2} - 1\right) = r^-\ (1.225-1) = 0.225r^-$$

$$\therefore \frac{r^+}{r^-} = .225$$

For square plannar crystal,

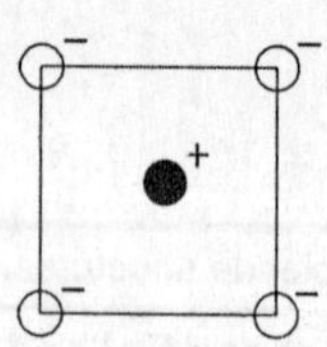

Fig. 6.8 Square plannar crystal.

$$r^+ + r^- = \frac{\sqrt{2}a}{2} = \frac{a}{\sqrt{2}} \quad \text{and} \quad 2r^- = a \quad \therefore\ r^+ + r^- = \frac{a}{\sqrt{2}} = \frac{2r^-}{\sqrt{2}} = \sqrt{2}r^-$$

$$r^+ = \sqrt{2}r^- - r^- = (\sqrt{2} - 1)\ r^- = 0.414r^- \qquad \therefore \frac{r^+}{r^-} = 0.414$$

The radius of Cl^- changes with co-ordination number.

Variation of ionic radii with co-ordination number:

Table 6.4 : Variation of ionic radii with coordination number

Ion	4	6	8	12
Li^+	0.73	0.90	1.06	
Na^+	1.13	1.16	1.32	1.53
K^+	1.51	1.52	1.65	1.78
Mg^{+2}	0.71	0.86		
Al^{+3}	0.53	0.68		
Ti^{+4}	0.56	0.75	0.88	
Fe^{+2}	0.77	0.92(HS), 0.75(LS)		
Fe^{+3}	0.63	0.79(HS), 0.69(LS)		

Thus ionic radius depends on the type of crystal arrangement or radius ratio. The table below shows the relation of radius ratio and crystal arrangement.

Table 6.5 : Relation of radius ratio and crystal arrangement

Coordination Number	Crystal arangements	Radius ratio
2	Linear	Up to 0.15
3	Triangular	0.15–0.22
4	Tetrahedral	0.22–0.41
	Square Planar	0.41–0.73
6	Octahedral	0.41–0.73
8	Cube	> 0.73

Table of ionic radii of a few ions are given below:

Table 6.6 : Ionic radii of a few ions

Ion	Goldschmidt empirical ionic radii	Pauling's calculated ionic radii	Ion	Goldschmidt empirical ionic radii	Pauling's calculated ionic radii
H^-	1.54	2.08	Mg^{2+}	0.78	0.65
Li^+	0.78	0.60	Al^{3+}	0.57	0.50
Be^{2+}	0.34	0.31	Si^{4+}	0.39	0.41
B^{3+}		0.20	P^{5+}	0.3-0.4	0.34
N^{3-}		1.71	P^{5-}		2.12
O^{2-}		1.40	S^{6+}	0.34	0.29
F^-		1.36	S^{2-}	1.74	1.84
Na^+	0.98	0.95	Cl^-	1.81	1.81

Remarkable difference in radius of H^- is observed in Goldschmidt value and Pauling's value. Theoretical calculation by Pauling assigned 2.08 Å for H^- that is not attained even by CsH. CsH is most ionic and this value is even high enough relative to He (0.93 Å), H (0.5 Å), Br (1.95 Å). The probable interpretation of this high value is (a) the nuclear charge is half the number of electrons, (b) electron-electron repulsion is strong and (c) screens each other from nuclear attraction. H^- radius of alkali metals decrease from CsH (H^- = 1.54 Å), NaH (H^- = 1.46 Å) and LiH (H^- = 1.26 Å). This decrease is due to increased covalence character of M–H bond.

The ionic radius also depends on the valence or oxidation number of the neighbouring ions. For correction the measured inter-ionic distance may be multiplied by $\sqrt[(n'-1)]{Z_1 Z_2}$, where Z_1 and Z_2 are ionic charges of the binary compound and n' is the Born exponent. Li and Sun proposed an empirical relation to calculate ionic radius. The relation is

$$R_x = 0.42\ Z^{-1} V^{1/3}\ n^3$$

where R_x is the ionic radius, Z is the atomic number, V is the oxidation number and n is the principal quantum number of outer electron. From Li to F, n^2 is used instead of n^3.

PERIODIC VARIATION

The ionic radius of an ion increases down a group as more and more quantum levels are involved. Along a period from left to right as the atomic or covalent radius decreases with higher effective nuclear charge, so also ionic radius if the oxidation number of the ion remains the same. The extent of this decrease in ionic radius is profound when oxidation number decreases with group number. So the change in ionic radius is more profound for normal or representative elements.

For transition metals the atomic radius and ionic radius decreases from left to right but this is less than that observed for representative elements. This is due to the fact that *ns* electrons are constant, added electrons enter the $(n-1)$ d-shell with increasing nuclear charge and screens effectively the ns electron as observed in the following table:

Table 6.7 : Ionic radius of transition metal ions with oxidation number

Ti^{2+}(0.90)	Fe^{2+}(0.76)	Ti^{3+}(0.76)	Fe^{3+}(0.64)
V^{2+}(0.88)	Co^{2+}(0.74)	V^{3+}(0.74)	Co^{3+}(0.63)
Cr^{2+}(0.84)	Ni^{2+}(0.72)	Cr^{3+}(0.69)	Ni^{3+}(0.62)
Mn^{2+}(0.80)	Cu^{2+}(0.74)	Mn^{3+}(0.66)	

The second series of the transition elements, which differ from its congener in the first series, is 18 electrons. The change in ionic radius is significant Ti^{4+} (0.68 Å) and Zr^{4+} (0.80 Å). But this increase is insignificant in case of 2nd and 3rd transition series, which differ from its congener by 32 electrons that occupy 4f-shell, valence shell being 6s. Hence for the trivalent ions of this inner transition series, increase in radius from that of the second which is less pronounced. This is known as *Lanthanide Contraction*. Among the lanthanides the ionic radius is more or less same, e.g., La^{3+} (1.22 Å), Lu^{3+} (0.99 Å). Thus, ionic radius being very close to each other, their properties are identical. this trend is also observed in actinide series as a result of *Actinide Contraction*.

The radius of polyatomic ions cannot be calculated easily, but can be done with difficulty from the heat of formation and the lattice energy of the ionic crystal. Such radius is known as Thermochemical Radius.

NH_4^+ (1.51), ClO_3^- (1.57), ClO_4^- (2.26), NO_2^- (1.78), NO_3^- (1.65), CO_3^{2-} (1.64), PO_4^{3-} (2.38), SO_4^{2-} (2.44), CrO_4^{2-} (2.42), MnO_4^- (2.15).

6.03 VAN DER WAALS' RADIUS

It is the shortest radius or half the distance between two non-bonded atoms of two adjacent molecules of non-metals before repulsion between their electron clouds comes into play. Non-metals in solid state exist as an aggregation of molecules, e.g., in solid xenon. As it is a non-bonded type, its inter-nuclear distance is 4.36 Å.

So, van der Waals' radius is 2.18 Å. The ionic radius and van der Waals' radius are numerically same and may have the same electronic arrangements away from the bonds, as do their negative ions.

van der Waals' radii of some non-metals are as follows:

N (1.5)	O (1.4)	F (1.35)
P (1.9)	S (1.85)	Cl (1.8)
As (2.0)	Se (2.00)	Br (1.95)

van der Waals' radius of a non-metal is useful in consideration of non-metal structure and in case of some organic compounds.

6.04 IONISATION POTENTIAL

The energy required to remove the most loosely bound electron, i.e., valence electron, from an isolated neutral gaseous atom of an element in its lowest energy state to convert it into a mono-positive ion is called the *ionisation energy* or *potential energy*. Obviously, the energy for the removal of first electron from the atom is the first ionisation potential energy; the removal of second electron from M^+ ion is the 2nd I.P. and so on. So, 1st I.P. < 2nd I.P. < 3rd I.P. and so on. It has been discussed earlier in Bohr's theory on hydrogen atom; the energy supplied to the atom promotes the electron to higher and higher energy orbital until removed completely.

The numerical values of ionisation potential depends on various factors:

1. *The charge upon the atomic nucleus*: If the elements have similar or same outer orbital with increased nuclear charge, the effective nuclear charge for an electron in the valence shell increases and so the first ionisation potential increases in general.
2. *The shielding effect of the inner electronic shell*: The nucleus attracts the valence electron; this attractive force is partly balanced by the repulsive force of the inner electrons, the inner electrons will screen or shield the outer electron from the nucleus and so the ionisation potential will decrease.
3. *Atomic radius*: With increase in atomic radius the valence electrons are away and so the ionisation energy will decrease.

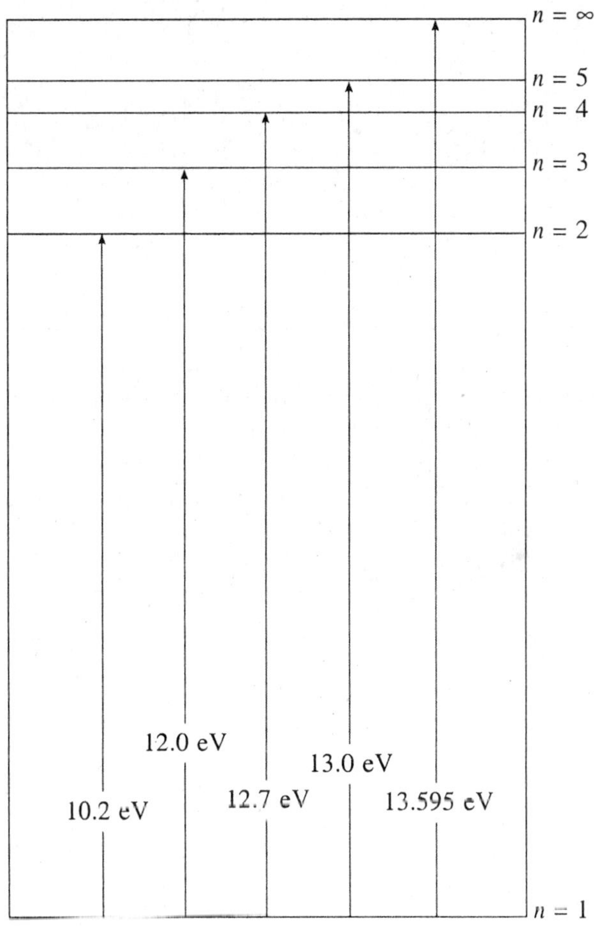

Fig. 6.9 Graphical representation of energy levels in the hydrogen atom.

4. The extent to which the outer electrons will penetrate the cloud of electronic charge of the inner electrons. The term, 'penetrate', means the extent to which an outer electron approaches more close to the nucleus. For the same quantum number n, n_l has different values and hence also the eccentricities of the orbital and so $s > p > d > f$ indicates the increasing penetrating effect of the orbital. It is evident that d and f electrons can be removed with greater ease and hence low ionisation energy. But I.P. of an element is dependent on the

combination of all these factors. The periodic variation of 1st ionisation potential can be graphically given as follows in Fig. 6.10:

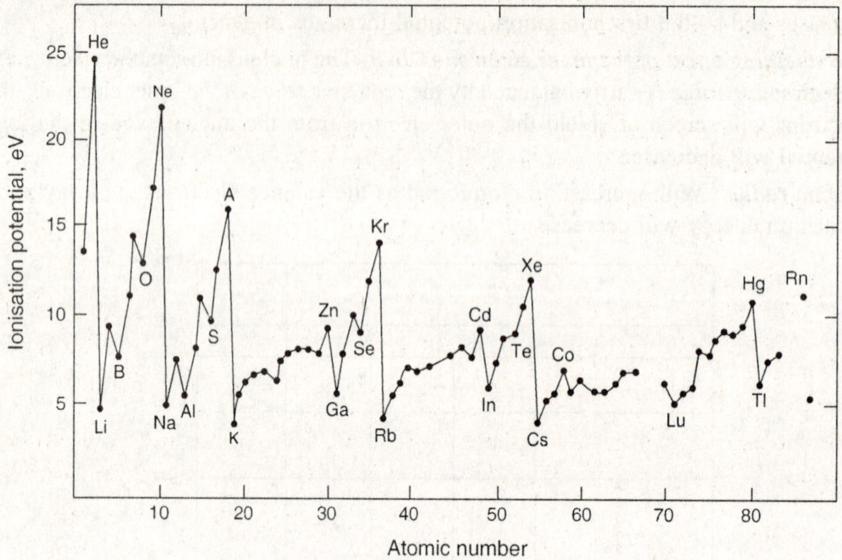

Fig. 6.10 Variations in first ionisation potentials with atomic number.

The ionisation potential is expressed in electron-volts. In general, chemical energy is insufficient to ionise an atom that required more than 15 eV. Hence simple cations of charges greater than 2 is uncommon. Observation of the graph reveals the following:

1. The inert gases have occupied the topmost position as they have complete ns^2p^6 configuration, which is most stable.
2. Next to inert gases, alkali metals have the lowest ionisation energy as the valence electrons, i.e., ns^1 are well screened by $(n-1)\ s^2p^6$ configuration. In these noble gases and the alkali metals, the ionisation energy decreases down the group as we pass on to the heavier numbers.
3. Along a period, as we pass from left to right, ionisation energy increases in general, excepting some anomalies. This is due to higher nuclear charge without expansion of the electronic orbit or, in other words, the effective nuclear charge for an electron increases.
4. For group III B elements, i.e., representative elements like B, Al, Ga, etc., there is significant drop in ionisation energy relative to the preceding members. This drop is maximum for B and Al, for others less. Here, the first np^1 appears after completion of ns^2 (being a duplet configuration) that is more stable and screens the np^1 effectively. Again, *p*-orbital is less penetrating than *s* and hence easier to remove. With higher elements Ga, In, etc., effect of this duplet configuration are more pronounced and have complex underlying electronic arrangement and hence drop in ionisation energy is less.
5. Stability for VI B, e.g., O, S, Se, etc., there is again drop in ionisation energy where these elements have s^2p^4 configuration. In the *p* level, the 4th electron is easier to remove in comparison to half-filled orbital, i.e., s^2p^3, which is more stable. Here also, this drop is pronounced in O and S and less for Se and Te, etc.
6. For N and P, etc., the ionisation energy values are higher compared with their neighbours. This is due to the stabilising influence of spin-spin interaction, which is maximum for filled and half-filled orbital.

7. For transition and inner transition series the electronic configuration of the valence shell is constant but nuclear charge increases with the addition of extra electron to the $(n-1)$ or $(n-2)$ shell. This results in slight decrease in atomic size and slight increase in ionisation energy and this increase in ionisation energy is still less pronounced in inner transition series. Zn, Cd and Hg have the highest value in the corresponding transition series as they are the last member of the series.
8. Hg has the highest ionisation energy value similar to noble gas Rn. This is because here *d* and *f* shells are filled and penetration effect in 6*s* electrons is larger.
9. The elements just following the lanthanide series have higher ionisation energy than the corresponding elements in the preceding series as lanthanide contraction brings about decrease in size.
10. Ionisation energy of *A* subgroup elements has low ionisation energy than the *B* subgroups of the periodic table because *B* subgroups have underlying 18 electrons arrangement that has more penetration effect than the underlying 8 electrons characteristics of *A* subgroup elements.

With the increase in oxidation states, the ionisation energy increases. If the difference of ionisation energy in two successive oxidation states is about 10–11 eV or less, then lower oxidation state is unstable. If this difference is about 16 eV, then lower state is stable. With gradually increasing oxidation state of a particular element, the electrovalent character of the compound formed gradually decreases. Ionisation energy data are very helpful in understanding the probability of formation of ionic compound, understanding reducing property of the free element, etc.

First Ionisation Potential (I.P.) of some elements in electron-volts (eV)

Table 6.8. Ionisation energies (in eV) for successive removal of electron-from atoms of the elements (1 eV = 8066 cm^{-1} = 23.06 kcal mol^{-1} = 96.49 kJ mol^{-1})[(a)]

Table 6.8 : First I.P. of elements

Z	Elements	1st	2nd	3rd	4th	5th	6th	7th	8th
1	H	13.598							
2	He	24.587	54.416						
3	Li	5.392	75.638	122.451					
4	Be	9.322	18.211	153.893	217.713				
5	B	8.298	25.154	37.930	259.368	340.217			
6	C	11.260	24.383	47.887	64.492	392.077	489.981		
7	N	14.534	29.601	47.448	77.472	97.888	552.057	667.029	
8	O	13.618	35.116	54.934	77.412	113.896	138.116	739.315	871.378
9	F	17.422	34.970	62.707	87.138	114.240	157.161	185.187	953.886
10	Ne	21.564	40.962	63.45	97.11	126.21	157.93	207.27	239.09
11	Na	5.139	47.286	71.64	98.91	138.39	172.15	208.47	264.18
12	Mg	7.646	15.035	80.143	109.24	141.26	186.50	224.43	265.90
13	Al	5.986	18.828	28.447	119.99	153.47	190.71	241.43	284.59
14	Si	8.151	16.345	33.492	45.141	166.77	205.05	246.52	303.17
15	P	10.486	19.725	30.18	51.37	65.023	220.43	263.22	309.41
16	S	10.360	23.33	34.83	47.30	72.68	88.049	280.93	328.23
17	Cl	12.967	23.81	39.61	53.46	67.8	97.03	114.193	348.28
18	Ar	15.759	27.629	40.74	59.81	75.02	91.007	124.319	143.456
19	K	4.341	31.625	45.72	60.91	82.66	100.0	117.56	154.86

Contd.

Z	Elements	1st	2nd	3rd	4th	5th	6th	7th	8th
20	Ca	6.113	11.871	50.908	67.10	84.41	108.78	127.7	147.24
21	Sc	6.54	12.80	24.76	73.47	91.66	111.1	138.0	158.7
22	Ti	6.82	13.58	27.491	43.266	99.22	119.36	140.8	168.5
23	V	6.74	14.65	29.310	46.707	65.23	128.12	150.17	173.7
24	Cr	6.766	16.50	30.96	49.1	69.3	90.56	161.1	184.7
25	Mn	7.435	15.640	33.667	51.2	72.4	95	119.27	196.46
26	Fe	7.870	16.18	30.651	54.8	75.0	99	125	151.06
27	Co	7.86	17.06	33.50	51.3	79.5	102	129	157
28	Ni	7.635	18.168	35.17	54.9	75.5	108	133	162
29	Cu	7.726	20.292	36.83	55.2	79.9	103	139	166
30	Zn	9.394	17.964	39.722	59.4	82.6	108	134	174
31	Ga	5.999	20.51	30.71	64				
32	Ge	7.899	15.934	34.22	45.71	93.5			
33	As	9.81	18.633	28.351	50.13	62.63	127.6		
34	Se	9.752	21.19	30.820	42.944	68.3	81.70	154.4	
35	Br	11.814	21.8	36	47.3	59.7	88.6	103.0	192.8
36	Kr	13.999	24.359	36.95	52.5	64.7	78.5	111.0	126
37	Rb	4.177	27.28	40	52.6	71.0	84.4	99.2	136
38	Sr	5.695	11.030	43.6	57	71.6	90.8	106	122.3
39	Y	6.38	12.24	20.52	61.8	77.0	93.0	116	129
40	Zr	6.84	13.13	22.99	34.34	81.5			
41	Nb	6.88	14.32	25.04	38.3	50.55	102.6	125	
42	Mo	7.099	16.15	27.16	46.4	61.2	68	126.8	153
43	Tc	7.28	15.26	29.54					
44	Ru	7.37	16.76	28.47					
45	Rh	7.46	18.08	31.06					
46	Pd	8.34	19.43	32.93					
47	Ag	7.576	21.49	34.83	52.0				
48	Cd	8.993	16.908	37.48					
49	In	5.786	18.869	28.03	54				
50	Sn	7.344	14.632	30.502	40.734	72.28			
51	Sb	8.641	16.53	25.3	44.2	56	108		
52	Te	9.009	18.6	27.96	37.41	58.75	70.7	137	
53	I	10.451	19.131	33					
54	Xe	12.130	21.21	32.1					
55	Cs	3.894	25.1						
56	Ba	5.212	10.004						
57	La	5.577	11.06	19.175					
58	Ce	5.47	10.85	20.20	36.72				
59	Pr	5.42	10.55	21.62	38.95	57.45			
60	Nd	5.49	10.72						
61	Pm	5.55	10.90						
62	Sm	5.63	11.07						
63	Eu	5.67	11.25						
64	Gd	6.14	12.1						
65	Tb	5.85	11.52						

Contd.

Z	Elements	1st	2nd	3rd	4th	5th	6th	7th	8th
66	Dy	5.93	11.67						
67	Ho	6.02	11.80						
68	Er	6.10	11.93						
69	Tm	6.18	12.05	23.71					
70	Yb	6.254	12.17	25.2					
71	Lu	5.426	13.9						
72	Hf	7.0	14.9	23.3	33.3				
73	Ta	7.89	16.2						
74	W	7.98	17.7						
75	Re	7.88	16.6						
76	Os	8.7	16.9						
77	Ir	9.1	16.0						
78	Pt	9.0	18.563						
79	Au	9.225	20.5	30.5	43.5				
80	Hg	10.437	18.756	34.2					
81	Tl	6.108	20.428	29.83					
82	Pb	7.416	15.032	31.937	42.32	68.8			
83	Bi	7.289	16.69	25.56	45.3	56.0	88.3		
84	Po	8.42							
85	At	9.2							
86	Rn	10.748							
87	Fr	3.98							
88	Ra	5.279	10.147						
89	Ac	5.1	12.1						

6.05 ELECTRON AFFINITY (E.A.)

Electron affinity of an element is the amount of energy released in bringing an electron from outside and to place it in the lowest energy state of a neutral gaseous atom to make it a uni-negative gaseous ion (preferably at 0 K). This process can be expressed as

$$E_{(g)} + e^- \longrightarrow E^-_{(g)} + \text{E.A.} \qquad \text{E. A.} = -\Delta H$$

Electron affinity of an element is measured indirectly and seldom with accuracy. In most cases it is determined using Born-Haber cycle. The electron affinity of an element can also be expressed as the energy required to remove the added electron of the uni-negative gaseous ion to form a neutral gaseous isolated atom that can be expressed as

$$E^-_{(g)} \longrightarrow E_{(g)} + e^- - \text{I.P.}$$

So, the electron affinity of an element is numerically equal to ionisation potential of the uni-negative ion of the same element with sign reversed. As in the case of ionisation potential, successive addition of electron will give rise to 2nd E.A., 3rd E.A. and so on. Addition of 2nd and 3rd electrons will feel electron-electron repulsion and so the energy is to be added to push the electron against repulsion and so EA_1, EA_2, etc. will be negative. Thus, electron affinity value is a measure of the affinity or attraction of an atom to hold the extra electron. If energy is released on adding electron to the atom, the anion is more stable than the neutral atom. The electron affinity is due to the pairing effect of the added electron with the singly occupied one, i.e., the unpaired electron already present in the atom.

The electron affinities of some elements are given in Table 6.9 in eV.

Table 6.9 : Electron affinity of some elements

Element	E.A.	Element	E.A.	Element	E.A.	Element	E.A.
H	0.75	F	3.6	Cl	3.62	Fe	0.16
He	−0.5	Ne	−1.2	Ar	−1.0	Co	0.66
Li	0.62	Na	0.55	K	0.5	Ni	1.16
Be	−0.5	Mg	−0.4	Ca	−0.3	Cu	1.23
B	0.3	Al	0.44	Sc	−0.3	Zn	−0.6
C	1.26	Si	1.4	Ti	0.08	Ga	0.3
N	−0.2	P	0.75	V	0.53	Ge	1.2
O	1.46	S	2.03	Cr	0.67	As	0.81
O^{-2}	−7.2	S^{-2}	−3.44	Mn	−0.5		

E.A. of a few elements are known accurately and many have been calculated theoretically. Hence it is quite difficult to make generalisation about the periodic trend but following observations can be noted:

(a) Factors affecting E.A. are the same as those affecting I.P. and so along a period from left to right it increases, i.e., more energy is released. This is due to the fact that nuclear charge increases and atomic size decreases.

(b) Along a group with increase in size, E.A. decreases in an irregular way in case of metals, but in case of non-metal the same trend is observed with some exceptions.

(c) Inert gases have closed electronic configuration and so non-bound state for added electron. Excess energy is needed to push the electron and hence negative value.

(d) Group II metals have negative value as they have either ns^2 or $(n-1)\,d^{10}ns^2$ configuration. They are reluctant to accept electron. Hg has –0.5 eV, but neighbouring elements Au (2.31 eV) and Pt (2.13 eV) have very high positive value.

(e) As expected the alkali metals have very low E.A. whereas halogens will have very high value. For halogens attainment of inert gas configuration and pairing effect plays the role. Among halogens Cl has the highest value; F<Cl>Br>I>At.This can be attributed to the fact that F has small size and no *d* orbital, here charge density is high and incoming electron will be repelled to some extent, hence low value relative to Cl. For Cl, 3*d* vacant orbital get mixed with 3*p* orbital and some electron cloud diffuses to the vacant *d* orbital thereby decreasing the charge density; this results in high electron affinity. For other members of this group, although *p* and *d* orbital mixed up, bigger size and higher screening effect of the inner electrons results in decrease in electron affinity.

(f) Being the first member and small size O^{-1} has low E.A. relative to S^{-1} state. Similar is the case with B and C relative to the other members of the family. Stable states of O^{2-} and S^{2-}, repulsion between the added electrons are more than the pairing effect and hence high negative value.

(g) For Group V elements half-filled shells are more stable. Formation of N^- will require more energy and hence negative value. The Born-Haber thermo-chemical cycle affords theoretical calculation of electron affinity. It is a thermodynamic cycle correlating heat of formation of a compound with other thermo-chemical quantities, and assumptions are made that the formation of an ionic crystal may occur either by direct combination of the elements or by successive steps evaporation of elements to gaseous state, then converted to ions and finally gaseous ions are combined to give the product. For example, let us consider the formation of the compound *MX*, this can be formulated as

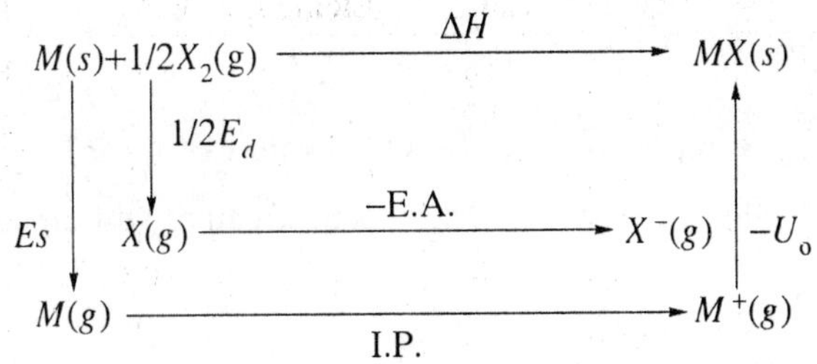

where *Es* is the sublimation energy, E_D is the dissociation energy, U_0 is the lattice energy and ΔH is the heat of formation. Other thermo-chemical data being known E.A. can be calculated. U_0 is calculated by Born-Lande equation,

$$U_0 = \frac{NAe^2Z^+Z^-}{4\pi\varepsilon_0 r}\left(1-\frac{1}{n}\right)$$

Here N = Avogadro's number, Z = ionic states, e = charge, r = inter-ionic distance, n = Born exponent or repulsion constant and A = Madelung constant depending on the geometry of the crystal. The value (n) can be derived theoretically from compressibility of the solid. Its value depends on the electronic configuration of the ions, e.g., for He = 5, for Ne = 7, Ar = 9, Kr = 10 and so on. If the ions have different electronic configuration, then mean value of n is taken.

6.06 ELECTRONEGATIVITY

It is not the inherent property of an atom. In order to study the polar character of the hetero-atomic covalent bond, this concept was introduced. The polar character of the hetero-molecules results from the unequal sharing of bonded pair electrons by them. This develops due to the difference in affinity towards the bonded electrons. This property is known as Electronegativity. Thus electronegativity may be defined as the tendency of an atom in a molecule to attract the bonded electrons towards themselves. The absolute value of electronegativity of an element is not fixed but differs according to the different theoretical approach.

Pauling scale: It is evident that atoms with small size of orbital with nearly filled electrons will attract electrons more than the atom with bigger size and of orbitals which are sparsely occupied. Let A_2 and B_2 be two molecules (purely covalent), they react to form the molecule *A—B*.

$$A—A + B—B = A—B + A—B$$

These reactions are exothermic, i.e., *A—B* molecule will be stable and the bond is stronger than *A—A* or *B—B* bond. If *A—B* be pure covalent, then bonded electron pair will be shared equally by *A* and *B*, and extra stability of *A—B* will not be obtained. Pauling in 1931 suggested that extra stability of *A—B* bond is obtained if the bonded electron pair is shared unequally, i.e., one atom will attract bonded electron more than the other, i.e., higher electronegativity of the former than the latter. This results in some ionic character to the *A—B* bond, i.e., polarity develops. Larger the difference of electronegetivity between the atoms *A* and *B* more polarity, or in otherworlds, more ionic character develops. It is found that when electronegativity difference between bonded atoms is 2.1, about 50% ionic character develops (although it is found that difference of 1.7 induces 50% ionic character).

Thus theoretical treatment of chemical bond shows that mixture of ionic and covalent nature will give rise to a stronger bond than the pure covalent or pure ionic bond. This additional bond energy is called the *ionic-covalent resonance energy* (Δ) and its relation is

$$\Delta = E_{A-B} - \sqrt{E_{A-A} \times E_{B-B}}$$

The term $\sqrt{E_{A-A} \times E_{B-B}}$ represents 100% covalent nature of *A—B* bond. E_{A-B} is the observed bond energy of *A—B*. Pauling suggested that the electronegativity difference of *A* and *B* in *A—B* is $\sqrt{\Delta}$, where Δ is in eV.

$X_A - X_B = \sqrt{\Delta}$ eV $= (\Delta / 23.06)^{1/2} = 0.208$ kcal/mole

(1 eV = 23.06 kcal/mole)

Thus $X_A - X_B = 0.208 \sqrt{E_{A-B} - (E_{A-A} \times E_{B-B})^{\frac{1}{2}}}$. Pauling thus calculated the value of $X_E - X_H$ for different elements and assumed $X_H = 0$. Thus X_E for different elements are obtained. This led to the negative value in several cases. Pauling changed his scale taking $X_H = 2.1$ and got positive value for all elements even for alkali metals. His calculated values are B = 2.0, C = 2.5, N = 3.0, Cl = 3.0, F = 4.0. Basic properties of elements are inversely related to the electronegativity. Along a period from left to right electronegativity increases and down a group decreases.

The electronegativity values of elements have wide applications:

(a) In studying the percent ionic character of a covalent bond

(b) In determining the bond energy

(c) In understanding thermal stability of compound

(d) In determining bond polarity and bond moment

Exact evaluation of ionic character in a polar bond is difficult but Pauling calculated by taking the ratio of exact dipole moment of the molecule to the moment considering the molecule to be ionic; the moment is the unit charge multiplied by bond distance. The percentage of ionic character when plotted against the electronegativity difference of the combining atoms gives the following curve (Fig 6.11):

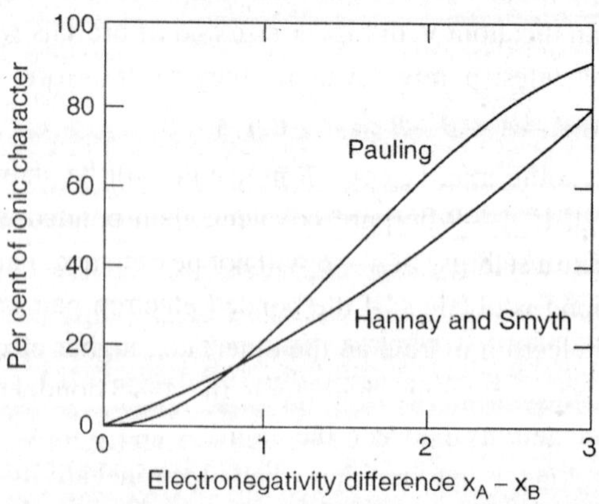

Fig. 6.11 Relation of amount of ionic character in a bond A-B to the electronegativities of the bonded atoms.

Pauling indicated 60% ionic character in HF but recent evaluation indicates 43%. So an empirical relation was established as follows:

Amount of ionic character = $0.16\ (X_A - X_B) + 0.035\ (X_A - X_B)^2$

The ionic character of a compound depends not only on the electronegativity difference but also on many other factors.

Mulliken's scale: In 1934, Mulliken proposed that the electronegativity of an element is the average of I.P and E.A. of the element, both expressed in eV. This is a straightforward relation only for univalent atom. When this average value is divided by 2.6, then electronegativity of an element has least deviation from Pauling's value. But this method has the limitation as the electron affinities of most of the elements are not known accurately. The linear relation between X_P and X_M is given as $X_P = 1.35(X_M)^{1/2} - 1.37$. Here X_P and X_M are the electronegativity of an element in Pauling's scale and Mulliken's scale respectively.

Allred and Rochow in 1958 proposed another method to evaluate the electronegativity of an element which is the electrostatic force, F acting on the electron at the distance equal to the covalent radius, r of the atom, $F = \dfrac{e^2 Z_{eff}}{r^2}$.

Thus, new electronegativity value can be related to that of Pauling's value by the relation

$$X_{AR} \approx X_P = 0.744 + 0.359\ Z_{eff} / r^2.$$

Here Z_{eff} is the effective nuclear charge. Here it is considered at the periphery and hence all valence electrons are considered unlike calculated previously. Thus, Allred and Rochow scale is based on the fundamental property of an atom and better agreement in many elements have been observed.

Electronegativity of an element is not constant but it depends on the two factors: (a) valence state and (b) atomic charge, i.e., chemical environment, e.g., S in SCl_2 will have less electronegativity than S in SF_6. Hydrogen in SiH_4 will have less electronegativity than that in CH_4, i.e., it depends on the nature of the other element with which it combines. Types of hybridization of the central atom in a compound change electronegativity of the element as the hybridized orbitals are of lower energy. As electron attracting power of *s* orbital is greater, so with greater *s* character of the hybridized orbital, it will be more electronegative, e.g., in case of nitrogen undergoing sp, sp^2, sp^3 hybridization X_P are 3.68, 3.26 and 3.04 respectively. In case of carbon undergoing sp, sp^2, sp^3 hybridization the respective electronegativities are 2.99, 2.66 and 2.48. *s* character of the orbitals of carbon decreases its electronegativity and increases the activity of H in the order $C_2H_2 > C_2H_4 > CH_4$ (50%, 33% and 25% respectively).

With change in oxidation number, the electronegativity changes are as follows:

Fe (II), $X_P = 1.83$; Fe (III), $X_P = 1.96$; Cu (I) $X_P = 1.9$; Cu (II) $X_P = 2$.

Higher cationic charged ions have higher electron attracting power, i.e., higher Lewis acidic character (Fajans' rule).

Group electronegativity: The inductive effect of substituents, overall charges, hybridization of bonding orbitals, all these together in a group is termed as group electronegativity. The group electronegativity is calculated directly from electronegativity of the constituent atoms in the group and then converted to Pauling scale. Thus, groups can be considered as super-atoms capable of absorbing large amount of charges, e.g., the reaction of CH_3I and CF_3I with alkali gives different products.

$CH_3I + OH^- \longrightarrow CH_3OH + I^-$ (C is less electronegative than I)

$CF_3I + OH^- \longrightarrow CF_3H + IO^-$

So nucleophillic attack of OH^- occurs at carbon in CH_3I. In case of CF_3I

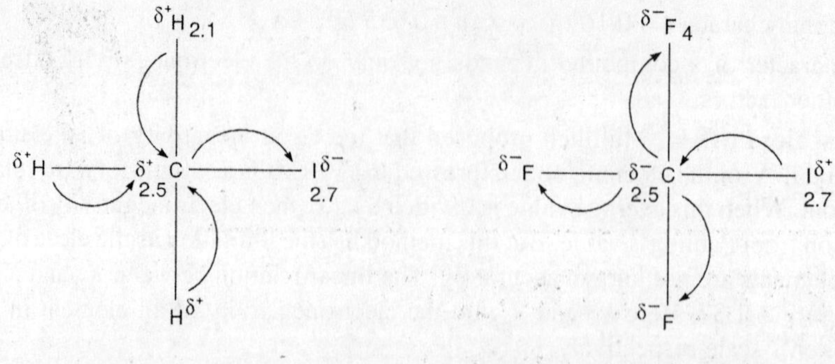

F increases the electronegativity of carbon greater than iodine by +δ and hence H attacks the carbon and so CF_3H is formed. The group electronegativities of CH_3 (2.31), CF_3 (3.47), CCl_3 (2.95), COOH (3.04), CN (3.32) have been calculated by S. G. Bratsch in 1985.

Table 6.10 : Electronegativity values in different scales

Atom	Pauling	Mulliken	Atom	Pauling	Mulliken
H	2.2	3.06	Cl	3.16	3.54
Li	0.98	1.28	K	0.82	1.03
Be	1.57	1.99	Ca	1.00	1.30
B	2.04	1.83	Ga	1.81	1.34
C	2.55	2.67	Ge	2.01	1.95
N	3.04	3.08	As	2.18	2.26
O	3.44	3.21	Br	2.96	3.24
F	3.98	4.42	In	1.78	1.30
Ne	—	4.60	Sn	1.96	1.83
Na	0.93	1.21	Sb	2.05	2.06
Mg	1.31	1.63	Te	2.10	2.34
Al	1.61	1.37	I	2.66	2.88
Si	1.90	2.03	Pb	2.33	—
P	2.19	2.39			

CHAPTER

7

EARLY CONCEPT OF VALENCY

7.01 EARLY CONCEPT OF VALENCY

In early days, study of compounds reveals that elements when combine to form compounds, cannot do so in a haphazard manner but follow some laws regarding their combining weights, although some are non-stoichiometric in nature. But it is clear that there is some guiding force that made the elements to combine among them. This guiding force was first termed as atomicity by Gerhardt and later termed as quantivalence by Hoffmann in 1865 and afterwards termed as valency or valence by Wickelhans in 1868. Electrolytic decomposition of substances led Davy and later Berzelius to think of the chemical or combining forces to be electrical in nature although discovery of organic compounds raised questions against this view. Later, discovery of electron gave support to this former idea.

In developing periodic table, Mendeleeff arranged elements in groups having chemical similarity and indicated relation among the valence number and the group in which the elements are placed. In 1904, Abegg expanded this concept and said that the maximum valence number both positive and negative together (without sign) will be eight and positive valence number will be the group number. Drude expanded Abegg's view in terms of modern terminology by saying that positive valence number v signifies the number of loosely bound electron in the atom and the negative valence v' is the number of electrons that can be removed from other element by that atom or can held firmly to itself.

The ordinary compounds and organic compounds have two types of valence forces—polar and non-polar. After the discovery of electron by J.J. Thomson, on the origin, nature, their number and the mode of formation of these compounds, several theories were developed. Out of all these theories, Kossel's (1916) theory of ordinary or polar or ionic compounds and Lewis (1916) theory of organic or non-polar compounds are worthy of mention. These two theories were based on the arrangement of electron in Bohr model and hence termed as *electronic theory of valency*. As the inert gases have

8 electrons in their outermost shell and are most stable, so these theories aimed at the distribution of electrons in the elements in a compound to attain inert gas configuration as the basis for these combining forces. This is known as *octet rule*. Thus the driving force to attain this inert gas configurations, i.e., octet rule or rule of eight, gives rise to different types of valencies—electro-valency, covalency and co-ordinate valency.

Kossel pointed out that elements just following the inert gases of closed electronic configuration have loosely bound electrons in their valence shell and hence during chemical combination they will loose these electrons and become positively charged ion. Similarly, the elements just preceding the inert gases have tendency to gain electron from others to attain closed configuration thereby forming negatively charged ions. Thus, the compound formed in this way, by loosing or gaining electrons equivalent to their valence number, have electrostatic attraction and the bond between them is called *electrovalent bond*. Electrovalent bonds not actually are bonds, have no directional nature and as the electrostatic force extends in all directions, large number of cations and anions come into close vicinity to form a three-dimensional arrangement called the *crystal*. They have high melting point, boiling point, soluble in polar solvents, and in solution or in molten state can conduct electricity, e.g., the NaCl, MgO are formed as follows:

$$Na(2,8,1)+Cl(2,8,7) \rightarrow Na^{+}(2,8)\,Cl^{-}(2,8,8) \text{ configuration of Ne and Ar}$$

$$Mg(2,8,2)+O(2,6) \rightarrow Mg^{+2}(2,8)\,O^{-2}(2,8) \text{ and so on.}$$

The energy consideration of such ionic bond will be discussed later. Some compounds are known where the inert gas configurations are closely approached. The same thing is observed for Nickel group of elements where 28 electrons are present (pseudo-insert gas or underlying 18-electron arrangement).

Lewis was of the opinion that in the compounds whether the atom will be electron donor or acceptor is difficult to ascertain; sharing of electrons between the atoms attains their inert gas arrangements. Each combining atom will give one electron to make a pair and then this pair of electrons commonly enjoyed by them forming a bond. This is called the *covalent bond*. When they share more than one pair, then it is called *multiple bond*. The formation of the following molecules are shown as follows:

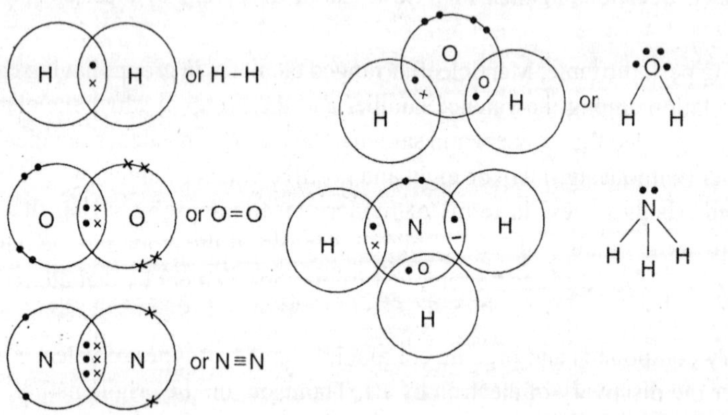

Fig. 7.1 Formation of some covalent molecules.

Thus, the covalent bonds are the real existing bond between atoms and they have directional character. As such compounds formed by covalent bond are non-polar in nature (very weak polarity in some cases) and soluble in non-polar solvent. They cannot conduct electricity. There is very weak attraction between the molecules and hence they have discrete nature, melting and boiling points are low. Some of these covalent molecules (e.g., SiO_2, BN, SiC) may form three-dimensional giant molecules, where covalent bonds extend over large number of atoms (that depend on the neutral electro-chemical nature of the central atom).

In some cases where one atom has more electrons than the other with which the former combines, the lone pair of electron is donated by one atom and shared by both. This type of valence is known as *co-ordinate co-valency*. Although the lone pair is donated by one and after bond formation, there exists no perceptible difference between covalent and co-ordinate bond. They are indistinguishable when they co-exist and an arrow pointing the direction of donation sometimes represents the co-ordinate bond. It is also indicated by a single line; with (+)*ve* and (–)*ve* signs.

STRUCTURE OF CO

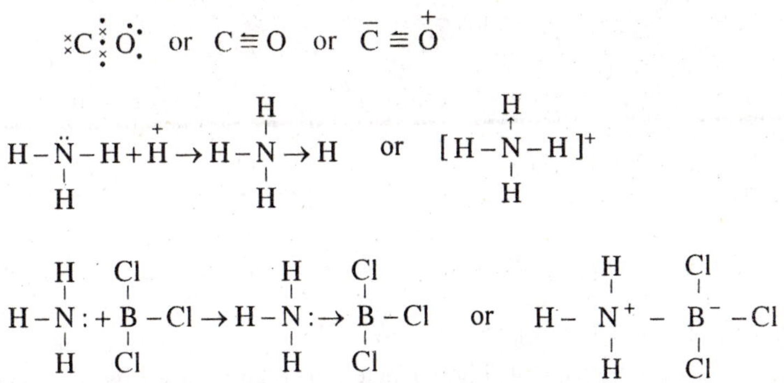

Fig. 7.2 Formation of co-ordinate bond of some compounds.

Although inert gas configuration attained by both cations and anions in electrovalent or covalent compounds are stable due to mutual cancellation of resultant spin momentum and orbital momentum, stable species are known where this rule is not followed. Again, octet rule is true to the neutral atom but not to the ions as electrostatic forces operate upon them. The species, which do not follow the rule, can be classified as follows (Foster):

A. *Ionic compounds with regular configurations*

(i) No electron species:	H^+, D^+, T^+
(ii) Species with two electrons:	H^-, Li^+, Be^{+2}
(iii) Species with 8 electrons:	N^{-3}, O^{-2}, $F^{-1} \rightarrow Ne \leftarrow Na^{+1}$, Mg^{+2}, Al^{+3}
	S^{-2}, $Cl^{-1} \rightarrow A \leftarrow K^+$, Ca^{+2}, Sc^{+3}
(iv) Species with 18 electrons:	Cu^+, Zn^{+2}, $Ga^{+3} \rightarrow Ni$
	Ag^{+1}, Cd^{+2}, $In^{+3} \rightarrow Pd$
	Au^{+1}, Hg^{+2}, $Tl^{+3} \rightarrow Pt$
(v) Species with (18+2) electrons:	Ga^{+1}, Ge^{+2}, $As^{+3} \rightarrow Zn$
	In^{+1}, Sn^{+2}, $Sb^{+3} \rightarrow Cd$
	Tl^{+1}, Pb^{+2}, $Bi^{+3} \rightarrow Hg$

The 2 electrons available are not involved in combination. These are the inert pair of Sidgwick.

B. *Ionic compounds with irregular configuration*

(i) Hg_2^{+2} has peculiar electron system.

(ii) Cations of transition and inner transition metals of varying cationic charge and varying number of electrons in the valence shell:

Cr^{+2} (2, 8, 12) Cr^{+3} (2, 8, 11)
Fe^{+2}(2, 8, 14) Fe^{+3} (2, 8, 13)
V^{+3}, V^{+2}, Nd^{+3}, Gd^{+3}, etc.

(iii) Anions of irregular configuration: SO_3^{-2} and SO_4^{-2}. Ō is surrounded by 8 electrons and S in SO_4^{-2} and SO_3^{-2} by 12 and 10 electrons respectively.

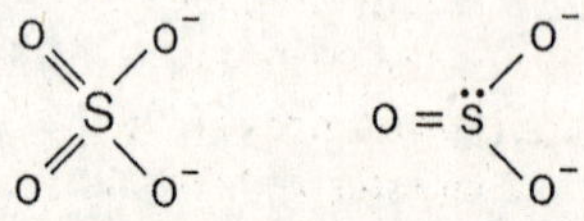

Fig. 7.3 Electron arrangements in SO_4^{-2} and SO_3^{-2} ions.

C. *Covalent compounds that do not follow the octet rule*

$BeCl_2$ (Be = 4, Cl = 8), BF_3 (B = 6, F = 8), PCl_5 (P = 10, Cl = 8), SF_6 (S = 12, F = 8), $AlCl_3$ (Al = 6, Cl = 8).

Similar species are WCl_6, OsF_8, metal carbonyls $Fe(CO)_5$, many co-ordination compounds, metal nitrosyls, etc. So it may be concluded that although octet rule is very useful, its application is limited, but in considering the chemical union between atoms, it is universal that electron pairing is must. Hence, it can be said, "The rule of two is far more fundamental than the rule of eight."

7.02 LIMITATIONS OF THE ELECTRONIC THEORY

The octet rule is empirical in nature. A large number of compounds do not follow this rule as just mentioned. In case of covalent or polar compounds the rule does not mention the geometry or shape of the molecule. Whether H_2O is linear or angular, or CH_4, NH_3, PCl_3, PCl_5 are planar or three-dimensional and the properties of the compounds are dependent on the structure of the molecule cannot be explained.

In case of ionic compounds the limitations are still more worthy to mention. The octet rule, applicable to atoms of the elements, explains their stability, but the ions having octet configuration are not stable. They enter into large number of electrostatic interactions; attraction between oppositely charged ions and repulsion between the ions of like charges in the close vicinity and also among ions away. This theory does not consider the net effect of these interactions. It also does not account for the energy change associated with the formations of ionic compound from solid metal and non-metal, their subsequent vaporisation, ionisation, dissociations, etc.

Thus, a compound formed whether electrovalent or covalent is always associated with energy change and the compound formed will have always less energy than the isolated atom, i.e., the energy is liberated. The same amount of energy is to be supplied to break the chemical bond in a compound in gaseous state. This is the bond dissociation energy. A chemical bond is the force with which the chemically distinct species are held together. The bond dissociation energy is dependent on the chemical environment and presence of other bonds in the compound. Another criteria for chemical bond is the bond length, which is the average distance between the nuclei of two atoms held by chemical bond. The bond length also depends on the multiplicity of the bond, lone pair-lone pair interaction, and also the presence of other atoms in the molecule.

7.03 IONIC BONDS AND LATTICE ENERGY

As mentioned earlier, ionic bond is formed between elements having low ionisation energy and high electron affinity. The electrostatic force holds the cation and anion to form an ionic compound, and as this electrostatic force extends in all directions, large number of cations and anions form an aggregate called *crystal*. In solid, liquid and gaseous state, these compounds exist as ions and hence instead of molecular weight formula weight is used. Thus, when these ions enter into the crystal lattice, impart, extra stability to the compound and is called *lattice energy*. Thus, lattice energy of an ionic compound is the energy released when a large number of positive and negative gaseous ions representing a molecule of the material, separated from each other by an infinite distance, are brought together into crystal lattice characterized by definite inter-ionic distance. Here, all attractive and repulsive forces in all directions and from nearer to far remote ions depending on the geometry of the crystal are to be considered. As a simple case, let us consider a simple ionic crystalline lattice to know the energy change. Here, ions are considered as point charge at infinite distance and then come closer, when force of attraction begin to play. They are at equilibrium distance. When they come more closer to each other, repulsion occurs, which is a short-range force.

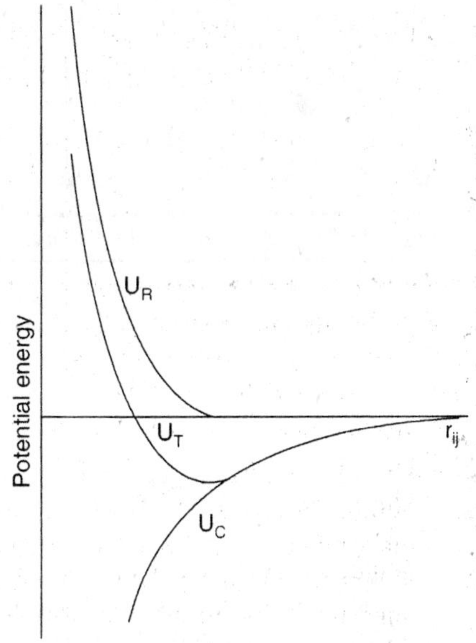

Fig. 7.4 Potential energy diagram for the formation of an ionic bond as a function of ionic separation.

$$U_n = \frac{B}{r_{ij}^n}, \text{ where } n \text{ is of the order 9.}$$

So, $U_T = U_C + U_n$

U_T is the total energy and U_C is the attractive force.

$$\therefore\ U_T = -\frac{ZiZje^2}{r_{ij}} + \frac{B}{r_{ij}^n}$$

At $r = r_0$, the equilibrium distance U_T will be zero. So we can write for uni-univalent ionic compound,

$$\left(\frac{\partial U_T}{\partial r}\right)_{r=r_0} = \frac{e^2}{r_0^2} - \frac{Bnr_0^{n-1}}{r_0^{2n}} = 0 \quad \text{or,} \quad \frac{e^2}{r_0^2} = \frac{nB}{r_0^{n+1}} \quad \text{or,} \quad B = \frac{e^2 r_0^{n-1}}{n}$$

$$\text{So,}\quad U_T = -\frac{e^2}{r_0^2} + \frac{e^2 r_0^{n-1}}{n r_0^n} = -\frac{e^2}{r_0^2} + \frac{e^2}{n r_0} = -\frac{e^2}{r_0}\left(1 - \frac{1}{n}\right)$$

This is the potential energy of single ionic system. If we want to calculate total energy in a crystal, we must multiply this with number of ions present together with attractive and repulsive forces arising out of second, third, fourth and so on layers of neighbouring ions. The second and onward energy term depends upon the particular geometry of the crystal that is termed as Madelung constant A, n is the Born exponent (discussed in the Article 6.02(*e*) of ionic radius)

$$U_T = -\frac{ANe^2}{r_0}\left(1 - \frac{1}{n}\right)$$

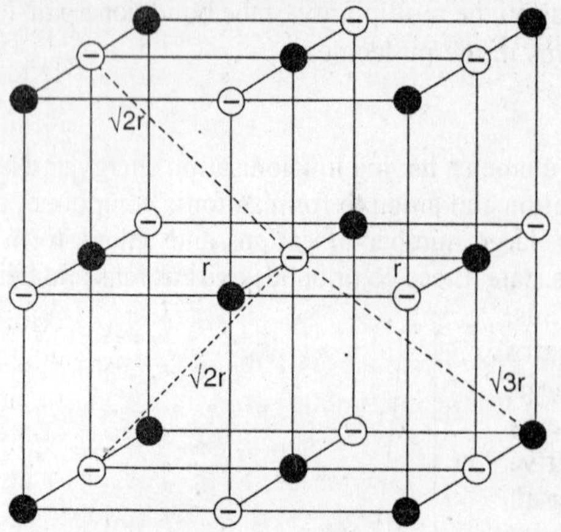

Fig 7.5 The sodium chloride structure.

Born exponent, n can be calculated from the compressibility of the crystal (κ) by the relation, $\kappa = \frac{18r_0^4}{Ae^2(n-1)}$ and Madelung constant can be calculated considering the simple 6 : 6 face-centered NaCl crystal. Each Na^+ is surrounded by 6Cl^- and each Cl^- by six Na^+, the Na^+-Cl^- distance is represented by r. The first layer of 6 Cl^- surrounds a Na^+ at a distance r. 12 Na^+ of second layer neighbour surrounds the Cl^- at $\sqrt{2}r$, 8 more distant Cl^- at $\sqrt{3}r$, 6 more Na^+ at $2r$, 24 more Cl^- at $\sqrt{5}r$ and so on.

So the electrostatic interaction energy of Na^+ with all these surrounding ions will be

$$E = U_T = \frac{6e^2}{r}(Z^{\pm})^2 + \frac{12e^2}{\sqrt{2}r}(Z^{+})^2 + \frac{8e^2}{\sqrt{3}r}(Z^{\pm})^2 + \frac{6e^2}{2r}(Z^{+})^2 + \frac{24e^2}{\sqrt{5}r}(Z^{\pm})^2 + \ldots$$

or, $$E = U_T = -\frac{e^2|Z|^2}{r}\left[6 - \frac{12}{\sqrt{2}} + \frac{8}{\sqrt{3}} - \frac{6}{2} + \frac{24}{\sqrt{5}} + \ldots\right]$$

The sum of this infinite series is called Madelung constant, A which is characteristic of the geometrical arrangement of ions, i.e., crystal structure and is independent of the nature of ions and their oxidation states. Madelung constants for the crystals are given as NaCl (1.747558), CsCl (1.762670), ZnS (1.63806). Wurtzite (Zinc Blende = 1.641), Fluorite (5.03878). The values of n determined theoretically and estimated from compressibility of the crystal are as given below:

Table 7.1 : Value of Born exponent for different compounds

Value of compound	: Experimental value n	Value of n : Nobel gas compound or ion	Theoretical value n
LiF	5.9	He	5
LiCl	8.0	Ne	7
LiBr	8.7	Ar	9
NaCl	9.1	Kr	10
NaBr	9.5	Xe	12

Born exponent n for a compound, say LiF, is the average of the noble gas configurations of the ions Li^+ (He = 5) and F^- (Ne = 7) which comes out to be 6, is close to the experimental value of 5.9. For higher ions the lattice energy calculated has 1–2% error.

In the theoretical consideration of lattice energy three correlation factors are to be introduced for accurate determination. These are:

1. van der Waals' forces operating between all atoms or ions or molecules arising out of the oscillating dipoles of the neighbours and this is very short range force of the order $\frac{1}{r^6}$.

2. Zero-point energy of the crystal. This is the vibrational energy of ions even at absolute zero.
3. Repulsion energy calculated from Bohr equation is not correct. Correct determination requires quantum mechanical approach. After introducing these corrections Mayer, Born and Helmholz gave the equation,

$$U_T = E = \frac{AZ_jZ_je^2N}{r} + NBe^{-kr} - \frac{NC}{r^6} + NE_0$$

where k is the compressibility, E_0 is the zero-point energy and C is another constant.

Using a thermodynamic cycle, known as Born-Haber cycle, the lattice energy is determined, which is in good agreement with the theoretical value. The cycle gives correlation between the heat of formation H_f of a compound with electron affinity, A; ionisation potential, I; heat of sublimation, S; dissociation energy, D and lattice energy, U. The construction of the cycle is based on the assumption that the formation of an ionic crystal may occur either by direct combination of elements or by the alternative method in which reactants are vaporized then converted to gaseous ions and the gaseous ions are converted to products. The cycle may be constructed as follows:

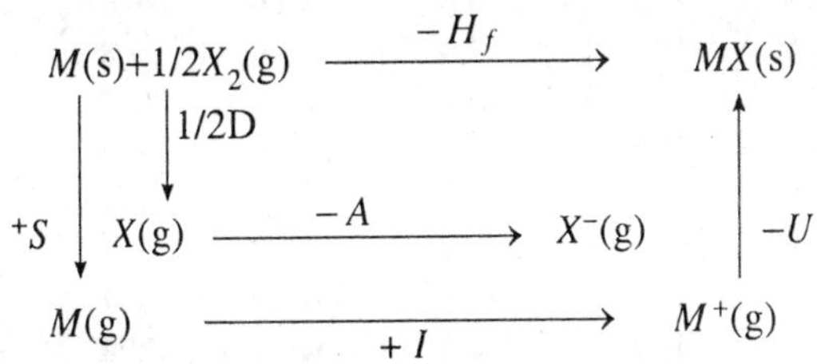

Fig. 7.6 Born-Haber cycle for the determination of lattice energy.

So, $-\Delta H_f = S + I + \frac{1}{2}D - A - U$

or, $U = \Delta H_f + S + I + \frac{1}{2}D - A$

Using this equation, lattice energy of ionic compounds can be calculated. As the electron affinities of elements are not known with certainty.

(a) This cycle is used to evaluate electron affinity from the theoretical lattice energy of the compound. The electron affinity of oxygen is measured from the ionic oxides using this cycle.

(b) Lattice energy data is helpful in understanding the stability of compounds. Heat of formation and lattice energy of alkali metal fluorides decrease with increasing size of metal ion but with other halide the reverse is true as observed from the table below. This is due to the closeness of approach of small metal ion and fluoride ion.

(c) Born-Haber cycle is also helpful to understand the instability of compounds of lower valent metal ions.

Table 7.2 : Born-Haber cycle data for certain ionic compounds

Compound	*H* kcal	*S* kcal	½*D* kcal	*I* kcal	*A* kcal	Theo. *U* kcal	Exp. *U* kcal
LiF	144.7	38.3	32.2	123.8	98.9	240.1	242.8
NaF	136.6	26.0	32.2	118.0	97.8	215.0	216.6
KF	134.5	21.7	32.2	99.6	97.6	190.4	191.8

contd.

Compound	*H* kcal	*S* kcal	½*D* kcal	*I* kcal	*A* kcal	Theo. *U* kcal	Exp. *U* kcal
RbF	132.8	19.9	32.2	95.9	99.0	181.8	184.6
CsF	131.5	19.1	32.2	89.4	99.4	172.8	176.0
LiCl	97.5	38.3	28.9	123.8	95.2	193.3	201.7
NaCl	98.2	26.0	28.9	118.0	90.7	180.4	193.9
KCl	104.9	21.7	28.9	99.6	90.7	164.4	168.3
RbCl	104.9	19.9	28.9	95.9	90.7	158.9	162.8
CsCl	106.6	19.1	28.9	89.4	95.1	148.9	157.2
LiBr	83.7	38.3	26.9	123.8	89.6	183.1	191.0
NaBr	86.3	26.0	26.9	118.0	85.6	171.7	175.5
KBr	94.2	21.7	26.9	99.6	84.6	157.8	160.7
RbBr	96.1	19.9	26.9	95.9	86.3	152.5	157.1
CsBr	97.5	19.1	26.9	89.4	89.4	143.5	151.2
LiI	65.0	38.3	25.4	123.8	81.8	170.7	178.4
NaI	69.5	26.0	25.4	118.0	78.1	160.8	164.8
KIr	78.9	21.7	25.4	99.6	76.6	149.0	151.5
RbI	80.8	19.9	25.4	95.9	77.8	144.2	147.9
CsI	83.9	19.1	25.4	89.4	81.7	136.1	143.7
MgO	146.0	36.5	59.2	520.6	−178	940.3	
CaO	151.7	47.6	59.2	412.0	−171	842.3	
SrO	141.0	39.7	59.2	383.8	−167	790.7	
BaO	133.0	49.1	59.2	349.0	−157	747.3	
MgS	82.2	36.5	66.6	520.6	−72.4	778.3	
CaS	114.0	47.5	66.6	412.9	−80.8	721.8	
SrS	113.0	39.7	66.6	383.8	−84.3	687.4	
BaS	111.0	49.1	66.6	349.0	−80.2	655.9	

7.04 LATTICE DEFECTS

In a real or ideal crystal of a compound, at absolute zero, every atom or ion is placed on an appropriate lattice point and each lattice point is occupied by right kind of ions or atoms. At temperature above this, the thermal vibration of ions or atoms causes the lattice defects that may be due to

(i) interchanging of ions or atoms from their lattice position,

(ii) the location of ions or atoms of any kind in interstitial positions, or

(iii) the existence of some vacant sites from which any ion or atom is missing.

The first kind is uncommon for energetic reason, and observed only in metallic phases. Therefore, the defects in crystals are either stoichiometric defects or non-stoichiometric defects.

Stoichiometric defects: Here the ratios of cations and anions represent their chemical formula. Two types of defects may occur:

A. *Schottky defect*: Here the ratio of $\frac{A^+}{B^-}$ is same but same number of cations and anions are missing from the crystal lattice creating equal number of vacancies or holes. This type of defect is generally found in highly ionic compounds of high co-ordination number—where size of cation and anion are equal, e.g., NaCl, CsCl, etc.

B. *Frenkel defect*: In this case an ion is shifted from its normal lattice position and occupies an interstitial position. This occurs in compounds, where A^+ and B^- differs widely in size and co-ordination number is low, e.g., AgBr, ZnS, etc. Here the cations are smaller, and so take up the interstices of bigger anions. Such compounds are polar or covalent due to greater polarisation. These compounds are feeble conductor of electricity as the cation moves from one lattice to the other leaving a hole, which is occupied by another creating a new hole and in this way conduction occurs.

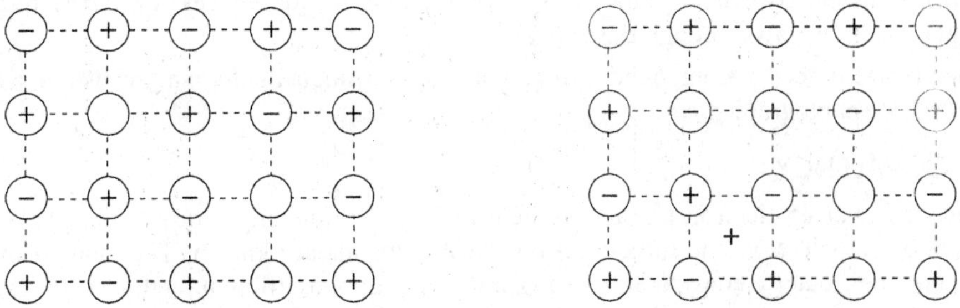

A Schottky defect B Frenkel defect

Fig. 7.7 A Schottky defect B Frenkel defect.

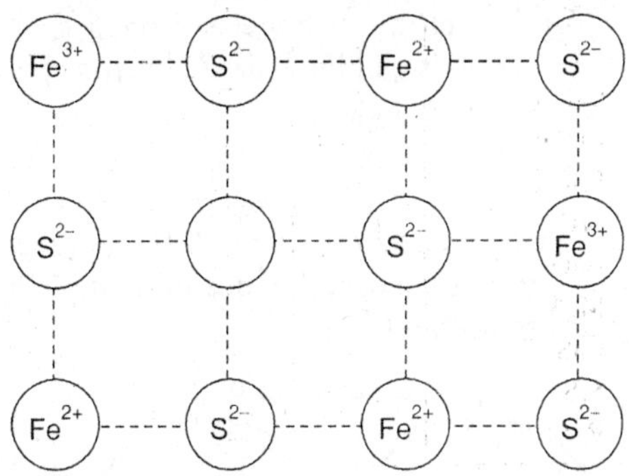

Fig. 7.8 Defects in FeS.

Non-stoichiometric defects: Here the ratio of cation and anion does not follow the law of constant proportion, i.e., they are non-stoichiometric or Berthollide compounds. These can be represented by the formula $M_{1\pm x}X$ where M and X is generally bivalent or monovalent and M has variable valency, e.g., FeS, FeO, NiO, CuO, CuI, etc., and x is the fraction of vacant cation sites per formula weight that has been oxidised to higher valence state to maintain the electrical neutrality. Thus, FeS may be represented by $Fe^{+2}_{1-3x}Fe^{+3}_{2x}S^{-2}$. This metal deficiency defect can be diagrammatically represented as in Fig. 7.8.

Sometimes metal deficiency crystal defect also occurs when an extra anion occupies the interstitial position, and charge is balanced by oxidation of cation to higher valence state. Non-stoichiometric defect may also occur with metal excess. Anion is missing from lattice site and an electron occupies that position, e.g., NaCl with sodium vapour forming a yellow $Na_{1+x}Cl$ compound or KCl forms $K_{1+x}Cl$ (lilac colour) and electron occupies the vacant site, as in the Fig. 7.9.

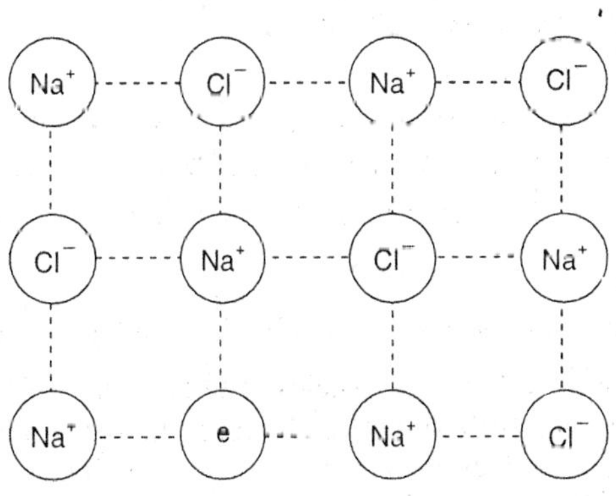

Fig. 7.9 Defects in NaCl lattice showing "tapped ***e***".

ZnO lattice contains Zn^{+2} excess having formula $Zn_{1+x}O$. When ZnO is heated, it looses some oxygen, leaving vacant anion sites where electron occupies that sites. Now electrons are excited to higher energy level with absorption of energy and assume yellow colour. In small crystals when other metal atom replaces one metal atom, crystal defects are observed. Presence of these metal impurities imparts some special properties to the crystal. Thus, Germanium metal containing some gallium (having less electron) leaves some electron holes and this gallium-doped germanium behaves as *p*-type semiconductor.

Similarly, arsenic(having more electron)-doped germanium gives *n*-type semiconductors as negative charge flows through the crystal lattice. So $Fe_{1-x}O$ may behave as *p*-type semiconductor as electron transfer between Fe^{+2} and Fe^{+3} takes place and $Zn_{1+x}O$ behaves as *n*-type semiconductor as trapped electron flows through the crystal.

Thus, lattice defective compounds are very useful as semiconductors in transistors, as catalyst, in colour T.V., photography, etc.

7.05 COVALENCY

Lewis concept of electron pairing in chemical bond formation is truly striking, although the existence of one-electron and three-electron covalent bonds are known. The method of electron pairing and their energy consideration was approached in two different methods: (a) the valence bond (V.B.) or atomic orbital (A.O.) method and (b) the molecular orbital (M.O.) method. The essential features of valence bond methods are mentioned below:

(i) The valence electrons of the combining atoms only interact to get paired.

(ii) The atomic orbitals retain their individuality.

(iii) The chemical bond is formed as a result of considerable degree of overlap of suitable orbitals and the paired electrons reside in the zone of overlap.

In the M.O. approach, it is considered that valence orbitals of the combining atoms first get mixed to form a set of molecular orbitals that include both the nuclei of the combining atoms and then the valence electrons occupy the M.O.s. For simplicity and ease of understanding and to interpret the chemical observation, the V.B. approach is more acceptable to chemists (although in simple cases). In this concept the idea of multiple bond formation and also the valence forces directed in space has been retained.

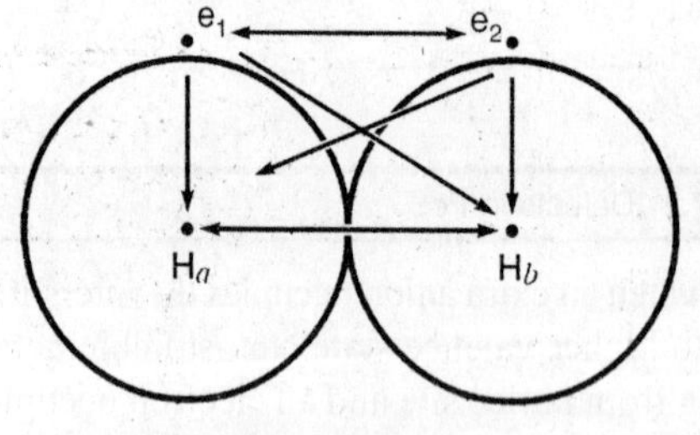

Fig. 7.10 Formation of H_2 molecule.

To understand the Valence Bond approach, let us take the formation of simple hydrogen molecule. We know that to dissociate the hydrogen molecule into atoms, same energy equivalent to bond dissociation energy is needed. So, during formation of molecule from two separated hydrogen atoms at a large distance, some energy is released and forms the stable molecule. Here we have to consider six electrostatic interactions—four are attractive in nature between electrons and protons and two are repulsive in nature between two electrons or two protons.

Total energy of the molecule will be the resultant of all these interactions. The magnitude of these interactions and hence the energy of the molecule can be worked out by classical method and quantum mechanical method which will be discussed later. Classically, hydrogen molecule can be represented as:

(i) $H_a^1\ H_b^2$ (ii) $H_a^2\ H_b^1$ (iii) $H_a^{-\delta 1,2}\ H_b^{+\delta}$ (iv) $H_a^{+\delta}\ H_b^{-\delta 1,2}$

The structures (i) and (ii) are most probable and here exchange of electrons occurs indicating the loss of individuality of electrons. Though structures (iii) and (iv) are less probable since the electro-negativity of the element is the same, yet their contribution to the resonance energy evaluation should be considered. So, H_2 molecule may be represented by all the canonical forms, i.e., the molecule is a resonance hybrid of the four, no one can express exactly the molecule. But the conclusion that can be drawn in that resonance is must to get a stable covalent bond and the stable bond will have some ionic character. The extent of this ionic character is dependent upon the electro-negativity difference primarily.

According to Pauling, the following conditions must be satisfied for resonance:

(a) The relative positions of the constituting atoms must be same in all canonical forms.

(b) The individual structures differ only in distribution of valence electrons.

(c) The different canonical forms will have nearly equal energy such that they can contribute to the resonance.

(d) All canonical forms will have same number of unpaired electrons.

(e) In the canonical forms, the negative charge, that is, the unpaired electron will be on the more electro-negative atom and the positive and negative charge will be placed as far away as possible.

(f) For effective resonance to permit, the molecule will be planar.

The energy consideration for the formation of stable hydrogen molecule can be represented as follows:

The curve indicates if the combining atoms have unpaired parallel spin, repulsion results and the molecule will be unstable. But if they have opposite spin, attraction results and forms the stable molecule. When the combining atoms approach more close to r_0 representing the equilibrium distance, nucleus-nucleus repulsion operates and the molecule will be unstable again. The first one is called the *anti-symmetric state* and the latter *symmetric state.* In quantum mechanical approach, this can be achieved as follows: The wave function of the two hydrogen atoms are $\Psi_a(1)$ and $\Psi_b(2)$ when they are separated. The combined molecule will have the wave functions for the structures:

(i) and (ii) $\Psi_I = \Psi_{ab} = \Psi_a(1)\Psi_b(2) + \Psi_a(2)\Psi_b(1)$ which indicate the resonance occurring there and wave function corresponding to structures (iii) and (iv) will be $\Psi_{II} = \Psi_a(1)\Psi_a(2) + \Psi_b(1)\Psi_b(2)$ which indicates ionic nature of the covalent bond. So overall wave function will be $\Psi_{VB} = \Psi_I + \Psi_{II} = \Psi_a(1)\Psi_b(2) + \Psi_a(2)\Psi_b(1) + \Psi_a(1)\Psi_a(2) + \Psi_b(1) + \Psi_b(2)$.

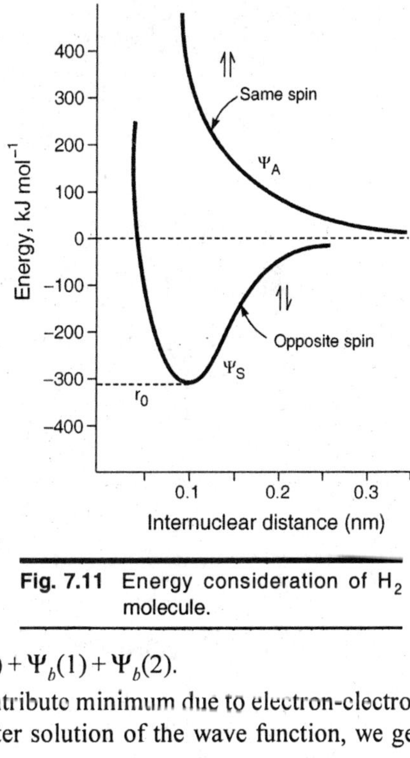

Fig. 7.11 Energy consideration of H_2 molecule.

The last two terms representing the ionic term will contribute minimum due to electron-electron repulsion. So, $\Psi_{VB} = \Psi_a(1)\ \Psi_b(2) + \Psi_a(2)\ \Psi_b(1)$ and after solution of the wave function, we get

$\Psi_S = \dfrac{1}{\sqrt{2+2S}}\left[\Psi_a(1)\Psi_b(2) + \Psi_a(2)\Psi_b(1)\right]$, $\Psi_A = \dfrac{1}{\sqrt{2-2S}}\left[\Psi_a(1)\Psi_b(2) - \Psi_a(2)\Psi_b(1)\right]$; Ψ_S is the symmetric function having lower energy, i.e., anti-parallel spin, and Ψ_A is the anti-symmetric function having higher energy due to parallel spin. Here S is the overlap integral and $S = \int \Psi_a(1)\Psi_b(2)\Psi_a(2)\Psi_b(1)\,d\tau_1 d\tau_2$. The anti-parallel spin character of Ψ_S can be understood from the following consideration: When two hydrogen atoms form the molecule, the electron density will increase in between the two nuclei, i.e., the electrons will spend more time there, or in other words the two electrons will be indistinguishable, i.e., $\Psi_a(1) \approx \Psi_b(2)$, and $\Psi_b(1) \approx \Psi_b(2)$ so we can write

$$\Psi_S = \frac{2}{\sqrt{2+2S}}\Psi_a(1)\Psi_b(2) \text{ and } \Psi_A = 0.$$

Here, Ψ_A represents no electron density in between the nuclei. So, only symmetric combination represents the bonding electron and according to Pauli's principle they will be paired. The plots of probability curves and wave function are as shown in Fig. 7.12.

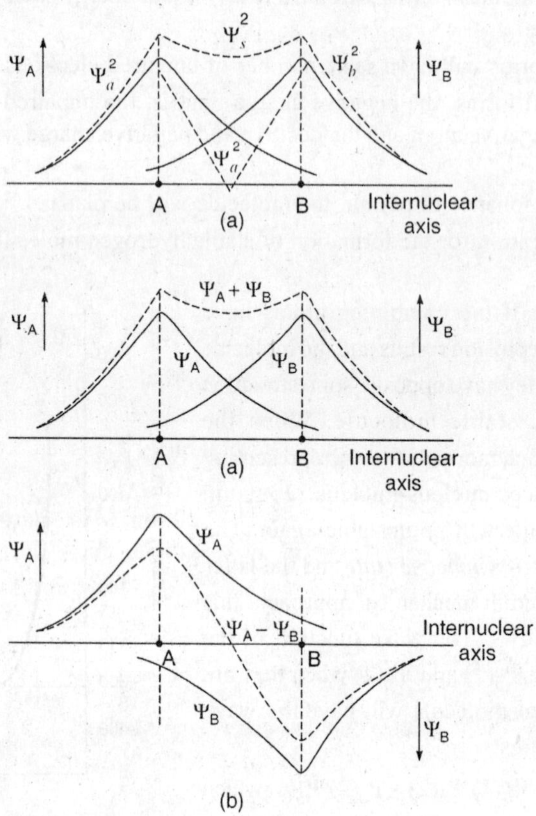

Fig. 7.12 (a) Symmetric and (b) anti-symmetric overlapping of two 1 *s* wave functions.

7.06 HEITLER AND LONDON CONCEPT (V.B. CONCEPT)

The concept of molecule formation adopts Lewis idea of electron pairing that comes from the combining atoms containing one or more unpaired electrons in them. The valency of the atom thus depends on the number of unpaired electron in valence shell. Already paired electron may involve in bond formation if they are made unpaired without expense of much energy that actually occurs when no change in principal quantum number takes place. For this reason NCl_3 is formed but not NCl_5 whereas PCl_3 and PCl_5 both are known. For nitrogen $2s^2 2p^3$, the $2s^2$ electron if be made unpaired, will go to the 3*s* orbital, i.e., change in principal quantum number *n* will take place, whereas for P_4, this transition occurs within the same quantum state.

Likewise, PH_3 is formed but not PH_5. Here transition of 3*s* electrons to 3*d* state will not take place as sufficient chemical energy is not available and no mixing of 3*d* and 3*s* orbital occurs. The three main features are available from the discussion about this concept:

1. After bond formation each combining atom attains closed electronic structure.
2. Each combining atom gives one unpaired electron for bond formation and the electrons will be paired.
3. The bonded paired electron will be localised in between the two atoms.

In many cases the first feature is not maintained as in the formation of PCl_5, the formation of co-ordinate bond is not accounted for.

The limitation of this concept can be shown as:

(a) NO being an odd electron molecule is fairly stable.
(b) Formation of B_2H_6 molecule cannot be explained.
(c) Oxygen molecule is highly paramagnetic.
(d) Perbromates are not formed.
(e) Stability of H_2^+ could not be explained.

All these indicate the modification of the concept.

Thus, simplified treatment of covalent bond in terms of orbital overlap and spin pairing is analogous to Lewis concept but as a result of overlap, bonds will be directional in nature in space and this was not accounted for in Lewis concept.

7.07 HUND AND MULLIKEN THEORY (M.O. CONCEPT)

In the valence bond concept it was accepted that the atomic orbitals remain unchanged when molecules are formed but this may not be the case as due to mutual electrostatic interactions, the atomic orbitals will be deformed and will form molecular orbital by the linear combination of atomic orbital (LCAO).

The concept of this combination can be developed as follows: When an electron moves round the nuclei of a molecule, at any instant of time it is closest to one nucleus where it is controlled by that nucleus rather than the one farther away. When it is nearer to the nucleus of H_a, its wave function will be Ψ_a and when it is closest to H_b, its wave function will be Ψ_b. The resultant wave function of the electron within the nuclear framework of the molecule will be $\Psi_{MO} = \Psi_a \pm \lambda\Psi_b$ for one-electron system.

For hydrogen molecule, i.e., 2-electron system, the wave function will be like Ψ_{VB} as described before. The difference being that here too much emphasis is given to ionic terms, the resonance term and the ionic term are given the same weight. The value of λ depends on the nature of interacting nuclei and is a measure of the ionic character of the bond. So, for homo-nuclear diatomic molecule $\lambda = 1$.

The molecular orbitals are formed by definite combinations of quantum numbers and possess some definite energy. After the M.O.s are formed, electrons are fed into these orbitals following Aufbau's principle and Pauli's exclusion principle but the nuclei will be at equilibrium distance. Here n, l, etc., will have little significance but new quantum number λ having values 0, ±1, ±2 are obtained; correspondingly, orbitals are designated as σ, π, δ, etc. This can be given in comparison with atomic orbitals as follows:

Table 7.3 : Different M.O.s and their λ values

Complete electron shell in atom	Value of l	Value of λ	Complete electron groups in molecule
s^2	0	0	σ^2
p^6	1	+1, 0, −1	σ^2, π^4
d^{10}	2	+2, +1, 0, −1, −2	$\sigma^2\pi^4\delta^4$

The combination of A.O.s to form the M.O.s can be diagrammatically represented as follows—different M.O.s are formed depending on the sign of the A.O.s. Ψ_A^* and Ψ_B are anti-bonding and bonding molecular orbitals. Thus, in bonding M.O.s electron density increases and hence stronger bond is formed. In the anti-bonding M.O.s the energy becomes higher, and there is no electron density in between the two and hence less stable bond is formed.

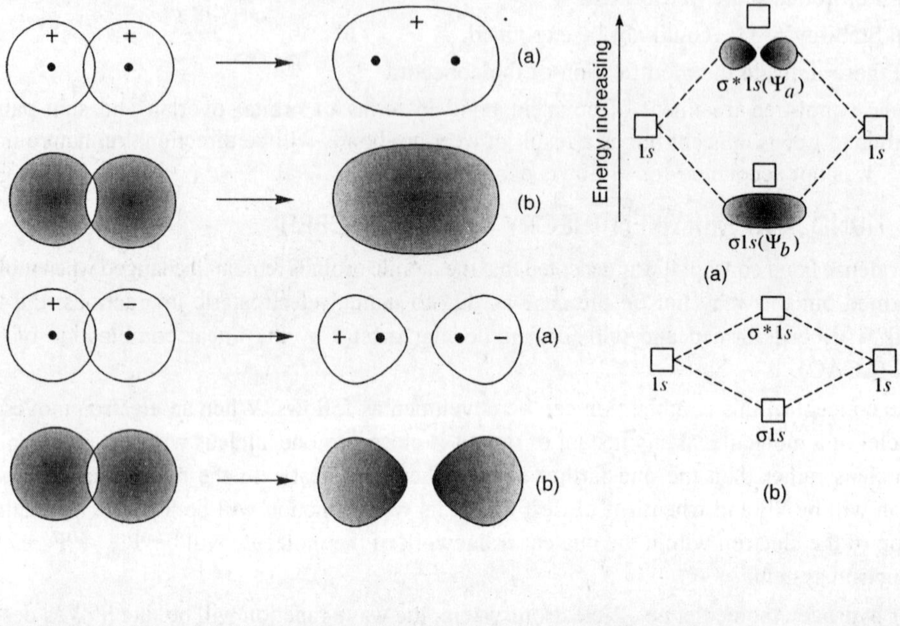

Fig. 7.13 Addition of two 1*s* A.O.s in tems of (a) boundary surfaces of the A.O.s and (b) corresponding charge clouds. The sign of the wave function appears in (a) but not in (b), since density of charge cloud is proportional to Ψ.
Combination of two 1*s* atomic orbitals having opposite sign of wave function in terms of (a) boundary surface of the atomic orbital and (b) charge density.

The difference between the V.B. theory and M.O. theory can be drawn in this way:

(i) Both the theories give same result in different method of approach.

(ii) V.B. approach is easier and involves localised bond and can be represented by diagram. The M.O. concept based on mathematics with gradual refinement is harder to realize and abstract in nature.

(iii) M.O. concept better explains the properties of molecules.

(iv) M.O. theory is applicable better to diatomic molecule and in some cases for triatomic molecules like HCN, etc., but not applicable to polyatomic and three-dimensional molecule.

The simple M.O. diagrams of some homo-nuclear and hetero-nuclear diatomic molecules are discussed below. The M.O. diagram for elements up to half-filled *p* orbital for homo-nuclear diatomic molecule π_{2p}^b becomes lower energy than the σ_{2p}^b and for elements having more than half-filled $2p$ orbitals the reverse case occurs. For hetero-nuclear diatomic molecules, depending upon the occupancy of the valence orbitals and the electro-negativity of the combining elements, M.O. becomes more complicated. Here energy consideration and symmetry of the orbitals are considered.

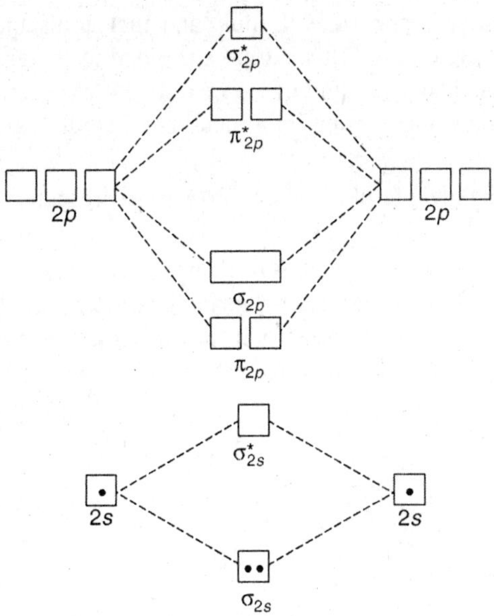

Fig. 7.14 M.O. diagram for Li_2.

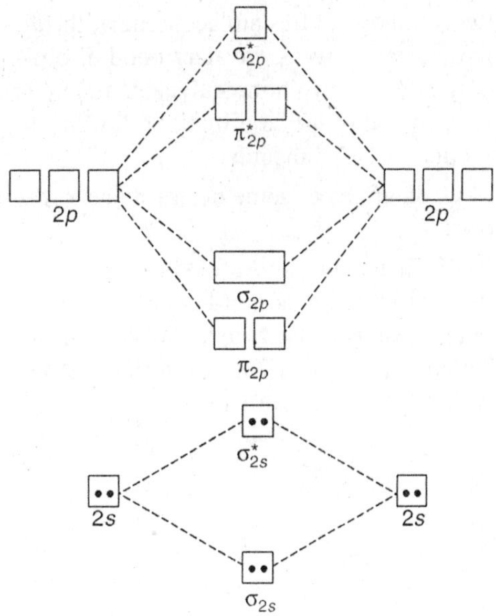

Fig. 7.14 M.O. diagram for Be_2.

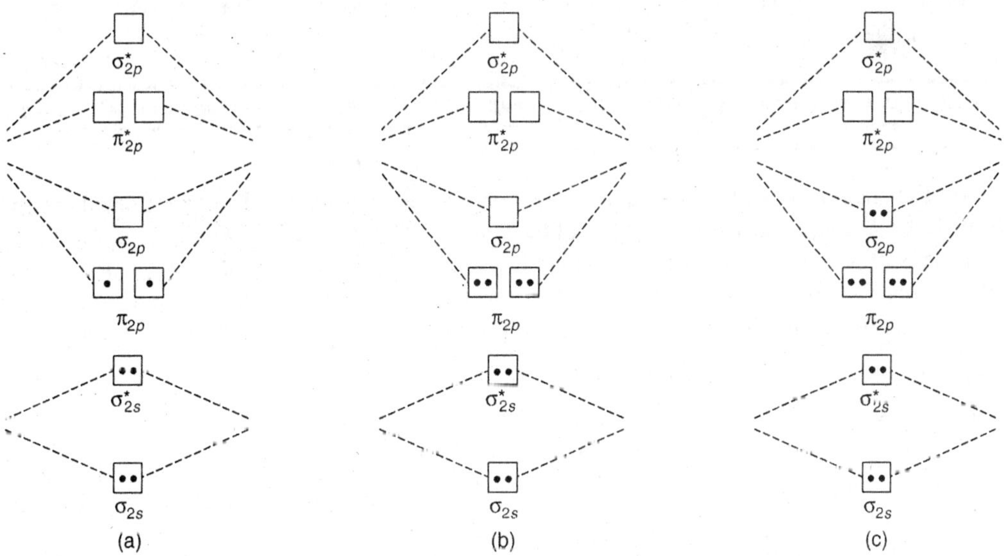

Fig. 7.14 M.O. diagram for (a) B_2, (b) C_2 and (c) N_2.

$Be_2\left(KK\sigma_{2s^2}\sigma^*_{2s^2}\right)$, bond order = ½ (2 – 2) = 0, i.e. Be_2 does not exist.

$N_2\left(KK\sigma_{2s^2}\sigma^*_{2s^2}\pi_{2p^4}\sigma_{2p^2}\right)$, bond order = ½ (8 – 2) = 3, so strong stable (diamagnetic) triple bond.

$O_2\left(KK\sigma_{2s^2}\sigma^*_{2s^2}\sigma_{2p^2}\pi_{2p^4}\pi^*_{2p^1}\pi^*_{2p^1}\right)$, bond order = ½(8 – 4) = 2, so it is O = O and paramagnetic equivalent 2 unpaired electrons.

$Ne_2\left(KK\sigma_{2s^2}\sigma^*_{2s^2}\pi_{2p^4}\pi^*_{2p^4}\sigma_{2p^4}\sigma^*_{2p^2}\right)$, bond order = ½ (8 – 8) = 0, so Ne_2 molecule does not exist.

Recent study* of Be_2 molecule shows the dissociation energy of 59 kJ mol^{-1} and nuclear distance 2.45 Å indicating weak covalent bond of order less than unity. This can be attributed to the anti-bonding effect of two non-equivalent atomic orbitals. The $2p$ orbital of Be lie only 3.4 eV energy higher than that of the $2s$ orbital. So, $2s$-$2p$ interaction is much more important here than the $1s$-$2p$ interaction of He_2 molecule.

M.O. diagrams of some hetero-nuclear diatomic molecule like HF, CO, NO are represented as follows:

In HF, F_2, having almost filled valence orbitals, will be in low energy state than the $1s$ of hydrogen. $2s$ of F_2 will remain as non-bonding and held the electrons that have fluorine character. Let the molecular axis is along X-axis. So, $1s$ of hydrogen and $2p_x$ of fluorine will form M.O.s. σ and σ* orbital and the $2p_y$ and $2p_z$ of fluorine will be unaffected and remain as non-bonding holding the electrons of fluorine character.

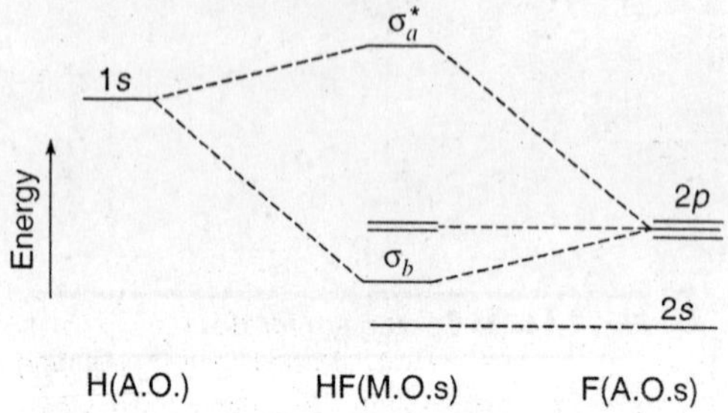

Fig. 7.15 M.O. energy level diagram for HF (not to scale).

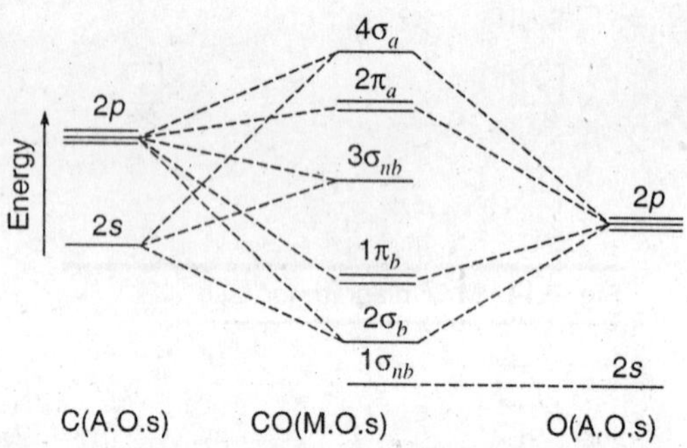

Fig. 7.15a M.O. energy level diagram for CO ($\sigma_{nb} = \sigma_s$).

The M.O. diagram of CO is given above. O_2 is of lower energy level and so $2s$ of O_2 will occupy σ_n orbital the $2s$ and $2p_x$ of carbon and $2p_x$ of oxygen then form the three M.O.s. σ, σ_n and σ_a^* with increasing energy, σ will hold electrons having oxygen character, σ_n will hold the electron having essentially carbon character (HOMO) and hence easily form metal carbonyls where co-ordination takes place through carbon. M ← CO. Both C and O have $2p_y$ and $2p_z$ orbital and form π and π* orbitals. So, the M.O. diagram is CO $\left[\sigma_n^2\sigma^2\pi^4\sigma_n^2\right]$ and bond order $= \frac{1}{2} \times 6 = 3$ of which two are conventional covalent bond and one is coordinate bond, the electron pair is donated by oxygen. :C ⥢ O: In N_2, the highest occupied molecular orbital (HOMO) is σ_{2p} bonding orbital, not available easily, hence being iso-electronic with CO, no M ← N_2 bond is formed ordinarily except in special situations.

In case of NO molecule, N_2 and O_2 having more or less identical energy, the M.O. diagram will be like that of CO, the HOMO will hold one electron. NO $\left[\sigma_n^2\sigma^2\pi^4\sigma_n^2\pi_a^{*1}\right]$, bond order is 2.5. Here, HOMO is π^{*1}, this electron can be easily removed to form NO^+. Again, the presence of one unpaired electron indicates the feeble tendency to dimerize to N_2O_2, which is observed at very low temperature in liquid state. The NO molecule is paramagnetic. This unpaired electron is delocalised in the molecule. It is written as

$$:\dot{N}::\ddot{O}: \leftrightarrow :\overset{-\partial}{\ddot{N}}::\overset{+\partial}{\dot{O}}: \leftrightarrow :N \equiv O:$$

*R.D.W. Smith; J. Chem., Ed. 77 (2000) 780

This 3(three)-electron bond indicates the bond order = 2.5.

σ and π M.O. formed as a result of mixing of different A.O.s can be given as follows as in Fig. 7.16.

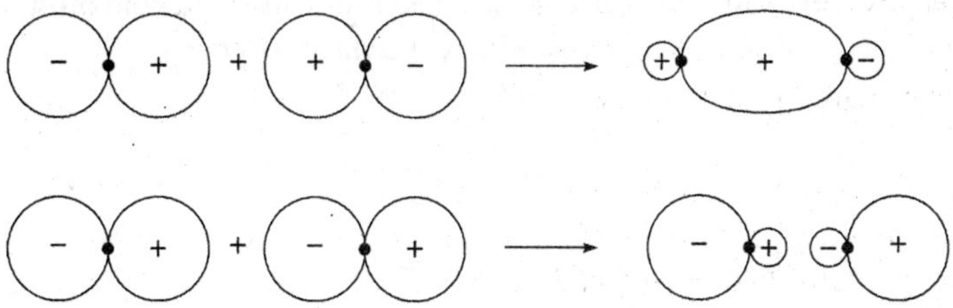

$p_x + p_x$
Combination of *p* atomic orbitals having lobes directed along the inter-nuclear axis forming σ m.o.'s.

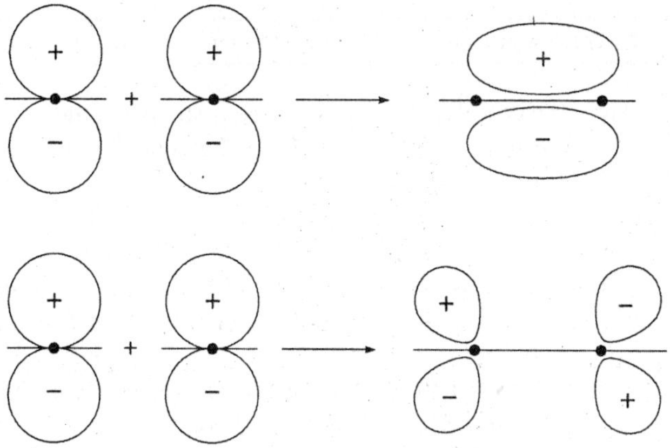

$p_y + p_y$
or
$p_z + p_z$
Combination of two *p* atomic orbitals to form bonding and anti-bonding π molecular orbitals.

s = s
σ and σ* molecular orbitals showing axis of symmetry passing through the nuclei.

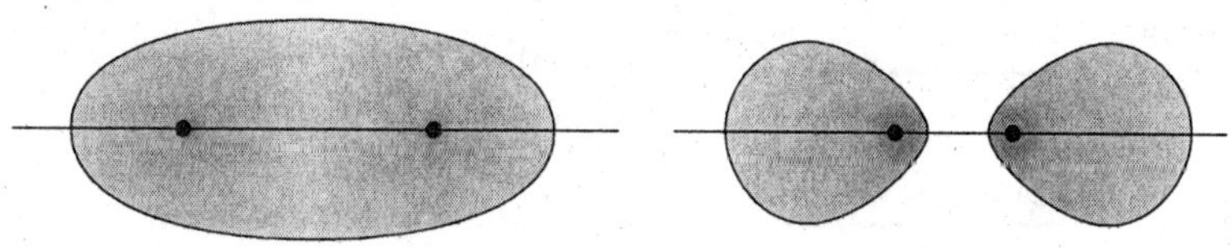

Fig. 7.16 Diagrammatic representation of s–s, p_x–p_x, p_y–p_y and p_z–p_z orbitals.

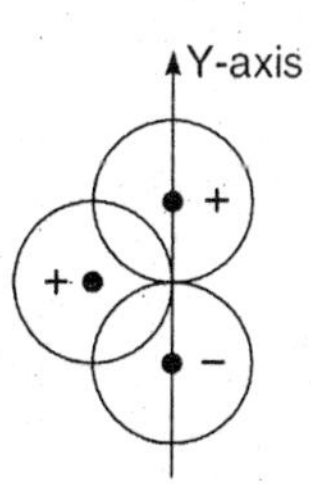

Fig. 7.17 Overlap of *s* and p_y atomic orbitals. (x-axis arbitrarily chosen as the molecular axis).

The overlap of *s* with p_y or p_z will not take place as their symmetry is different. If that somehow takes place then due to opposite sign of p_y or p_z orbitals, overlap of one lobe will be balanced by the other.

From the foregoing discussion, it is evident that for effective combination of A.O.s to form M.O.s the conditions are as follows:

(i) The A.O.s will have comparable energies with same symmetry with respect to molecular axis.

(ii) To form an effective bond the overlapping will be appreciable. The following chart will show how the atomic orbitals give the M.O.s of different types. *s* with $\left(s, p_x, d_{x^2-y^2}, d_{z^2}\right)$ will give σ M.O. p_y with (p_y, d_{xy}) and p_z (d_{xz}, p_z) give π M.O. d_{yz} with d_{xy} gives δ M.O.

7.08 PAULING-SLATTER THEORY (HYBRIDISATION THEORY)

The foregoing discussions of localised M.O. when applied to the polyatomic molecules, we met with some discrepancies. Say, to explain the formation of H_2O, NH_3 and CH_4 molecules, we see the central atom, O, N and carbon respectively, has the electronic configurations and available orbitals for overlap as mentioned below. The geometrical figure can be represented as below:

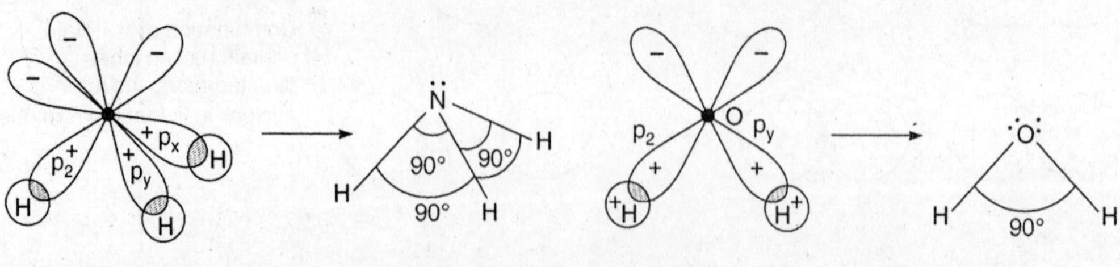

Fig. 7.18 Overlap of p orbitals of nitrogen and oxygen with s orbital of hydrogen to form NH_3 and H_2O.

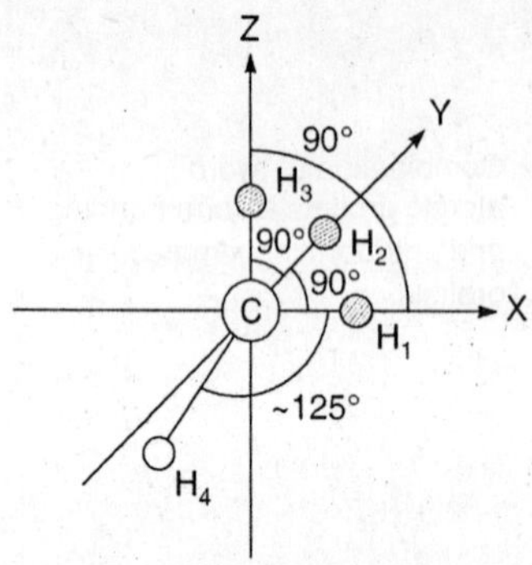

Fig. 7.19 Overlap of 2s and 2p orbitals of C with 1s orbital of H_2 to form CH_4.

In H_2O, the two O–H bonds will be equivalent and the bond angle in between the bonds will be 90°. The bonds are formed by the overlap of p_y and p_z with 1s of H_2. So the relative strengths are $1 \times \sqrt{3} = \sqrt{3}$, the molecule will be in Y-Z plane. The single O–H bond has heat of formation 103 kcal mol^{-1} and in H_2O it is, 221 kcal mol^{-1}, which is higher than double the value of single O–H bond. The observed bond angle is 104.5°. First, it was attributed to be due to the coulombic repulsion between the two polar $H^{+\delta}$ ends and also due to the intervention of $2s^2$ electrons. But this explanation is inadequate. In NH_3, the structure will be pyramidal with N atom at the apex, and will be 90° as overlap of 1s of H takes place with $2p_x$, $2p_y$ and $2p_z$ of nitrogen, $2s^2$ being lone pair. The observed H–N–H angle is 107.2° and each N–H bond is stronger than $\sqrt{3}(1 \times \sqrt{3})$. In case of CH_4, one electron from 2s is shifted to 2p, making 4 electrons unpaired. Now overlapping of 1s of hydrogen four bonds (s-s, 3 s-p) are formed. The strength of s-s overlap is 1 and that of s-p is $\sqrt{3}$. The three s-p bonds will be at 90° with each other, and s-s bond will be any where in space with less energy. Thus, three C–H bonds will be equivalent and one C–H different. Experimental results show that 4 C–H bonds are equivalent with relative strength 2 and H–C–H bond angles are 109°28′ where the hydrogen atoms occupy the corners of a regular tetrahedron. The bond length of all the four C–H is equal. How this can be achieved as a result of mixing of 3p and 1s of different character? The answer is that the four completely equivalent orbitals must have same fraction of s character or $\frac{1}{4}$ s character and $\frac{3}{4}$ p character. Let us consider the first equivalent orbital along Z-axis where p_x and p_y will have no contribution, so it consists of only s and p_z. We can, therefore, write,

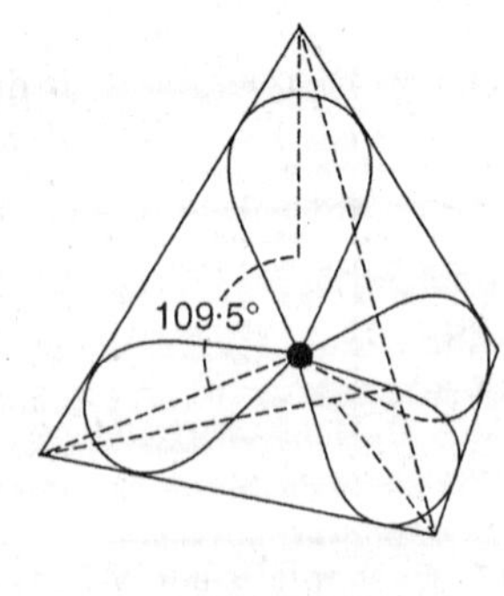

Fig. 7.19a sp^3 hybridisation of carbon.

$$\Psi_{sp^3}(1) = \frac{1}{2}\Psi_s + \sqrt{\frac{3}{2}}\Psi_{p_z} \quad \text{[From quantum mechanical calculation]}$$

$$\Psi_{sp^3}(2) = \frac{1}{2}\Psi_s - \frac{1}{2\sqrt{3}}\Psi_{p_x} + \sqrt{\frac{2}{3}}\Psi_{p_z}$$

$$\Psi_{sp^3}(3) = \frac{1}{2}\Psi_s - \frac{1}{2\sqrt{3}}\Psi_{p_x} + \frac{1}{\sqrt{2}}\Psi_{p_y} - \frac{1}{\sqrt{6}}\Psi_{p_z}$$

$$\Psi_{sp^3}(4) = \frac{1}{2}\Psi_s - \frac{1}{2\sqrt{3}}\Psi_{p_x} - \frac{1}{\sqrt{2}}\Psi_{p_y} - \frac{1}{\sqrt{6}}\Psi_{p_z}$$

The angle of Ψ_2, Ψ_3, Ψ_4 with respect to Ψ_1 has been calculated to be 109°28′. The relative strength of each of these orbitals will be

$$\Psi_{sp^3} = -\left(-\frac{1}{2\sqrt{3}}\right)\sqrt{3}\left(1+\frac{1}{3}\right)+\sqrt{3-12\left(\frac{1}{2\sqrt{3}}\right)^2\left(\frac{2\sqrt{2}}{3}\right)} = 2$$

Thus, the relative strength of the new orbital called *hybrid orbital* is greater than thats of pure *s* or *p* atomic orbital and thus the bonds formed with the combining orbital through overlap are of greater strength. So it can be concluded that in atoms or ions, when several outer electronic orbitals have more or less identical energies, they can get mixed or combined to form corresponding number of equivalent bonding orbitals with greater electron density in particular directions different from the atomic orbitals from which they are constructed. The hybrid orbitals can better overlap with the orbitals of other atoms along the directions and thus the molecule assumes certain geometrical configurations. Thus, CH_4 has tetrahedral configuration with bond energy 396 kcal mol^{-1}.

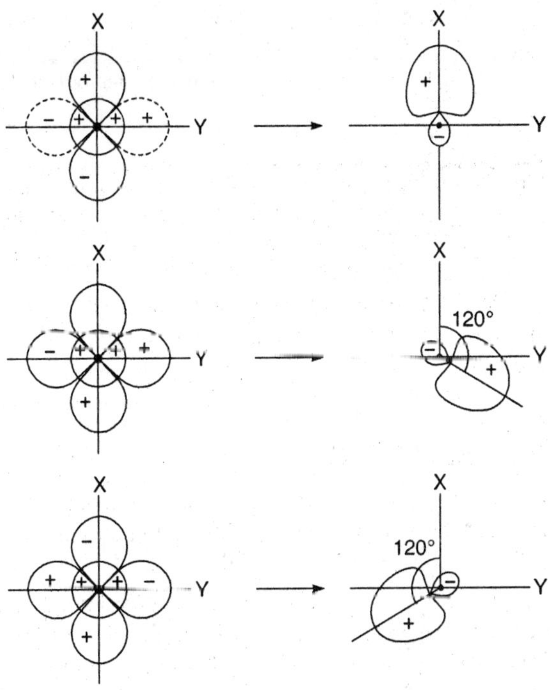

Fig. 7.20 Formation of sp^2 hybrid orbitals.

(i) Hybrid orbital formed by the combination $2s+2p_x$. The $2p_y$ orbital does not contribute as its two lobes make "equal" but "opposite" overlap with the *s* orbital (shown in dotted line).

$$\Psi_{sp^2}(1) = \frac{1}{\sqrt{3}}\Psi_{2s} + \sqrt{\frac{2}{3}}\Psi_{2p_x}$$

(ii) Hybrid orbital formed by the combination $2s-2p_x+2p_y$.

$$\Psi_{sp^2}(2) = \frac{1}{\sqrt{3}}\Psi_{2s} - \frac{1}{\sqrt{6}}\Psi_{2p_x} + \frac{1}{\sqrt{2}}\Psi_{2p_y}$$

(iii) Hybrid orbital formed by the combination $2s-2p_x-2p_y$.

$$\Psi_{sp^2}(3) = \frac{1}{\sqrt{3}}\Psi_{2s} - \frac{1}{\sqrt{6}}\Psi_{2p_x} - \frac{1}{\sqrt{2}}\Psi_{2p_y}$$

The non-equivalent hybrid orbitals also result if the four bonds of tetrahedral carbon are satisfied with different atoms having different electro-negativity. Thus, in CH_3Cl and $CHCl_3$ where angles differ as *p*-character of the hybrid orbitals slightly increases. Thus, non-equivalent hybrid orbital is obtained when

1. more electro-negative atom bound to the central atom will attract more electron density.
2. orbital with greater *p*-character will have less electro-negativity.
3. more electro-negative atom will attach the hybrid orbital with greater *p*-character and less electro-negative atom with greater *s*-character.

The *s-p* type hybrid orbitals are sp^3 or tetrahedral hybridization, sp^2 or triangular hybridization and *sp* or linear hybridization (discussed above) The latter two types can form multiple bond that can be shown later. In sp^2 type, 1*s* and 2*p* orbitals, say, p_x p_y can mix together and form three equivalent bonding orbitals having Ψ functions obtained considering the one orbital along *X*-axis.

The unhybridized p_z orbital of two adjacent carbon atoms can overlap sidewise to form the π bond. The *s*-character of such hybrid orbital is 33%. The hybrid orbital and the orbital of incoming atom may overlap to form σ bond (e.g., in C_2H_4).

In *sp*-hybridization or linear hybridization, the *s*-orbital and one *p*, say, p_z mix together to give 2 equivalent orbitals as shown below (Fig 7.21):

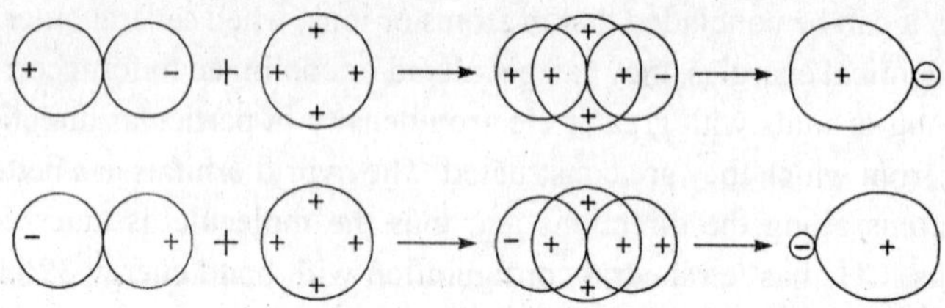

Fig. 7.21 Combination of s and p atomic orbital forming two sp hybrid orbitals, The wave functions are

$$\Psi_{sp}(1) = (1/\sqrt{2})(\Psi_{2s} + \Psi_{2p_z})$$

$$\Psi_{sp}(2) = (1/\sqrt{2})(\Psi_{2s} - \Psi_{2p_z})$$

Here *s*-character is 50%. The unhybridized p_x and p_y can form two bonds with adjacent C-atom; e.g., in C_2H_2, p_x and p_y form 2π bonds with adjacent C-atom and with 1*s* of hydrogen forms σ bond. When *d* orbital of penultimate shell (inner orbital type) or of outer shell (outer orbital type) are involved in hybridization, it is of special interest. Three types of hybridization may occur and where square planar, octahedral and trigonal bi-pyramidal configurations are obtained.

Table 7.4 : The types of orbitals used, their relative strength, geometry of the molecule with examples

dsp^2 or sp^2d	Square planar	$d_{x^2-y^2}, s, p_x, p_y$	$AuCl_4^-$, $PtCl_4^-$, $Pt(NH_3)_4^{+2}$ $[Cu(py)_4]^{+2}$
d^2sp^3 or sp^3d^2	Octahedral	$d_{x^2-y^2}, d_{z^2}, s, p_x, p_y, p_z$	$Cr(CO)_6$, $[Co(NH_3)_6]^{+3}$ SF_6, PF_6^-
dsp^3	Trigonal bipyramid	$d_{z^2}, s, p_x, p_y, p_z$	$Ni(CN)_5^{-3}$, $Fe(CO)_5$, PCl_5
dsp^3	Square pyramid	$d_{x^2-y^2}, s, p_x, p_y, p_z$	$MnCl_5^{-3}$, $Ni(CN)_5^{-3}$

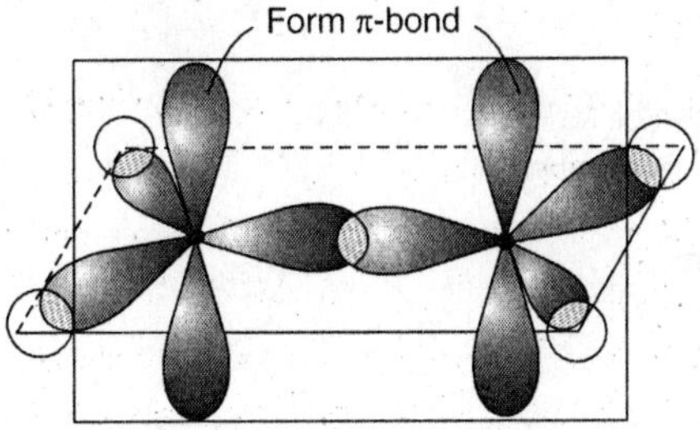

Formation of σ and π bonds in ehylene

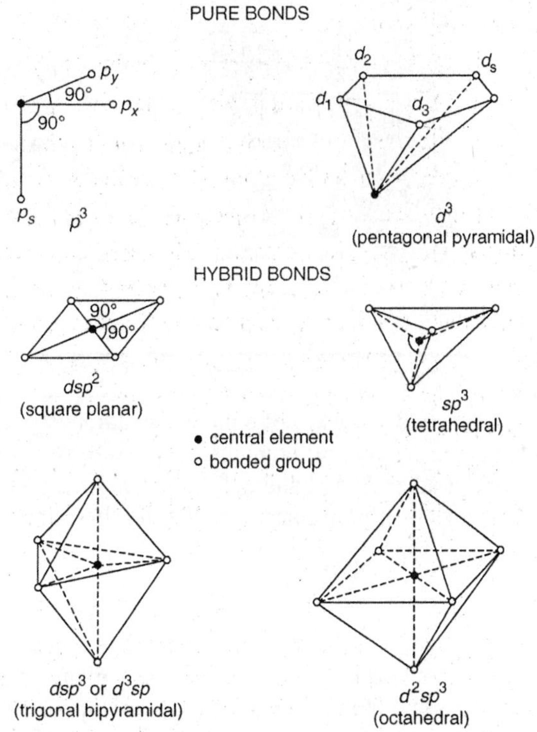

Fig. 7.22 Directional characteristics of various types of covalent bond.

Table 7.5 : Directional characteristics of different covalent bonds and their relation strength

Bond type	Number of bonds	Geometry	Relative strength
s	1	Non-directional	1
p	3	Right angle	$\sqrt{3} = 1.73$
d	5	Pentagonal (pyramid)	2.23
sp^3	4	Tetrahedral	2.00
dsp^2	4	Square planar	2.69
d^2sp^3	6	Octahedral	2.93
dsp^3 or d^3sp	5	Trigonal bipyramidal	—

The structure of H_2O and NH_3 can be explained on the basis of hybridized model. Both N and O undergo sp^3 hybridization with one lone pair of N in one hybrid orbital and 2 lone pairs in two hybrid orbitals of O, the remaining hybrid orbitals having one electron each overlapping with 1s orbital of H to form σ-bonds.

The one lone pair on N pushes the N–H bonds close and hence ∠H–N–H angle is 107°45′. Similarly, the two lone pairs of oxygen repel each other and pushes the O–H bonds more. So ∠H–O–H bond is 104°27′ (Fig 7.23).

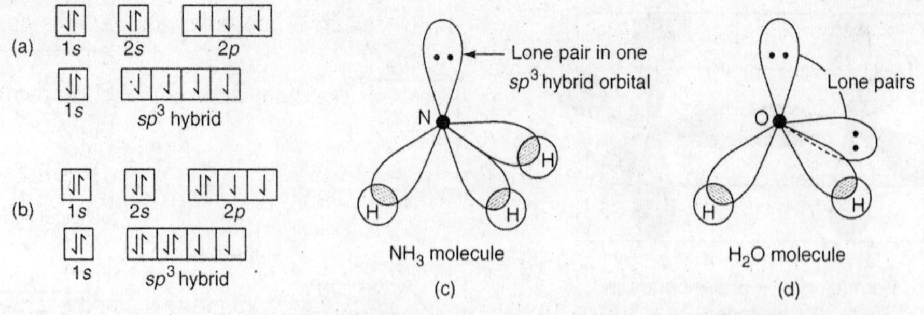

Fig. 7.23 Hybrid orbitals of (a) nitrogen and (b) oxygen and the geometry of (c) ammonia and (d) water molecule.

The actual geometry of a molecule although primarily depends on the hybridization of orbitals of the central atoms different interaction resulting from lone pair bond pair electrons, resonance and ionic character of the bond deform the molecule from actual geometry. So, before going to discuss the geometry of molecules, we consider these factors in short.

7.09 RESONANCE

Resonance is the mechanism developed theoretically to explain the stability and other properties of a compound with the help of different hypothetical electronic structures; no single structure can explain it and the properties can best be explained by considering different arrangements of electron over the same atomic skeleton. The structure of the molecule will be a combination of all such canonical forms. All the canonical forms cannot contribute equally to the resonance. The requisite condition for resonance to occur has been discussed earlier in Valence Bond Concept (Article 7.05).

The energy of the resonance hybrid is lower than the energy of any canonical forms and the difference of experimental heat of formation and the calculated heat of formation is called the *resonance energy*. The resonance structures of some compounds (or ions) are given as follows:

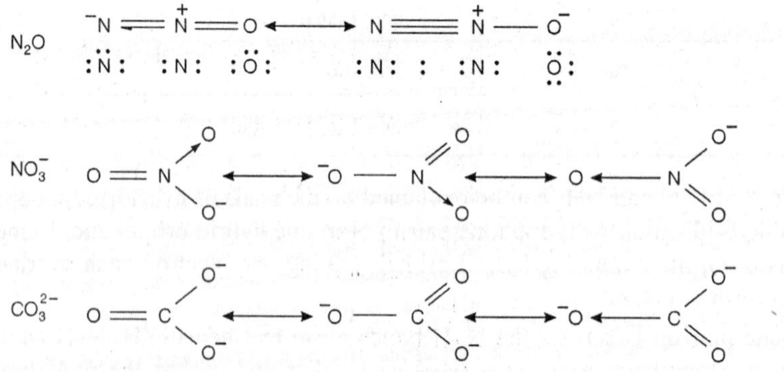

Fig. 7.24 Resonance structure of CO, CO_2, N_2O, NO_3^-, CO_3^{2-}.

Resonance differs from tautomerism as under:

(i) Resonance has hypothetical canonical forms, whereas tautomerism contains some real isomeric forms of the molecules.

(ii) In resonance the canonical forms have same atomic skeleton, but in tautomerism the isomeric structures have different atomic skeletons.

(iii) In resonance electron shift occurs from one atom to another, in tautomeric forms atom shift occurs from one position to the other.

7.10 PARTIAL IONIC CHARACTER AND POLARITY

In discussing the V.B. concept earlier in Article 7.05, it has been found that ionic character of a covalent bond is a natural outcome and represented by the wave function $\Psi_{11} = \Psi_a(1)\,\Psi_a(2) + \Psi_b(1)\,\Psi_b(2)$. The bond between like atoms has mainly covalent character, but slight ionic nature is present, e.g., even in H_2 molecule, 2% ionic character is present. Bonds between fluorine with 1st short period elements of the periodic table are less than 50% ionic. Lithium and beryllium compounds except with F and O, are covalent in nature. Thus, the extent of ionic or covalent nature of the compounds depends on the electro-negativity difference between the combining elements as has been discussed in the Chapter of Electro-negativity, Article 6.06. Electro-negativity difference of 2.1 in general imparts 50% ionic character to the compounds.

How the electro-negativity difference between the combining elements imparts the ionic or covalent nature, can be understood considering the distribution of electron cloud over them. Thus, the distribution of electron cloud over like atom and unlike atoms can be represented as Fig 7.25:

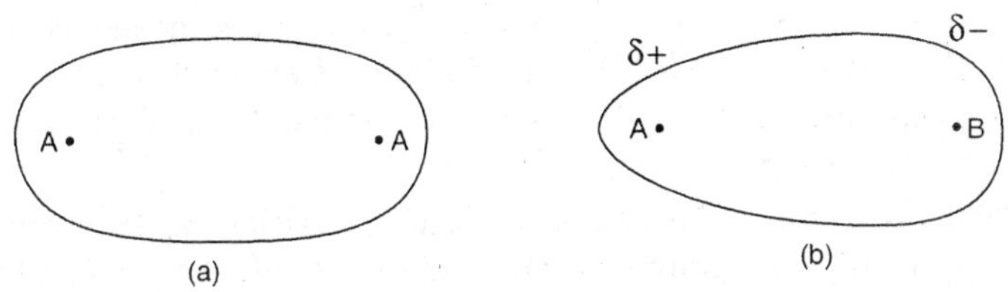

Fig. 7.25 Symmetrical and unsymmetrical distribution of charge cloud between two (a) like atoms, (b) unlike atoms. E.N. of B > E.N. of A (assumed).

Here, *B* is more electro-negative than *A* and draws more electron cloud towards itself, i.e., unsymmetrical distribution of charge and the bond formed is polar. In ionic compounds the cation has the greater attraction towards the electron cloud over the anion, and simultaneous repulsion of the nucleus results in deformation, distortion or polarisation of the anion. The cation will also be similarly polarised and greater the degree of polarisation of the ions, the greater will be the degree of sharing of electron, and hence greater the covalent character, of course, cations are less polarised than the anions.

Ion polarisation depends on a number of factors which is represented by Fajans' rule (1924).

(a) *High charge upon either cation or anion*

Ions with higher charges cause greater polarisation and hence higher covalent character. Comparison of the melting points of the anhydrous halides having same anion but different charge on cations illustrates the point.

$NaCl$ ($r^+ = 0.95$ Å) m.p. = 800°C $\qquad$ $CaCl_2$ ($r^+ = 0.99$ Å) m.p. = 772°C

$MgCl_2$ ($r^+ = 0.65$ Å) m.p. = 712°C $\qquad$ $AlCl_3$($r^+ = 0.50$ Å) m.p. = Sublimes

(b) *Small size of cation and large size of anion*

Small cations have greater concentration of positive charge and large anions have greater concentration of negative charge. Hence, small cations with large anion will give a covalent compound; for example, LiI is a purely covalent compound. Thus, comparison of anhydrous chlorides of group II elements shows the covalent character of small cations with same anion.

Be^{+2} $r^+ = 0.31$	$BeCl_2$ m.p. = 405°C
Mg^{+2} $r^+ = 0.65$	$MgCl_2$ m.p. = 712°C
Ca^{+2} $r^+ = 0.99$	$CaCl_2$ m.p. = 772°C
Sr^{+2} $r^+ = 1.13$	$SrCl_2$ m.p. = 872°C
Ba^{+2} $r^+ = 1.35$	$BaCl_2$ m.p. = 960°C

Similarly, the same cation with increasing anion size shows the higher covalent nature.

CaF	$r_{F^-} = 1.36$	m.p. 1392° C
$CaCl_2$	$r_{Cl^-} = 1.81$	m.p. 772° C
$CaBr_2$	$r_{Br^-} = 1.95$	m.p. 730° C
CaI_2	$r_{I^-} = 2.16$	m.p. 575° C

The charge and size of a cation have some effect in polarising an anion and Cartledge (1928–30) correlated the two in terms of ionic potential ϕ, where $\phi = \dfrac{\text{cationic charge}}{\text{cation radius}}$.

Higher the ionic potential, higher is the polarising capacity and greater the covalent character. It has been found that when $\sqrt{\phi} < 2.2$, the compounds are ionic and conduct electricity on molten state and when $\sqrt{\phi} > 2.2$, the compounds are purely covalent and does not conduct current.

(c) Cations with 18-electron structures have greater polarising capacity compared to normal 8- electron systems, as *d*-electron cannot effectively shield the nucleus. Hence, higher the effective nuclear charge, greater is the polarising power.

KCl (m.p. 776°C)	NaCl (m.p. 800°C)	$CaCl_2$ (m.p. 772°C)
AgCl (m.p. 455°C)	CuCl (m.p. 422°C)	$HgCl_2$ (m.p. 270°C)

With greater polarisation and hence greater covalent nature, the decrease in solubility in polar solvent is also observed, e.g.,

AgF > AgCl > AgBr > AgI and

$HgF_2 > HgCl_2 > HgBr_2 > HgI_2$.

Similarly, with gradual increase in covalent character colour intensifies.

Colourless	$HgCl_2$	AgCl	$PbCl_2$	$SnCl_4$
Colour	HgI_2(red)	AgI(yellow)	PbI_2(yellow)	SnI_4(red)

CHAPTER

8

GEOMETRICAL STRUCTURE OF COVALENT MOLECULES

8.01 INTRODUCTION

As discussed before, localized covalent bonds are directional in nature and as a result the molecules assume some definite geometrical shape. The hybridization scheme, dependent on the available valence shell orbital, first determines the approximate special distribution of orbital of the central atom (Helferich rule). The presence of lone pair electrons, the interaction between lone pair and bond pair electrons (VSEPR rule) and the unsymmetrical substitution of the bonds of the central atom, i.e., non-equivalent hybrid orbital (Bent's rule) ultimately determine the geometry of the molecule. All these rules are considered gradually as follows:

HELFERICH RULE

In ascertaining the shape of some molecules, Helferich (1946) gave some more or less empirical generalization known as Helferich rule. The rules are as follows:

(a) *The molecule AX_2 type or AX_2^- (X is same or different)*

It will have linear structure if the central atom *A* has no lone pair electron, and if one or two lone pair is present, the molecule assumes angular structure. Thus, CO_2, CS_2, N_2O, N_3^- are linear, whereas H_2O, H_2S, NO_2^- are angular.

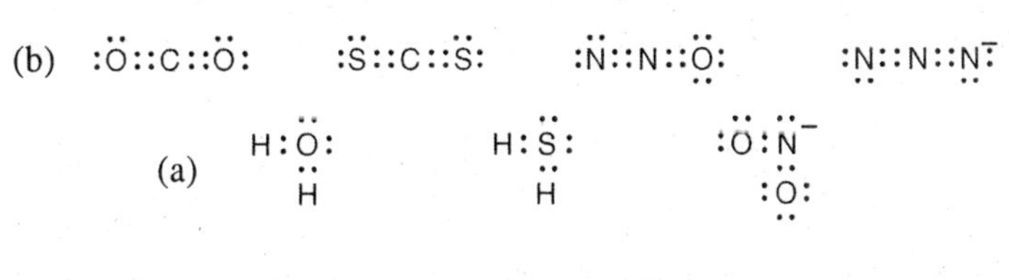

Fig. 8.1 Structure of AX_2 type molecule (a) with and (b) without lone pair electron.

(b) *The molecule* AX_3 *type or* AX_3^- *ion (X is same or different)*

It will be planar if the central atom *A* has no lone pair of electron, or pyramidal with *A* at the apex, if *A* has lone pair of electron. Thus, CO_3^{-2}, NO_3^-, BF_3^- are planar while NH_3, SO_3^{-2}, ClO_3^- are pyramidal.

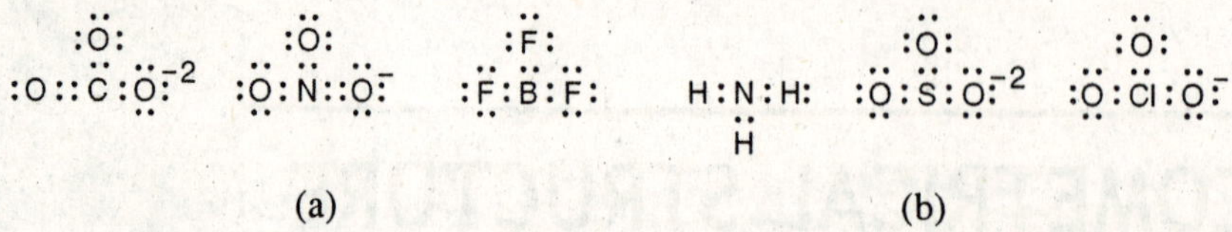

Fig. 8.2 (a) Planar: No unshared electron on central atom (b) Pyramidal: The central atom contains unshared pair(s) of electron.

(c) *Molecule or ion of the type* AX_4

It will have tetrahedral structure if the central atom *A* has no unshared pair or lone pair of electron in the outer shell and the central atom has noble gas arrangement but if the central atom has non-inert gas configuration it assumes planar structure.

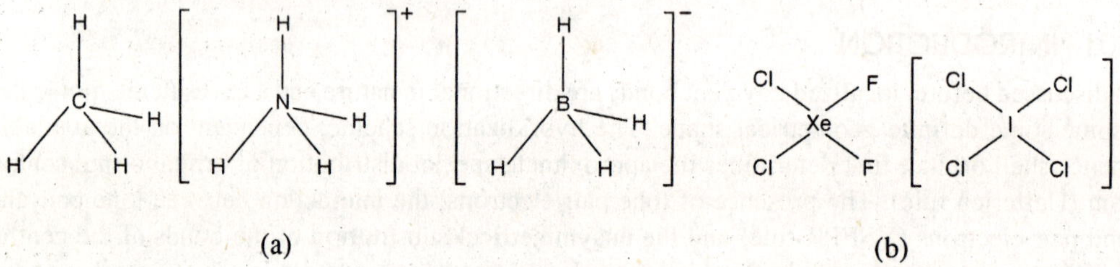

Fig. 8.3 (a) Tetrahedral: The valence shell of the central atom contains eight electrons in each and conform to noble gas configuration, (b) Planar: The valance shell of the central atom in each has four bond pairs and two lone pairs—a total of 12 electrons.

On the basis of this Helferich rule, geometry of some molecules, their number of electrons around central atom, hybridization and distribution of orbital, shape of the molecule with examples are listed below:

See Table 8.1.

8.02 BENT'S RULE

For non-equivalent hybrid orbital, i.e., where all the atoms or groups linked to the central atom are not identical, the symmetry of the geometrical configuration is lost. This is explained by Bent's rule.

The deviation depends upon the nature of bonding involved and varies from molecule to molecule although the gross spatial distribution of hybrid orbital remains the same. For example, in CH_3Cl, $\angle$H–C–H is 110° and $\angle$H–C–Cl is less. Similarly, in $CHCl_3$, $\angle$Cl–C–Cl is 110° and $\angle$H–C–Cl is different. This can be attributed to the fact that H and Cl have different electron affinity, Cl having higher electron affinity will tend to form the bond where higher electron density is available. We know that *p*-orbital has higher electron density. So the hybrid orbital having higher *p* character will form bond with more electro-negative element. Thus, C–Cl or C–F bond have less than 25% *s*-character or 75% higher *p*-character, and C–H bond will have > 25% *s*-character and <75% *p*-character. The percentage of *s*-character or *p*-character is calculated from the observed bond angle θ as follows:

$$\cos\theta = \frac{s}{s-1} = \frac{p-1}{p}$$

In CH_4, $\theta = 109.5°$ $s = 0.25$ and $p = 0.75$.

In NH_3, $\theta = 107.2°$ $s = 0.22$ and $p = 0.78$

In H_2O, $\theta = 104.5°$ $s = 0.20$ and $p = 0.80$

Similarly, in $BeCl_2$, the two hybrid orbitals have 50% *s*-character, but in BeClBr, the Be–Br bond has greater than 50% *s*-character, and Be–Cl bond will have greater than 50% *p*-character.

The *sp* hybrid orbital has $\theta = 180°$ ($s = 50\%$), for sp^2 (33.3% *s*-character) $\theta = 120°$, for sp^3 (25% *s*-character) $\theta = 109°28'$, i.e., with increasing *s*-character, the inter-valence bond angle increases. The bond lengths also vary. The hybrid orbital with higher *s*-character will have shorter bond length and higher bond angle and is more acidic in nature. If the molecule has lone pair over the central atom, they will occupy the hybrid orbital with higher *s*-character, i.e., the bond will swell up, i.e., greater angular volume. This may push the other bonds more closely. "The *p*-character or conversely *s*-character concentrates in h.a.o.s directed toward the electro-negative (conversely, electro-positive) substituents."

Thus, Bent's rule can be summarised as follows:

(i) The greater *p*-character of a hybrid orbital has less electro-negativity.

(ii) The more electro-negative atom will bound to the central atom having higher *p*-character of the hybrid orbital.

(iii) The less electro-negative atom will bound to the central atom using the hybrid orbital having higher *s*-character.

Table 8.1 : Geometry of some typical molecules

Type of molecule	Number of valence shell electron pairs on A			Hybridization and distribution of orbitals	Shape of molecule	Example
	Bond pair	Lone pair	Total			
AX_2	2	0	2	*sp* linear	linear	Cl—Be—Cl
	2	1	3	sp^2 trigonal planar	bent (V-shaped)	$PbCl_3$
	2	2	4	sp^3 tetrahedral	bent (V-shaped)	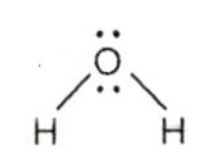
	2	3	5	sp^3d trigonal bipyramidal	linear	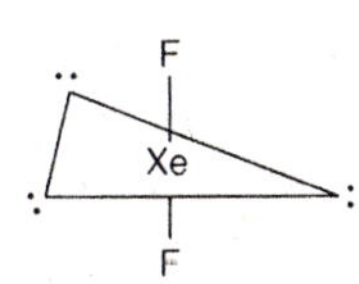
AX_3	3	0	3	sp^2 trigonal planar	triangular	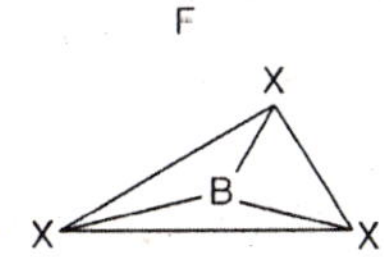

contd.

Type of molecule	Number of valence shell electron pairs on A			Hybridization and distribution of orbitals	Shape of molecule	Example
	Bond pair	Lone pair	Total			
	3	1	4	sp^3 tetrahedral	trigonal pyramidal	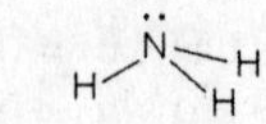
	3	2	5	sp^3d trigonal bipyramidal	T-shaped	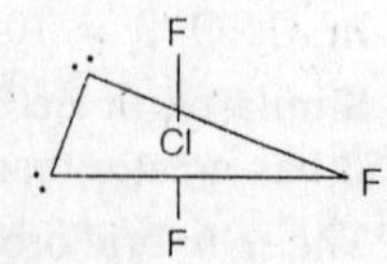
AX_4	4	0	4	sp^3 tetrahedral	tetrahedral	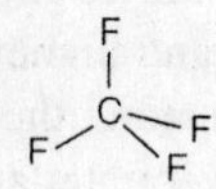
	4	1	5	sp^3d trigonal bipyramidal	tetrahedron (irregular)	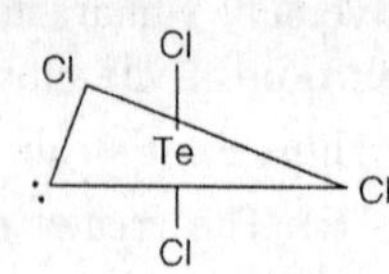
	4	2	6	sp^3d^2 octahedral	square planar	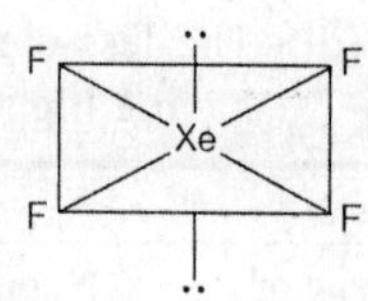
AX_5	5	0	5	sp^3d trigonal bipyramidal	trigonal bipyramidal	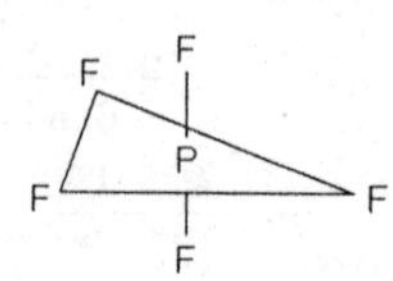
	5	1	6	sp^3d^2 octahedral	square pyramidal	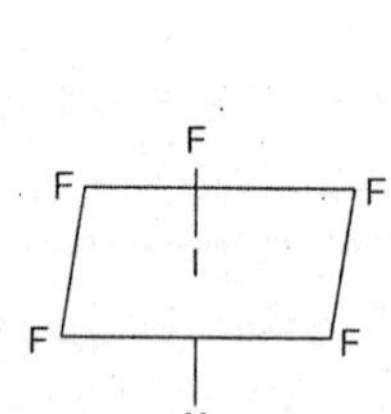
AX_6	6	0	6	sp^3d^2 octahedral	octahedral	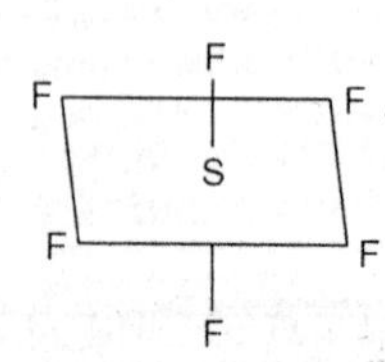

8.03 VALENCE SHELL ELECTRON PAIR REPULSION (VSEPR) THEORY

The basic postulate of this concept is that the valence shell electrons occupy the localized orbitals with opposite spins. The shape of the molecule depends upon the number of such electron pairs occupying the localized orbitals around the central atom. The occupied orbitals again repel each other so that they orient themselves round the central atom in such a way to minimize the repulsive interactions. This idea of valence shell electron repulsion was first introduced by Sidgwick and Powell (1940) and then developed by Gillespee and Nyholm (1957). The generalizations drawn are given below:

1. The lone pairs of electrons repel more strongly than the bond pair electrons,

   lp-lp > *lp-bp* > *bp-bp*.

It is natural as the lone pair is concentrated more toward the central atom while the bond pairs are to some extent dragged towards the bonded atom. So, lone pair electron has greater repulsion.

This rule can be illustrated considering the bond angles of CH_4, NH_3, H_2O (respective bond angles are 109°28′, 107°20′ and 104°30′). CH_4 has no lone pair, the *bp-bp* interactions minimized at the equilibrium bond angle 109°28′. NH_3 has one and H_2O has two lone pairs. The *lp-bp* for and *lp-lp* interaction in H_2O reduces the bond angles. In NH_3 it is 107°20′ (having one lone pair) and in H_2O it is 104°30′ (having 2 lone pairs), CO_2 (180°) and SO_2 (119°30′). This is also evident from CO_2(180°) which has no lone pair, but SO_2(119°30′) which has lone pair, hence bent.

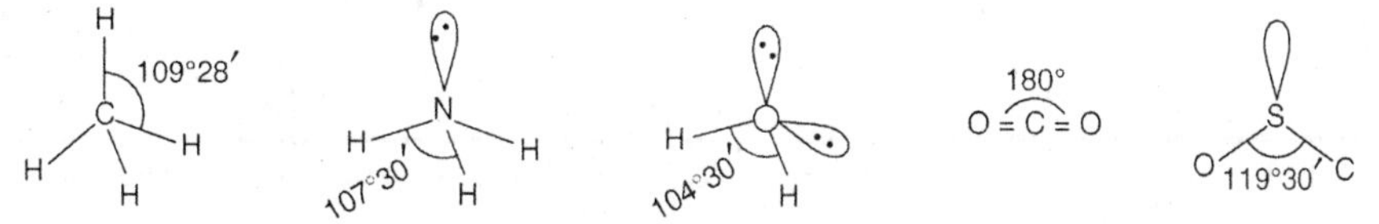

Fig. 8.4 Effect of lone pair electron on inter-valence bond angle of different molecules.

In CO_2, the *bp-bp* interaction is reduced when the angle between them is 180°, while in SO_2, the lone pair of sulphur pushes the S–O bonds closer.

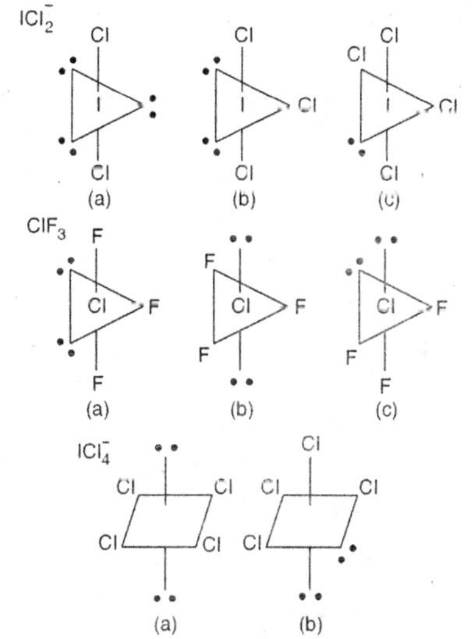

Fig. 8.5 Possible structures of ICl_2^-, ClF_3 and ICl_4^-.

In trigonal bipyramid, there are three equatorial and two axial groups. The angle between equatorial groups is 120° and the angle between axial and equatorial groups is 90°. The repulsion between groups of lone pair decreases as the angle between them increases. Hence, the lone pair occupies the equatorial position as in ICl_2^-, ClF_3 or ICl_4^-. In case of ClF_3, (a) is the most stable structure, as it has 4 *lp-bp* + 2 *bp-bp* interaction whereas structure (b) possesses 6 *lp-bp* and structure (c) possesses 1 *lp-lp* + 3 *lp-bp* + 2 *bp-bp* interactions. In all the cases total interactions are 6. In case of ICl_2^- structure (a) is the most stable as here all the lone pairs of electrons are in equatorial positions.

Similarly, the stable structure of $TeCl_4$ will be structure (a) that has only two *lp-bp* interactions at 90° but (b) has three such interactions. XeF_6 has octahedral structure with distortion with one lone pair projected either through face or through the edge of the distorted octahedron as shown in the figure 8.7(B). Actually, XeF_6 is rapidly inter-convertible between the two (known as fluxional).

2. Repulsion by a bond pair decreases if the bonded atom has higher electro-negativity because it will pull the electron cloud towards itself thereby decreasing the *bp-bp* repulsion and the lone pair will have longer angular volume. The bond angles as observed are NH_3(107°20′), NF_3(102°), H_2O(104°30′), OF_2(103°15′).

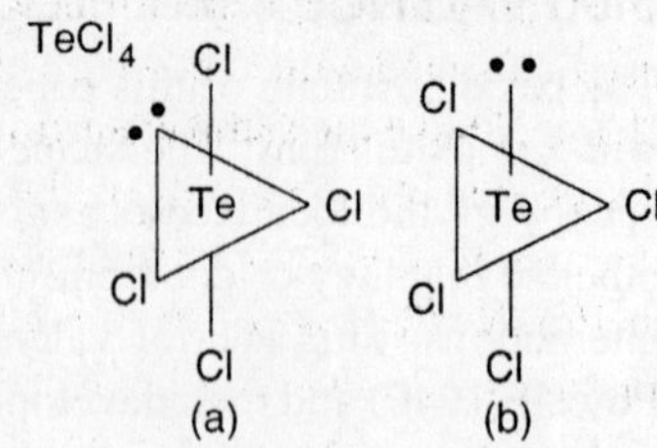

Fig 8.6 Possible structures of $TeCl_4$.

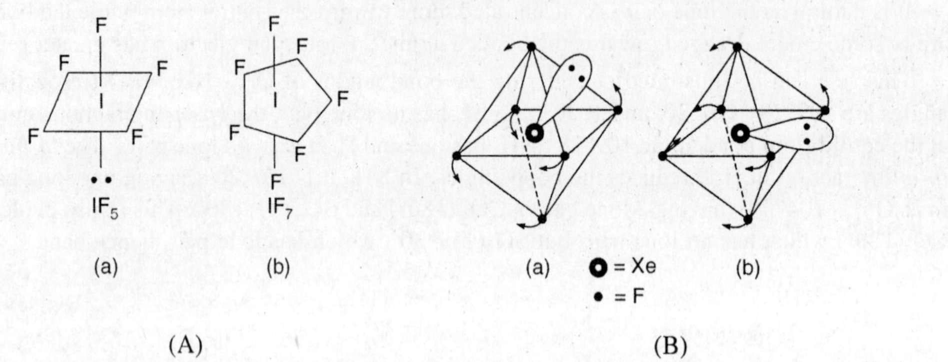

Fig. 8.7 (A) Structures of IF_5 and IF_7, (B) Possible structures of XeF_6 : lone pair (a) through surface, (b) through edge.

3. The lone pair from one atom having filled orbital may sometimes gets shifted to the unfilled orbital of the bonded atom, e.g., PH_3(93°50′), PF_3(97°50′) and AsH_3(91°50′), AsF_3(96°12′). In PF_3, the vacate orbital of P mixed with filled orbital of F, forming *dπ–pπ* partial double bond and transfer some charge density of F (high) to the *d* orbital of P, thereby shortening of P–F bond [contrary to rule (2)]. Thus reduces the *bp-bp* repulsion and increases the ∠F–P–F relative to ∠H–P–H where H has no filled orbital and no possibility of formation of bond. Similar is the case with AsF_3. Cl_2O(110°50′), H_2O(104°30′), F_2O(103°15′).

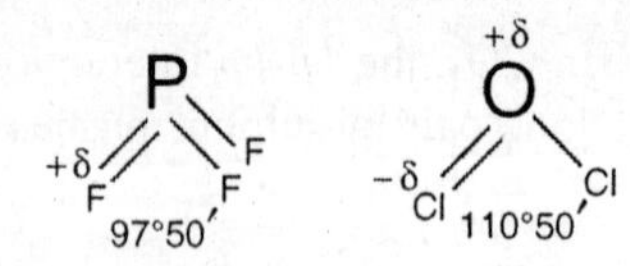

Fig 8.8 PF_3 and OCl_2 showing d_π-p_π partial double bond.

8.04 HYDROGEN BOND AND HYDROGEN BRIDGE BOND

The hydrogen atom with $1s^1$ can form one covalent bond with an electro-negative atom but simultaneously can establish an extra weak bond to another electro-negative atom in the same or different molecule or with a polarisable double bond or with aromatic ring system with polar group like O–H. This secondary bond formed is called the *hydrogen bond* and denoted by broken longer line compared to continuous shorter line covalent bond, X–H-----Y. The existence of such bond was first suggested by T.S. Moor and T.F. Winmill (1912) between N and O in $(CH_3)_3NHOH$ and between H and > C = O in ortho position in benzonoid compounds, later developed by Latimer and Rodebush (1920). This bond is purely electrostatic in nature. The higher electro-negative atom X pulls the bonded electron towards itself leaving the bare proton of hydrogen with positive(+) charge that can easily form electrostatic bond with the nearby negative ion of the same or neighbouring molecule.

Hydrogen bonds are formed with F, O, Cl and N mainly having high electro-negativity. It can also be formed with C in which electronagetive atoms are attached, e.g., $CHCl_3$, HCN, etc. Although Cl and N have the same electro-negativity, Cl cannot form stronger bond due to bigger size and low charge density. Strength of the hydrogen bond depends on the electro-negativity of atom. The average bond energy of hydrogen bond is 2-10 kcal/mole and that for a covalent bond is 100-300 kcal/mole. Because of its low bond energy and low activation energy involved in the formation or rupture, this bond may be formed at normal temperature.

Experimental studies of ice show that the distance O–H–O is 2.76 Å. By neutron diffraction experiment, one O–H is found to 1 Å, so the other O...H is 1.76 Å. Thus, the hydrogen bond is a longer bond.

The formation of hydrogen bond can be explained in this way: With its 1*s* orbital hydrogen can form one covalent bond. Due to high energy difference it cannot use 2*s* or 2*p* orbital for the formation of second bond. So the electro-negative atom can polarise the X–H bond to $X^{-\delta} - H^{+\delta}$. The nucleus of hydrogen becomes bare and can induce ion-dipole or dipole-dipole interaction and can attach to Y^- of the neighbouring atom through electrostatic interaction, which indicates the linear structure for maximum interaction. Some covalent character can be assigned to it due to resonance which may be represented by the canonical forms, e.g., with O–H bond with contribution to the resonance are given:

$$\mathrm{O-H : O \leftrightarrow O : \overset{+}{H}\ O \leftrightarrow O : H-O}$$

(A) 64% (B) 31% (C)5%

Hydrogen bond cannot be linear always, e.g., in HF it is zig-zag.

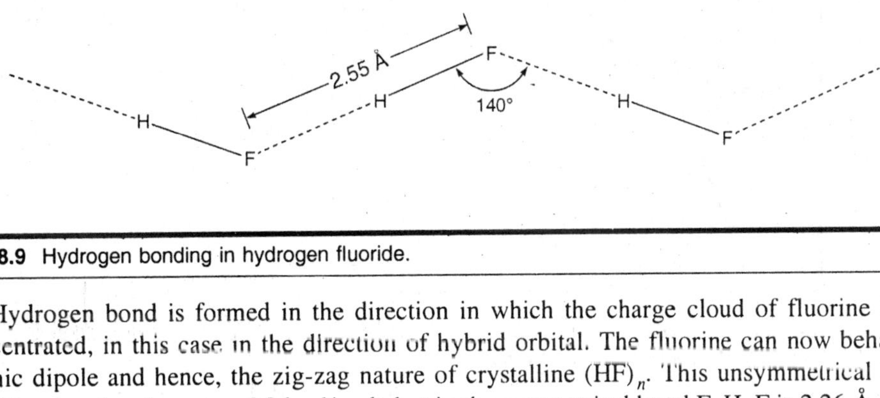

Fig. 8.9 Hydrogen bonding in hydrogen fluoride.

Hydrogen bond is formed in the direction in which the charge cloud of fluorine is mostly concentrated, in this case in the direction of hybrid orbital. The fluorine can now behave as an atomic dipole and hence, the zig-zag nature of crystalline $(HF)_n$. This unsymmetrical hydrogen bond has less bond energy, 6.8 kcal/mole but in the symmetrical bond F–H–F is 2.26 Å and bond energy is 50 kcal/mole.

Hydrogen bond may be of two types: intermolecular and intramolecular hydrogen bonds.

Intermolecular hydrogen bond: Intermolecular hydrogen bond is present in the hydrides of elements of periodic group (IV) to group (VII). It shows that the melting point and boiling point gradually increases with increasing molecular weight with the exception of the first member of each group. As their electro-negativity is high and intermolecular hydrogen bond causes the association of the molecules, thereby increasing the effective molecular weight. From second member onwards, the molecules exist in discrete form and hence increase in melting point and boiling point is observed with the increase in molecular weights of the hydrides. For group (IV) elements, no hydrogen bond is formed and hence gradual increase is noticed in melting point and boiling point with molecular weights (Fig 8.10).

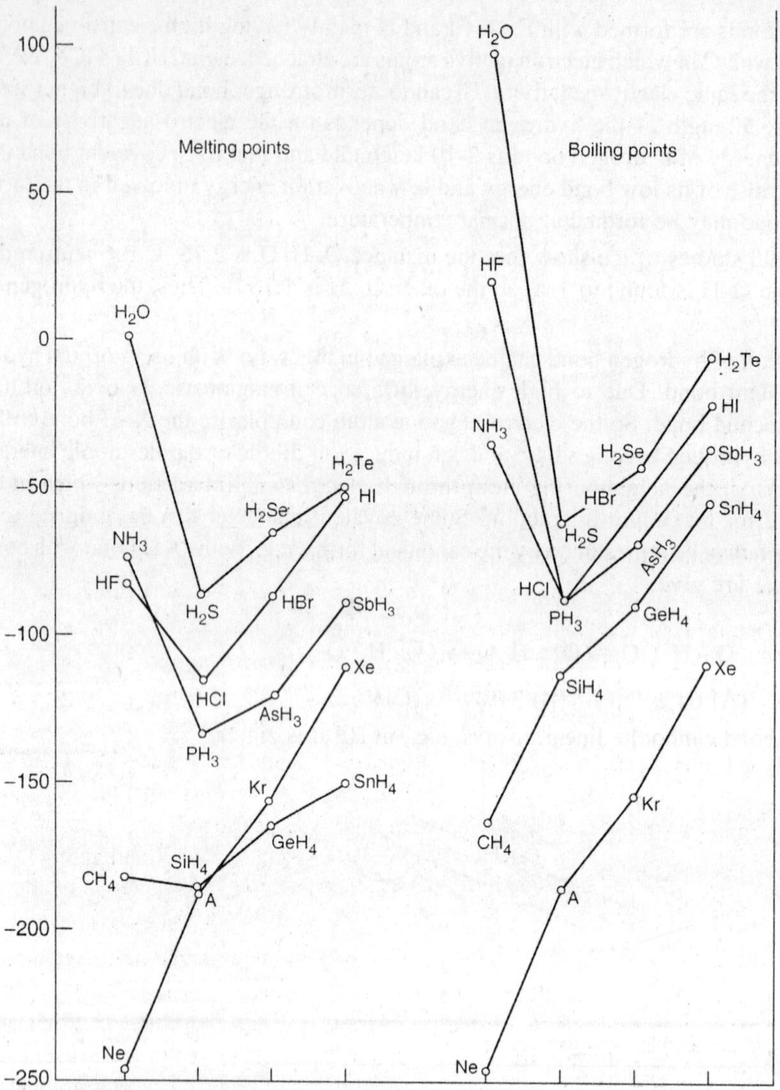

Fig. 8.10 Melting and boiling points of simple hydrides.

Well-known and clear-cut molecular association is known for dimeric acetic acid. Here, heat of dissolution is 13.8 kcal/mole.

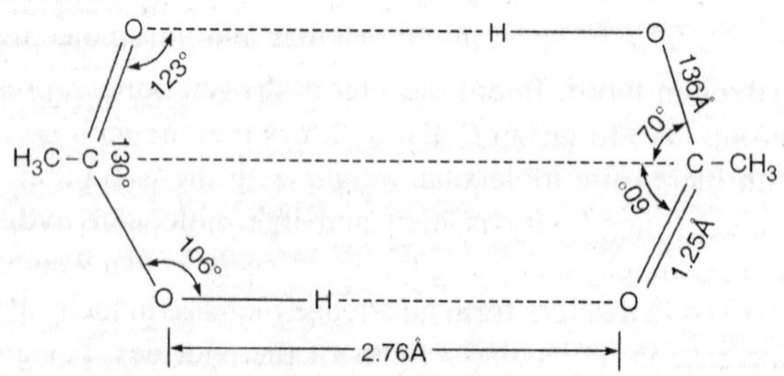

Fig. 8.11 Dimeric form of acetic acid.

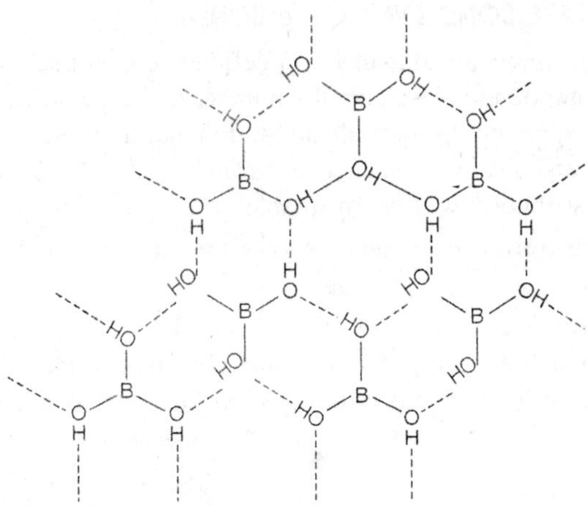

Fig. 8.12 Molecular association due to hydrogen bonding in crystals of boric acid.

When 5- or 6-member ring is formed, that becomes more stable.

X-ray and neutron diffraction studies show the pattern of hydrogen bond in H_3BO_3 (Fig 8.12).

The intermolecular association in compounds like *p*-nitrophenol increases its boiling point relative to *o*-nitrophenol where **intramolecular hydrogen bond** occurs as the two groups, –OH and $-NO_2$ are in adjacent position (Fig 8.13).

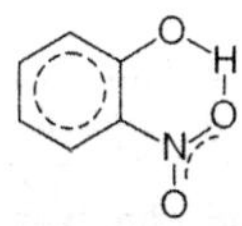

Fig. 8.13 Hydrogen bond in *o*-nitrophenol.

In ice, the H_2O molecules are associated tetrahedrally in such a way that each O-atom attached to two H-atom at a distance of 0.99 Å indicating O–H and two H-atom at a distance 1.77 Å indicating O....H bond (Fig 8.14).

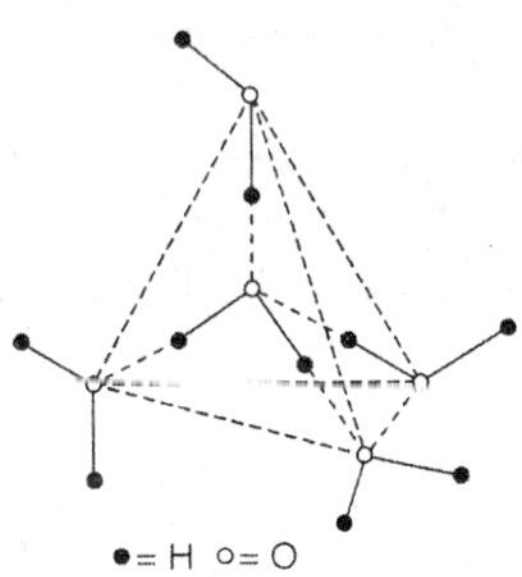

Fig. 8.14 Hydrogen bonding in crystals of ice

Hydrogen bond is also observed intramolecularly between suitably located O–H and polarisable double bond as in C_2H_4 or delocalised electron cloud in aromatic ring. In proteins also hydrogen bond is observed. The formation of hydrogen bond with solute containing oxygen or nitrogen functional groups in solvents like $CHCl_3$ or $CHCl_2F$ increases the solubility relative to the solvents CCl_4 or CCl_3F. The boiling point of catechol, resorcinol and quinol are 245°, 277° and 285°C respectively, which is evident from their structures forming hydrogen bond.

Pyrogallol has the structure corresponding to I.R. study.

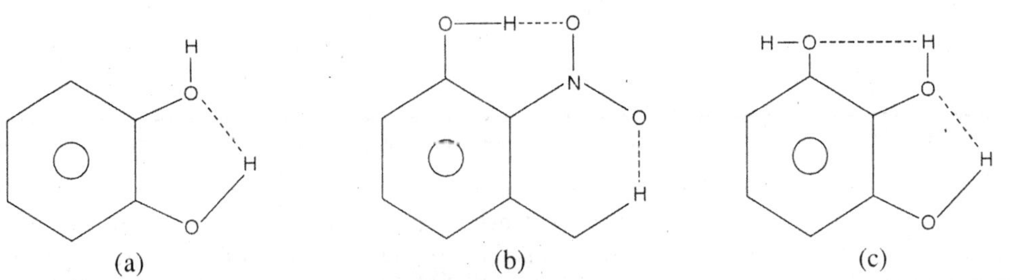

Fig. 8.15 Hydrogen bond in (a) Catechol, (b) 2-Nitroresorcinol, (c) Pyrogallol.

8.04(a) HYDROGEN BRIDGE BOND OR (3C 2e BOND)

Hydrogen bridge bonds are observed in electron-deficient compounds, i.e., the total number of valence electrons in the compound is less than the number of available valence orbitals. In B_2H_6, both the boron has $2s$, $2p_x$, $2p_y$ and $2p_z$ orbitals and individual hydrogen has $1s$ orbital, and, thus making total is 14 orbitals and available valence electrons are $(2 \times 3 + 1 \times 6) = 12$. So, to fit them usual bonding, electron distribution will be unsuitable.

Spectroscopic and other physical evidence shows the structure of B_2H_6 to be as shown in Fig 8.16.

The four-terminal hydrogen have ligancy 1 and B–H = 1.187 Å, i.e., higher than single-bond value of 1.13 Å and the bridging hydrogen has ligancy 2 of B–H bond 1.33 Å, i.e., higher value B–B = 1.77 Å. So the bond order of each B–H bond is less than one. This bridge-structure for B_2H_6 is evident that the bridged hydrogen is not methylated unless broken to two units giving $B(Me)_3$.

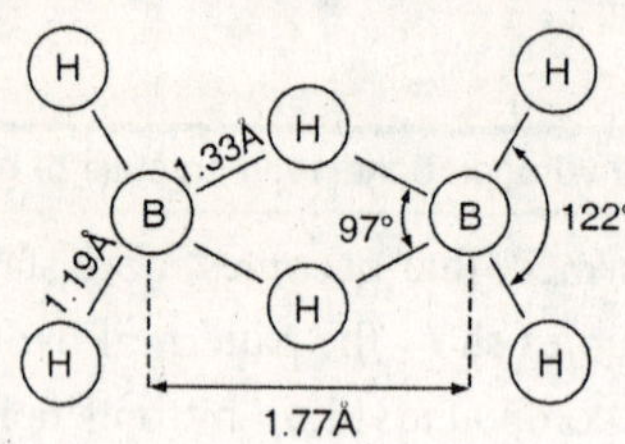

Fig. 8.16 Structure of B_2H_6.

The electron diffraction data indicates that the four-terminal hydrogens and the two boron atoms are coplanar and the bridging hydrogens are perpendicular to the plane one above and the other below the plane. Thus, the two BH_3 tetrahedra joined to each other by edge with the two bridging hydrogens common to both. The bridging hydrogen binds the two boron atoms making the three centres two electrons (3C 2*e*) bond unlike the single (2C 2*e*) bond. The 3C 2*e* bond uses $1s^1$ of hydrogen and one electron of *sp* hybrid orbital of one boron and the vacant *sp* hybrid orbital of the other boron atom, thus the three A.O.s will form three M.O.s as shown in the diagram. In the same way another 3C 2*e* bond will be formed.

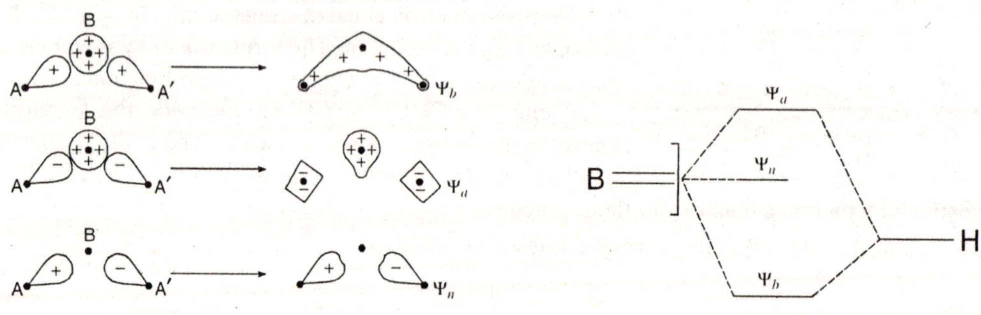

(a) (b)

Fig. 8.17 (a) Combination of atomic orbitals to give molecular orbitals. (b) Energy level diagram for 3-centre molecular orbitals.

This is the open 3C 2*e* bond. The energy consideration shows the greater stability of 3C 2*e* bond compared to 2C 2*e* bond. The two electrons (one from BH_2 and other from H) occupy the Ψ_b. The semi-topological diagram of B_2H_6 is given as in the diagram [Fig 8.18(a)]. Similarly, closed 3C 2*e* bond can be diagrammatically represented as given in Fig 8.18(b).

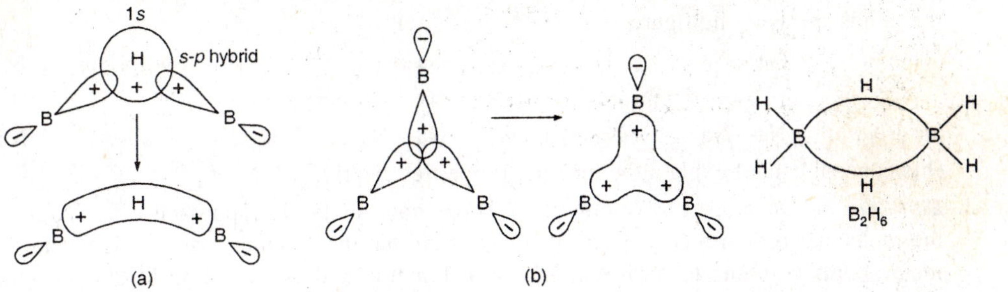

Fig. 8.18 Formation of three-centre bonds of (a) open and (b) closed types.

Neutron diffraction study of some organometallic complexes indicates the presence of some sort of C–H–M interaction of 3C 2*e* type bond. This is called as **AGOSTIC** hydrogen bond. In such interaction the metal will have vacant co-ordination site and the electron density from C–H bond is shifted to this vacant orbital making C–H–M bridge, e.g., $Os_3(H)\ CH_3(CO)_{10}$.

8.05 METALLIC BOND

The metals have fewer electrons in their valence shell and show electropositive character. They have some characteristic properties as given below:

(a) The metals have high electrical conductivity—highest for univalent metals and decreases with higher valence metals, i.e., periodic function of atomic numbers. The conductivity decreases with temperature. The metals also show thermal conductivity.

(b) The metals can emit electrons when heated to certain temperature (thermoionic emission) or when radiation of short wavelength falls on it (photoelectric effect). The photoelectric emission is temperature-independent up to some point, then thermoionic emission occurs. For this emission threshold energy is needed.

(c) Metals have metallic lustre, malleability and ductility. The metals can absorb light of all frequencies and immediately radiate it and hence the lustre. Malleability and ductility are due to the gliding of adjacent atomic planes over each other.

(d) All metals have high melting and boiling points and constant atomic heat. They are also hard indicating strong cohesive forces among atoms. Strong non-directional and non-rigid bonds hold atoms. Thus, the metals have close-pack structures with high co-ordination numbers.

There are some elements, which have borderline properties, i.e., they have some metallic character and some non-metallic characters, e.g., germanium, antimony, tellurium, etc. The metallic bond will be of such nature as to permit the transfer of free electron from one atom to other metal atoms and will thus hold 8-12 neighbours.

The theories of metallic bonds are:

1. *The electron gas model*: Drude (1900) and Lorentz (1923) developed the idea that the valence electrons of the metal atom leave the parent atom. The positive core of the atom enters into the joint crystal lattice and the free electrons move freely throughout the lattice like gas, thus the positive core are embedded in this electron gas. The attractive force between the electron gas and the positive ion stabilised the metal. Though this theory explains the conduction phenomena, but cannot explain the constancy in atomic heat.

2. *Valence bond theory*: According to this theory, metal atoms are held together by electron pair bond and these bonds resonate between the neighbouring atoms, e.g., in sodium metal the bonds are as in the figure:

Na—Na Na—Na ⟷ Na—Na Na—Na ⟷ Na^+ Na—Na Na

Na—Na Na—Na   Na—Na Na—Na Na^+ Na^-—Na Na

This concept cannot explain the metallic properties clearly.

3. *Band theory*: Developed by Sommerfeld, Block and others, this approach is analogues to the molecular orbital theory where it is supposed that the *N* number of atomic orbitals of metal atoms combine to form *N* numbers of bonding and anti-bonding M.O.s. Bonding orbitals are called *energy bands*. The magnitude of the splitting depends on the inter-nuclear distance and the nature of atomic orbitals. Greater the overlap of the A.O.s of neighbouring atoms, wider will be the energy bands and then the valence electrons of the atoms will occupy these bands in accordance with Pauli's Exclusion Principle. These bands are known as *Brilloin Zones*. These successive bands may have some forbidden zone or band gaps or may overlap to some extent. In some cases before the first Brilloin zone is filled, the electron flows to the second zone if the two bands overlap. This can be shown as follows:

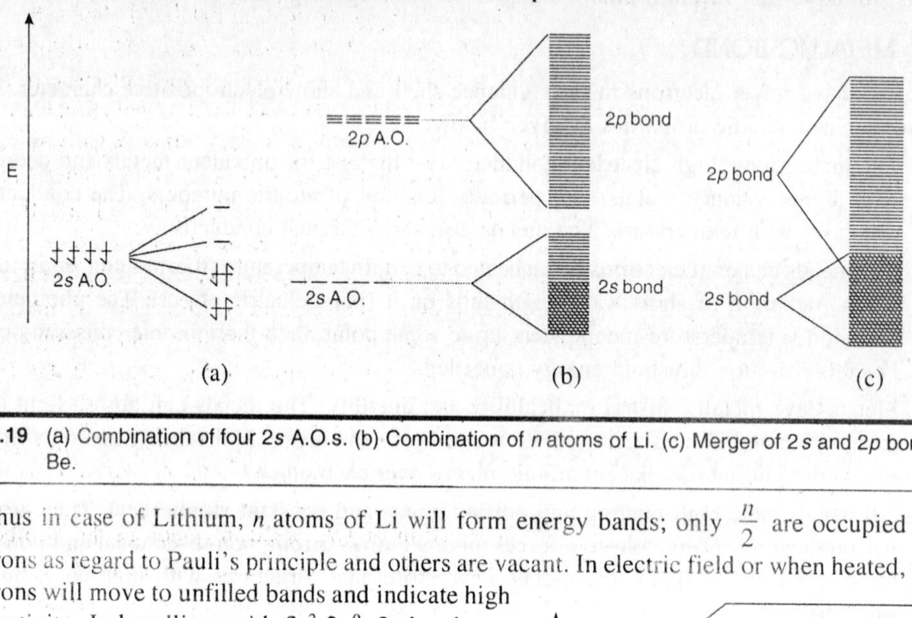

Fig. 8.19 (a) Combination of four 2*s* A.O.s. (b) Combination of *n* atoms of Li. (c) Merger of 2*s* and 2*p* bonds of Be.

Thus in case of Lithium, *n* atoms of Li will form energy bands; only $\frac{n}{2}$ are occupied by *n* electrons as regard to Pauli's principle and others are vacant. In electric field or when heated, these electrons will move to unfilled bands and indicate high conductivity. In beryllium with $2s^2\ 2p^0$, 2*s* bands are completely filled here, 2*p* band merges with 2*s* filled bands and under excitation electron from 2*s* will move to 2*p* bands and hence the metal shows conductivity. Here 2*p* bands are called conductor bands. In all cases of metallic conductors, the conductivity decreases with temperature.

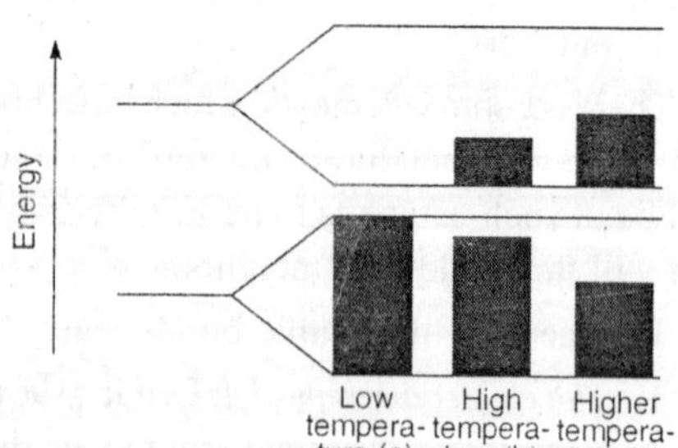

Fig. 8.20 (a) At very low temp. (b) At high temp. (c) At very high temp.

In case of insulators, the energy gaps between the energy band/valence band and the conduction band is high and so transition of electron from the first to the second does not take place.But there are some elements where the energy gap between the valence band and the conduction band is low and some electronic movement from the first to the second may take place on heating and shows conductivity to some extent. This property increases with temperature. These are called *semiconductors*. So, at sufficiently low temperature, semiconductor may behave as an insulator.

If the energy gap between filled valence band and the empty conduction band is of the magnitude of the optical spectrum, then by absorption of light energy, some electron may move from the valence band to the conduction band causing conductivity on illumination, called *photo-conductivity*. This type of semiconductors is called *intrinsic semiconductors*. Pure germanium, silicon and grey tin are intrinsic semiconductors. For insulators, this energy gap ranges from 3.0 to 5.5 eV, for semiconductors it ranges from 0.4 to 1.4 eV. The specific conductivity of semiconductors is about 10^{10} to 10^{16} times higher than the insulators.

The conductivity of semiconductors can be increased by adding impurities (i.e., doping), like atoms of neighbouring group elements of the periodic group. This type of semiconductivity is known as *extrinsic semiconductivity*. These types of semiconductors are of two types: (1) n-type semiconductors and (2) p-type semiconductors.

1. *n-type semiconductors*: Si, Ge, etc., are semiconductors. If some As (having 5 valence electrons) is added (1:10^9 atoms ratio to Si) the As replacing the Si as host material has one electron in excess and is localized in a band called the *donor band*. This donor band is localized above the filled energy band and near the empty conduction band. If temperature is increased some electron from the donor band will move to the empty conduction band and consequently some electron from the filled energy band will move to the donor band to fill the vacancy. As a result conduction continues and electron flow occurs throughout the crystal lattice of the host element. Since negative electron causes the conductivity it is called n-type semiconduction and the material as n-type semiconductor.

2. *p-type semiconductors*: Here the doped material has one electron less (Ga, In) than the host semiconductor Si or Ge. The empty band of Ga, called the *acceptor band* (as it accepts electron from the host) is localized near the filled energy band of host Si or Ge crystal lattice. On heating, some electrons from the energy band move to acceptor band creating electron holes in the energy band, then electron from the neighbouring energy band moves to occupy the hole making another positive hole. So, as positive holes move through the crystal lattice of the host Si or Ge lattice causing conduction, this is called the p-type semiconductor.

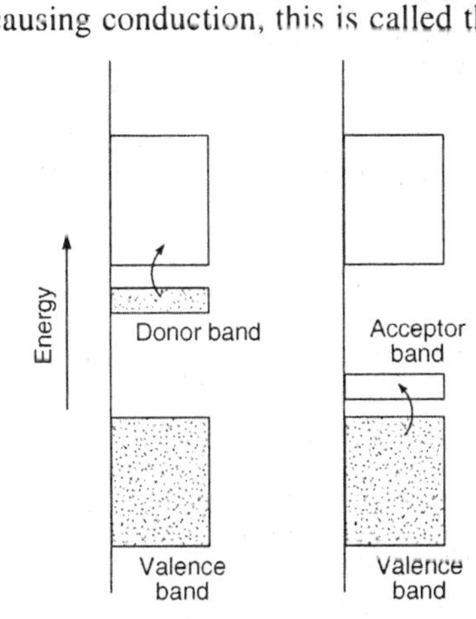

Fig 8.21 Band structure for doped extrinsic semiconductors (a) n-type, (b) p-type.

Pauling's theory: In a metal lattice having high co-ordination number the neighbouring like atoms at equal distances linked to the central atom are not bonded by one electron or two electron bonds as the number of valence electrons are insufficient in number. Hence Pauling introduced Heitler-London's valence bond concept and exchange forces. According to this approach, the bonding orbitals of the metal atoms in the crystal are hybridized and these hybrid orbitals resonate between the pairs of atoms.

Co-ordination number being high, hybridization that occurs is of the type d^2sp^3, some will be of non-bonding and some will be vacant metallic d-orbital. The orbitals containing one electron is responsible for paramagnetic behavior, the vacant orbitals account for the conduction, resonance of these bonds gives stability and non-directional character and so high co-ordination number.

8.06 MOLECULES WITH ELECTRON PAIRS BEYOND SIX

Molecules with seven electron pairs can have three possible geometry: pentagonal bipyramid, capped octahedral and capped trigonal prism which has almost identical energies and can be interconverted. The most familiar is the "*pbp*" and can be illustrated with a typical example, IF_7. Another interesting example is XeF_6 which contains 6 *bp*'s and 1 *lp*. Expectedly, its structure deviates from the regular octahedron. Determination of the exact structure of gaseous XeF_6 appears to be unexpectedly difficult because the lone pair as well as the molecule itself is not rigid and thus provides an example of "stereochemically non-rigid" (fluxional) molecule. Electron diffraction study indicates that it adapts a slightly distorted geometry, in which the lone pair extends either through a face or through an edge but the former is the lowest energy conformation. As in Fig 8.7(a) and (b).

The molecule is highly dynamic and probably passes through several conformers. This model is also consistent with (i) the dipole moment measurement which reveals that no measurable dipole moment is associated with XeF_6 (g) and thus rules out large static distortion and (ii) ^{19}F NMR where all six F-atoms appear to be equivalent.

In the solid state, XeF_6 ionises as [XeF_5^+] [F^-] ions, the former ion has 5 *bp*'s and 1 *lp*, and adapts a distorted square pyramid geometry (recall BrF_5) with a lone pair occupying one of the six equivalent sites of a regular octahedron.

Molecules having 5 *bp*'s and 2 *lp*'s such as XeF_5, the only example of this category so far known has perfect pentagonal structure with 2 *lp*'s positioned at the axial sites. Apicophilicity of *lp*'s can be accounted by both VSEPR and Bent's model.

According to VSEPR theory, *lp*'s seek more space in comparison to *bp*'s (because unlike *bp*'s, *lp*'s are essentially localised on the central atom) and hence preferentially occupy less crowded axial site. The same conclusion can be obtained from Bent's rule which states more electro-negative substituents (here F-atoms) prefer h.a.o.s. Xe having less *s*-character XeF_5^- have seven electron pairs (5 *bp*'s and 2 *lp*'s) and undergo sp^3d^3 hybridisation, for accommodating them. Importantly, all h.a.o.s are not equivalent, rather a combination of two different sets of h.a.o.s. More electro-negative F-atoms occupy equatorial sites having p^2d^3 hybridization (no *s*-character) leaving *sp* hybridized axial sites for the *lp*'s. This results perfect pentagonal structure.

CHAPTER

9

OXIDATION-REDUCTION REACTION

9.01 OXIDATION-REDUCTION REACTION

Oxidation-reduction reactions are the type of chemical reactions where electron transfer takes place from an atom or ion (i.e., itself is oxidised and hence reducing agent or reductant, i.e., electron donor) to other atom or ion (i.e., accepting electron means reducing and hence electron acceptor and is the oxidising agent or oxidant) as the existence of free electron is not possible. As a result of their electron transfer, the positive valence state of the reductant increases or negative valence state decreases and reverse is the case with oxidant. As for example, the reaction of $FeCl_3$ with $SnCl_2$, $2Fe^{3+} + Sn^{2+} \rightarrow 2Fe^{2+} + Sn^{4+}$.

This total reaction can be expressed separately

$Fe^{3+} + e^- \rightarrow Fe^{2+}$ (Reduction and Fe^{3+} is oxidant).

$Sn^{2+} - 2e^- \rightarrow Sn^{4+}$ (Oxidation and Sn^{2+} is reductant).

After electron balance the above equation is obtained.

The equivalent weight of oxidant and reductant can be determined as the weight of reagent that contains or reacts with 1.008 gm of available hydrogen or 8.00 gm of available oxygen. The amount of available oxygen in potassium permanganate and potassium dichromate can be determined from the hypothetical equation as given below:

$2KMnO_4 = K_2O + 2MnO + 5O$

$K_2Cr_2O_7 = K_2O + Cr_2O_3 + 3O$

The equivalent weights are

$2 \times$ molecular weight of $KMnO_4 / 2\times5$ and molecular weight of $K_2Cr_2O_7/3\times2$ respectively.

The equivalent weight of the oxidant or reductant is also obtained from the number of electron change in their half-reactions, e.g.,

$MnO_4^- + 8H^+ + 5e^- \rightarrow Mn^{2+} + 4H_2O$, $\therefore$ equivalent weight = molecular weight/5

$Cr_2O_7^{2-} + 14H^+ + 6e^- \rightarrow 2Cr^{3+} + 7H_2O$, $\therefore$ equivalent weight = molecular weight/6

The equivalent weights are also determined by oxidation number (O.N.) method. The equivalent weight of an element is determined by dividing the atomic weight by the number of electron change of the atom taking part in the redox reaction, i.e., the number of change in the oxidation number of the element. In case of a complex molecule it is determined by dividing the molecular weight by the number of change in oxidation number of the reacting atom of oxidant or reductant. For example, the equivalent weight of copper in the reaction with HNO_3 is $Cu^0 - 2e^- \rightarrow Cu^{2+}$ [ox]; 1/2 the atomic weight of copper is the equivalent weight. In $KMnO_4$, the reacting species is Mn^{7+}, which after reacting with reductant changes to Mn^{2+}, i.e., change in oxidation number is 5 units and so equivalent weight is 1/5 of the molecular weight of potassium permanganate.

Moreover, the equivalent weight of a substance depends on the type of the reaction in which it is used. Thus, copper sulphate solution when reacts with potassium iodide, $2Cu^{2+} + 2I^- = I_2 + 2Cu^+$ the equivalent weight of $CuSO_4$ is its molecular weight. But in the reaction $CuSO_4 + Fe = FeSO_4 + Cu$; its equivalent weight is 1/2 the molecular weight of copper sulphate.

In writing a redox reaction as the reaction takes place in aqueous solution, and water being a weak electrolyte, H^+, OH^- or H_2O can be utilized in the half-reaction for atom/mass balance. As oxidant and reductant are strong electrolytes, they ionise completely and only the ions taking part in the reaction or produced after reactions are to be mentioned in the reaction. Weak electrolytes do not ionise completely and sparingly soluble salts are present in undissociated form, their formulae are written in the equation.

The ion-electron equation of the reaction between oxalic acid and potassium permanganate can be written as

$H_2C_2O_4 \rightleftharpoons 2H^+ + C_2O_4^{2-}$ and

$KMnO_4 \rightleftharpoons K^+ + MnO_4^-$.

The active ions are MnO_4^{2-} and $C_2O_4^{2-}$. The half-reactions are

$C_2O_4^{2-} - 2e^- \rightarrow 2CO_2$ [Ox]

$MnO4^- + 8H^+ + 5e^- \rightarrow Mn^{2+} + 4H_2O$ [Red]

The balanced equation is

$2MnO_4^- + 16H^+ + 5C_2O_4^{2-} = 2Mn^{2+} + 10CO_2 + 8H_2O$

The reaction of $K_2Cr_2O_7$ and KI in acidic solution can be written as

$2I^- - 2e^- = I_2$ [Ox]

$Cr_2O_7^- + 14H^+ + 6e^- \rightarrow 2Cr^{3+} + 7H_2O$ [Red]

So, the balanced equation is

$Cr_2O_7^{2-} + 14H^+ + 6I^- = 2Cr^{3+} + 3I_2 + 7H_2O$

The strength of reducing agents or oxidising agents, i.e., chemical activity differs among themselves. The strong oxidising agent can oxidise reducing agents and even weak oxidising agents. The strengths of the oxidising or reducing agents are indicated by oxidation potentials *E*. For understanding oxidation potential or redox potential the following experiment is done as in the diagram 9.1 and its explanations are to be noted.

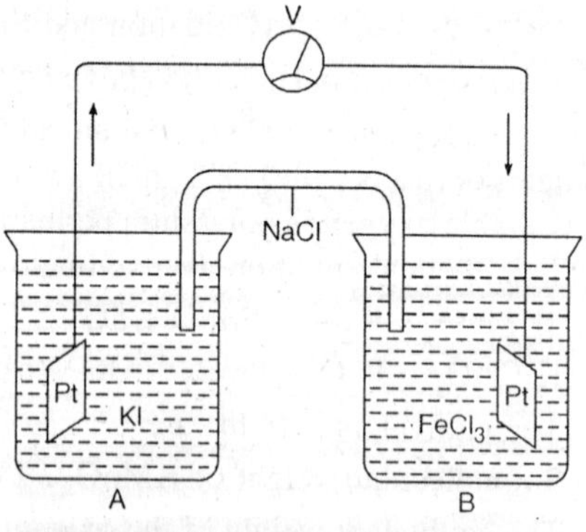

Fig. 9.1 Experimental determination of potential.

Electric current flows and the yellow colour due to liberation of iodine develops around platinum electrode in KI solution and Fe^{2+} ion is collected in the $FeCl_3$ solution. The electron produced (here KI solution with Pt) in the cell is negative electrode or cathode and the cell where reduction takes place is the positive electrode i.e., anode, here $FeCl_3$ solution with Pt electrode. Absolutely, pure oxidising or reducing agents are never met with; they are always admixed with oxidation or reduction products. Hence the redox potentials of the redox systems are written as $I_2/I^- = 0.54$ V, $Fe^{3+}/Fe^{2+} = 0.77$ V, etc., the higher valence states are the oxidised states and the lower valence states are the reduced states.

The flow of current in the above experiment is due to *e.m.f.* produced by the chemical change in the redox reaction. If a metal plate is immersed in a solution containing its own ion, say Zn metal in $ZnSO_4$ solution, a potential difference is developed between the metal and its solution. The electrode reaction is $M - ne^- \rightleftharpoons M^{n+}$ and the potential for this reaction as given by *Nernst* equation as $E = E_0 + \frac{0.0591}{n}\log C_M^{n+}$ or more accurately $E = E_0 + \frac{.0591}{n}\log\frac{[\text{Ox}]}{[\text{Red}]}$ where E is the experimental potential, E_0 is the standard potential, n is the number of electron change in the electrode reaction.

The redox potential of a redox system is always measured with reference to standard hydrogen electrode. This is a piece of platinum foil, coated electrically with platinum black, immersed in hydrochloric acid of unit activity of H^+ that corresponds to 1.8 M hydrochloric acid at 25°C. Hydrogen gas at one atmospheric pressure is bubbled through a tube round the platinum foil. By convention potential of the hydrogen electrode is taken as zero. Connecting this electrode to a metal electrode through salt bridge, the standard potential of the metal electrode is determined and the combined cell is written as

$$\text{Pt, } H_2 \mid H^+(a = 1) \parallel M^{n+}\ (a = 1) \mid M$$

Here, the activity of the M^{n+} is unity, i.e., $a = 1$ and the potential observed is called the standard potential E_0. The elements occupying higher position in the electrochemical series are more ready to lose electron, i.e., good reducing agent and behave as negative electrode. If the activity of hydrogen ion is reduced to 10^{-7} M and pressure of hydrogen gas is one atmosphere, the resulting potential is $E = E_0 + \frac{0.0591}{1}\log\frac{pH_2}{10^{-7}} = 0 + 0.414$ volt. So in the alkali region, the potential becomes more positive. Thus the potential of a system in alkaline medium is more positive than that in acidic medium.

When we write a cell on paper, the convention is that the negative electrode is written on the left and the positive electrode on the right so that the electron flows from the left to the right through the external circuit indicating the spontaneity of the redox reaction. Thus, the Daniel cell is written as

$$\text{Zn} \mid ZnSO_4 \parallel CuSO_4 \parallel \text{Cu; not as Cu} \mid CuSO_4 \parallel ZnSO_4 \mid \text{Zn}$$

The positive and the negative electrode in a cell can be ascertained from the value of cell e.m.f. Suppose in a cell Zn | $ZnSO_4$ ($a = 1$) and Pb | $PbSO_4$ ($a = 1$), if we consider Pb | $PbSO_4$ as negative electrode, then the cell is Pb | $PbSO_4$ ($a = 1$) || $ZnSO_4$ ($a = 1$) | Zn.

The e.m.f. of Pb | $PbSO_4$ ($a = 1$) is

$$E = E_{Pb \to Pb^{2+}} + \frac{0.0591}{2}\log\left[Pb^{2+}\right] = -0.13 \text{ volt}$$

The electrode potential for Zn / $ZnSO_4$ ($a = 1$) is

$$E = E_{Zn \to Zn^{2+}} + \frac{0.591}{2}\log\left[Zn^{2+}\right] = -0.76 \text{ volt}$$

Hence, the cell e.m.f. of the cell written above is

$$E = E_{Zn^{2+} \to Zn} - E_{0\,Pb \to Pb^{2+}} = -0.76 - (-0.13) = -0.63 \text{ volt}$$

The negative sign indicates the non-spontaneity of the reaction and hence Pb/$PbSO_4$ cannot be negative plate and the cell will be written as

$$Zn \mid ZnSO_4\ (a = 1) \parallel PbSO_4\ (a = 1) \mid Pb$$

$$E = E_{0\,Pb \to Pb^{2+}} - E_{0\,Zn \to Zn^{2+}} = -0.13 - (-0.76) = 0.63 \text{ volt}$$

In a half-reaction, oxidation is written in the left with requisite number of electron to be added to get the resultant, $MnO_4^- + 8H^+ + 5e^- \rightarrow Mn^{2+} + 4H_2O$.

Regarding redox potential, the following points are to be remembered:

1. In a redox system, the higher oxidation state is the oxidant and the lower oxidation state is the reductant. Higher the value of the potential, the powerful oxidising agent will be the oxidant and the reductant will be the weaker reducing agent, e.g., MnO_4^- / Mn^{2+} (E = 1.51 volt). MnO_4^- is a powerful oxidant and Mn^{2+} is a weak reductant.
2. Positive potential value indicates the powerful reducing agent and the reductant will liberate H_2 gas from acids, e.g.,

   $Mg - 2e^- \rightarrow Mg^{2+}\ (E_{0\,Mg \rightarrow Mg^{+2}} = +2.34)$, so $Mg + 2H^+ \rightarrow Mg^{2+} + H_2$.

   Negative potential value indicates the powerful oxidant and will oxidise H_2 gas to H^+, e.g., $Cl_2 \mid Cl^-$, the Cl_2 is a powerful oxidant. $Cl^- - e^- \rightarrow \frac{1}{2}Cl_2$ ($E_0 = -1.36$ volt).
3. In a half-reaction where H^+ or OH^- is not involved, their redox systems are independent of the concentration of H^+ or OH^-, e.g., E_0 for Na/Na^+ is independent on the entire pH range. In $Fe^{2+} - Fe^{3+}$ system, E^0 remains constant only over that pH range in which Fe $(OH)_2$ or $Fe(OH)_3$ is not precipitated. But once this pH is exceeded both $Fe(OH)_2$ and $Fe(OH)_3$ precipitate and role of H^+ or OH^- becomes important, due to different basicities, there is no continuous transition in potential from highly acidic to highly alkaline range.
4. In a half reaction where H^+ is used up or produced the E_0 for the system are dependent on pH and the oxidant will be the good oxidising agent in strong acid solution, i.e., in 1-2 (N) acid solution, e.g., $MnO_4^- + 8H^+ + 5e^- \rightarrow Mn^{2+} + 4H_2O$, ($E_0$= 1.51 volts) behaves as a good oxidising agent in 1-2 (N) acid solution.
5. If the two redox systems are combined, the stronger of the oxidant will gain electrons from the stronger reductant with the formation of weaker reductant or oxidant, e.g., Cl^-/Cl, (E_0= −1.36 volt) and $Fe^{3+} \mid Fe^{2+}$, (E_0 = 0.77 volt) are combined, Cl_2 will oxidize Fe^{2+} and the equation is $\frac{1}{2}Cl_2 + Fe^{2+} \rightarrow Cl^- + Fe^{3+}$.

Thus, an oxidant of higher potential will oxidise a reductant with lower potential and reductant of a lower potential will reduce the oxidant of higher potential. With the help of this standard redox potential data we can predict the direction of redox reaction, choose suitable oxidant or reductant, suitable indicator and solve many other important problems in analytical chemistry.

Table 9.1 : Standard oxidation potentials in acidic solution

Couple	Equation	E^0_{296} volts
Li(0)–Li(I)	$Li \rightleftharpoons Li^+ + e^-$	+ 3.02
Cs(0)–Cs(I)	$Cs \rightleftharpoons Cs^+ + e^-$	3.02
Rb(0)–Rb(I)	$Rb \rightleftharpoons Rb^+ + e^-$	2.99

Contd.

Couple	Equation	E^0_{296} volts
K(0)–K(I)	$K \rightleftharpoons K^+ + e^-$	2.922
Ba(0)–Ba(II)	$Ba \rightleftharpoons Ba^{+2} + 2e^-$	2.90
Sr(0)–Sr(II)	$Sr \rightleftharpoons Sr^{+2} + 2e^-$	2.89
Ca(0)–Ca(II)	$Ca \rightleftharpoons Ca^{+2} + 2e^-$	2.87
Na(0)–Na(I)	$Na \rightleftharpoons Na^+ + e^-$	2.712
La(0)–La(III)	$La \rightleftharpoons La^{+3} + 3e^-$	2.37
Mg(0)–Mg(II)	$Mg \rightleftharpoons Mg^{+2} + 2e^-$	2.34
H(–I)–H(0)	$H^- \rightleftharpoons \frac{1}{2}H_2 + e^-$	2.23
Th(0)–Th(IV)	$Th + 2H_2O \rightleftharpoons ThO_2 + 4H^+ + 4e^-$	1.80
Be(0)–Be(II)	$Be \rightleftharpoons Be^{+2} + 2e^-$	1.70
Hf(0)–Hf(IV)	$Hf + H_2O \rightleftharpoons HfO^{+2} + 2H^+ + 4e^-$	1.68
Al(0)–Al(III)	$Al \rightleftharpoons Al^{+3} + 3e^-$	1.67
Zr(0)–Zr(IV)	$Zr + 2H_2O \rightleftharpoons ZrO_2 + 4H^+ + 4e^-$	1.43
Mn(0)–Mn(II)	$Mn \rightleftharpoons Mn^{+2} + 2e^-$	1.05
Ti(0)–Ti(IV)	$Ti + 2H_2O \rightleftharpoons TiO_2 + 4H^+ + 4e^-$	0.95
Si(0)–Si(IV)	$Si + 2H_2O \rightleftharpoons SiO_2 + 4H^+ + 4e^-$	0.84
U(0)–U(VI)	$U + 2H_2O \rightleftharpoons UO_2^{+2} + 4H^+ + 6e^-$	0.82
Zn(0)–Zn(II)	$Zn \rightleftharpoons Zn^{+2} + 2e^-$	0.7620
B(0)–B(III)	$B + 3H_2O \rightleftharpoons H_3BO_3 + 3H^+ + 3e^-$	0.73
Ta(0)–Ta(V)	$2Ta + 5H_2O \rightleftharpoons Ta_2O_5 + 10H^+ + 10e^-$	0.71
Cr(0)–Cr(III)	$Cr \rightleftharpoons Cr^{+3} + 3e^-$	0.71
Te(–II)–Te(0)	$H_2Te(aq) \rightleftharpoons Te + 2H^+ + 2e^-$	0.69
Nb(0)–Nb(V)	$2Nb + 5H_2O \rightleftharpoons Nb_2O_5 + 10H^+ + 10e^-$	0.62
P(I)–P(III)	$H_3PO_2 + H_2O \rightleftharpoons H_3PO_3 + 2H^+ + 2e^-$	0.59
As(–III)–As(0)	$AsH_3 \rightleftharpoons As + 3H^+ + 3e^-$	0.54
Ga(0)–Ga(III)	$Ga \rightleftharpoons Ga^{+3} + 3e^-$	0.52
P(0)–P(III)	$P + 3H_2O \rightleftharpoons H_3PO_3 + 3H^+ + 3e^-$	0.49
Fe(0)–Fe(II)	$Fe \rightleftharpoons Fe^{+2} + 2e^-$	0.440
Eu(II)–Eu(III)	$Eu^{+2} \rightleftharpoons Eu^{+3} + e^-$	0.43
Cr(II)–Cr(III)	$Cr^{+2} \rightleftharpoons Cr^{+3} + e^-$	0.41
Cd(0)–Cd(II)	$Cd \rightleftharpoons Cd^{+2} + 2e^-$	0.402
Se(–II)–Se(0)	$H_2Se(aq) \rightleftharpoons Se + 2H^+ + 2e^-$	0.36
In(0)–In(III)	$In \rightleftharpoons In^{+3} + 3e^-$	0.340
TI(0)–TI(I)	$TI \rightleftharpoons TI^+ + e^-$	0.3363
P(0)–P(I)	$P + 2H_2O \rightleftharpoons H_3PO_2 + H^+ + e^-$	0.29
Co(0)–Co(II)	$Co \rightleftharpoons Co^{+2} + 2e^-$	0.277
Ni(0)–Ni(II)	$Ni \rightleftharpoons Ni^{+2} + 2e^-$	0.250
V(II)–V(III)	$V^{+2} \rightleftharpoons V^{+3} + e^-$	0.20
P(III)–P(V)	$H_3PO_3 + H_2O \rightleftharpoons H_3PO_4 + 2H^+ + 2e^-$	0.20
Sn(0)–Sn(II)	$Sn \rightleftharpoons Sn^{+2} + 2e^-$	0.136
Pb(0)–Pb(II)	$Pb \rightleftharpoons Pb^{+2} + 2e^-$	0.126
P(–III)–P(0)	$PH_3(g) \rightleftharpoons P + 3H^+ + 3e^-$	0.04

Contd.

Couple	Equation	E^0_{296} volts
Fe(0)–Fe(III)	$Fe \rightleftharpoons Fe^{+3} + 3e^-$	0.036
D(0)–D(I)	$\frac{1}{2}D_2 \rightleftharpoons D^+ + e^-$	0.0034
H(0)–H(I)	$\frac{1}{2}H_2 \rightleftharpoons H^+ + e^-$	0.0000
S(–II)–S(0)	$H_2S(aq) \rightleftharpoons S + 2H^+ + 2e^-$	–1.141
Re(0)–Re(VII)	$Re + 4H_2O \rightleftharpoons ReO_4^- + 8H^+ + 7e^-$	–0.15
Sn(II)–Sn(IV)	$Sn^{+2} \rightleftharpoons Sn^{+4} + 2e^-$	–0.15
Cu(I)–Cu(II)	$Cu^+ \rightleftharpoons Cu^{+2} + e^-$	–0.167
S(IV)–S(VI)	$H_2SO_3 + H_2O \rightleftharpoons SO_4^{-2} + 4H^+ + 2e^-$	–0.20
Sb(0)–Sb(III)	$Sb + H_2O \rightleftharpoons SbO^+ + 2H^+ + 3e^-$	–0.212
As(0)–As(III)	$As + 2H_2O \rightleftharpoons HAsO_2 + 3H^+ + 3e^-$	–0.2475
V(III)–V(IV)	$V^{+3} + H_2O \rightleftharpoons VO^{+2} + 2H^+ + e^-$	–0.314
Bi(0)–Bi(III)	$Bi + H_2O \rightleftharpoons BiO^+ + 2H^+ + 3e^-$	–0.32
Cu(0)–Cu(II)	$Cu \rightleftharpoons Cu^{+2} + 2e^-$	–0.3448
S(0)–S(IV)	$S + 3H_2O \rightleftharpoons H_2SO_3 + 4H^+ + 4e^-$	–0.45
Cu(0)–Cu(I)	$Cu \rightleftharpoons Cu^+ + e^-$	–0.522
Te(0)–Te(IV)	$Te + 2H_2O \rightleftharpoons TeO_2(s) + 4H^+ + 4e^-$	–0.529
I(–I)–I(0)	$I^- \rightleftharpoons \frac{1}{2}I_2 + e^-$	–0.5345
As(III)–As(V)	$HAsO_2 + 2H_2O \rightleftharpoons H_3AsO_4 + 2H^+ + 2e^-$	–0.559
Sb(III)–Sb(V)	$2SbO^+ + 3H_2O \rightleftharpoons Sb_2O_5(s) + 6H^+ + 4e^-$	–0.64
O(–I)–O(0)	$H_2O_2 \rightleftharpoons O_2 + 2H^+ + 2e^-$	–0.682
Se(0)–Se(IV)	$Se + 3H_2O \rightleftharpoons H_2SeO_3 + 4H^+ + 4e^-$	–0.740
Fe(II)–Fe(III)	$Fe^{+2} \rightleftharpoons Fe^{+3} + e^-$	–0.771
Hg(0)–Hg(I)	$Hg \rightleftharpoons \frac{1}{2}Hg_2^{+2} + e^-$	–0.7986
Ag(0)–Ag(I)	$Ag \rightleftharpoons Ag^+ + e^-$	–0.7995
N(I)–N(III)	$H_2N_2O_2 + 2H_2O \rightleftharpoons 2HNO_2 + 4H^+ + 4e^-$	–0.80
N(IV)–N(V)	$N_2O_4 + 2H_2O \rightleftharpoons 2NO_3^- + 4H^+ + 2e^-$	–0.81
O(–II)–O(0)	$H_2O \rightleftharpoons \frac{1}{2}O_2 + 2H^+(10^{-7}M) + 2e^-$	–0.815
Pd(0)–Pd(II)	$Pd \rightleftharpoons Pd^{+2} + 2e^-$	–0.83
Os(0)–Os(VIII)	$Os + 4H_2O \rightleftharpoons OsO_4 + 8H^+ + 8e^-$	–0.85
Hg(0)–Hg(II)	$Hg \rightleftharpoons Hg^{+2} + 2e^-$	–0.854
Hg(I)–Hg(II)	$Hg_2^{+2} \rightleftharpoons 2Hg^{+2} + 2e^-$	–0.910
N(III)–N(V)	$HNO_2 + H_2O \rightleftharpoons NO_3^- + 3H^+ + 2e^-$	–0.94
N(II)–N(V)	$NO + 2H_2O \rightleftharpoons NO_3^- + 4H^+ + 3e^-$	–0.96
N(II)–N(III)	$NO + H_2O \rightleftharpoons HNO_2 + H^+ + e^-$	–0.99
I(–I)–I(I)	$I^- + H_2O \rightleftharpoons HIO + H^+ + 2e^-$	–0.99
V(IV)–V(V)	$VO^{+2} + 3H_2O \rightleftharpoons V(OH)_4^+ + 2H^+ + e^-$	–1.00
Cl(V)–Cl(VII)	$ClO_3^- + H_2O \rightleftharpoons ClO_4^- + 2H^+ + 2e^-$	–1.00
Te(IV)–Te(VI)	$TeO_2(s) + 4H_2O \rightleftharpoons H_6TeO_6(s) + 2H^+ + 2e^-$	–1.02
N(II)–N(IV)	$2NO + 2H_2O \rightleftharpoons N_2O_4 + 4H^+ + 4e^-$	–1.03
Br(–1)–Br(0)	$2Br^- \rightleftharpoons Br_2(l) + 2e^-$	–1.0652
N(III)–N(IV)	$2HNO_2 \rightleftharpoons N_2O_4 + 2H^+ + 2e^-$	–1.07
I(–I)–I(V)	$I^- + 3H_2O \rightleftharpoons IO_3^- + 6H^+ + 6e^-$	–1.085
Br(–I)–Br(0)	$2Br^- \rightleftharpoons Br_2(aq) + 2e^-$	–1.087

Contd.

Couple	Equation	E^0_{296} volts
Se(IV)–Se(VI)	$H_2SeO_3 + H_2O \rightleftharpoons SeO_4^{-2} + 4H^+ + 2e^-$	–1.15
I(0)–I(V)	$\frac{1}{2}I_2 + 3H_2O \rightleftharpoons IO_3^- + 6H^+ + 5e^-$	–1.195
Pt(0)–Pt(II)	$Pt \rightleftharpoons Pt^{+2} + 2e^-$	ca. –1.2
O(–II)–O(0)	$H_2O \rightleftharpoons \frac{1}{2}O_2 + 2H^+ + 2e^-$	–1.229
Cl(III)–Cl(V)	$HClO_2 + H_2O \rightleftharpoons ClO_3^- + 3H^+ + 2e^-$	–1.23
Tl(I)–Tl(III)	$Tl^+ \rightleftharpoons Tl^{+3} + 2e^-$	–1.25
Mn(II)–Mn(IV)	$Mn^{+2} + 2H_2O \rightleftharpoons MnO_2 + 4H^+ + 2e^-$	–1.28
N(I)–N(III)	$N_2O + 3H_2O \rightleftharpoons 2HNO_2 + 4H^+ + 4e^-$	–1.29
Au(I)–Au(III)	$Au^+ \rightleftharpoons Au^{+3} + 2e^-$	ca. –1.29
Br(–I)–Br(I)	$Br^- + H_2O \rightleftharpoons HBrO + H^+ + 2e^-$	–1.33
Cl(0)–Cl(VII)	$\frac{1}{2}Cl_2 + 4H_2O \rightleftharpoons ClO_4^- + 8H^+ + 7e^-$	–1.34
Cl(–I)–Cl(0)	$Cl^- \rightleftharpoons \frac{1}{2}Cl_2 + e^-$	–1.3583
Cr(III)–Cr(VI)	$2Cr^{+3} + 7H_2O \rightleftharpoons Cr_2O_7^{-2} + 14H^+ + 6e^-$	–1.36
Au(0)–Au(III)	$Au \rightleftharpoons Au^{+3} + 3e^-$	–1.42
Br(–I)–Br(V)	$Br^- + 3H_2O \rightleftharpoons BrO_3^- + 6H^+ + 6e^-$	–1.44
I(0)–I(I)	$\frac{1}{2}I_2 + H_2O \rightleftharpoons HIO + H^+ + e^-$	–1.45
Cl(–I)–Cl(V)	$Cl^- + 3H_2O \rightleftharpoons ClO_3^- + 6H^+ + 6e^-$	–1.45
Pb(II)–Pb(IV)	$Pb^{+2} + 2H_2O \rightleftharpoons PbO_2 + 4H^+ + 2e^-$	–1.456
Cl(0)–Cl(V)	$\frac{1}{2}Cl_2 + 3H_2O \rightleftharpoons ClO_3^- + 6H^+ + 5e^-$	–1.47
Cl(–I)–Cl(I)	$Cl^- + H_2O \rightleftharpoons HClO + H^+ + 2e^-$	–1.49
Mn(II)–Mn(III)	$Mn^{+2} \rightleftharpoons Mn^{+3} + e^-$	–1.51
Br(0)–Br(V)	$\frac{1}{2}Br_2 + 3H_2O \rightleftharpoons BrO_3^- + 6H^+ + 5e^-$	–1.52
Mn(II)–Mn(VII)	$Mn^{+2} + 4H_2O \rightleftharpoons MnO_4^- + 8H^+ + 5e^-$	–1.52
Cl(–I)–Cl(III)	$Cl^- + 2H_2O \rightleftharpoons HClO_2 + 3H^+ + 4e^-$	–1.56
Br(0)–Br(I)	$\frac{1}{2}Br_2 + H_2O \rightleftharpoons HBrO + H^+ + e^-$	–1.59
N(I)–N(II)	$N_2O + H_2O \rightleftharpoons 2NO + 2H^+ + 2e^-$	–1.59
Ce(III)–Ce(IV)	$Ce^{+3} \rightleftharpoons Ce^{+4} + e^-$	–1.61
Cl(0)–Cl(I)	$\frac{1}{2}Cl_2 + H_2O \rightleftharpoons HClO + H^+ + e^-$	–1.63
Cl(0)–Cl(III)	$\frac{1}{2}Cl_2 + 2H_2O \rightleftharpoons HClO_2 + 3H^+ + 3e^-$	–1.63
Mn(IV)–Mn(VII)	$MnO_2 + 2H_2O \rightleftharpoons MnO_4^- + 4H^+ + 3e^-$	–1.67
Au(0)–Au(I)	$Au \rightleftharpoons Au^+ + e^-$	–1.68
I(V)–I(VII)	$IO_3^- + 3H_2O \rightleftharpoons H_5IO_6 + H^+ + 2e^-$	ca. –1.7
Ni(II)–Ni(IV)	$N^{+2} + 2H_2O \rightleftharpoons NiO_2 + 4H^+ + 2e^-$	–1.75
O(–II)–O(–I)	$2H_2O \rightleftharpoons H_2O_2 + 2H^+ + 2e^-$	–1.77
N(0)–N(I)	$N_2 + H_2O \rightleftharpoons N_2O + 2H^+ + 2e^-$	–1.77
Co(II)–Co(III)	$Co^{+2} \rightleftharpoons Co^{+3} + e^-$	–1.84
Ag(I)–Ag(II)	$Ag^+ \rightleftharpoons Ag^{+2} + e^-$	–1.98
	$2SO_4^{-2} \rightleftharpoons S_2O_8^{-2} + 2e^-$	–2.05
	$O_2 + H_2O \rightleftharpoons O_3 + 2H^+ + 2e^-$	–2.07
O(–II)–O(II)	$H_2O + 2F^- \rightleftharpoons F_2O + 2H^+ + 4e^-$	–2.1
O(–II)–O(0)	$H_2O \rightleftharpoons \frac{1}{2}O_2(g) + 2H^+ + 2e^-$	–2.42
F(–I)–F(0)	$F^- \rightleftharpoons \frac{1}{2}F_2 + e^-$	–2.85
F(–I)–F(0)	$HF \rightleftharpoons \frac{1}{2}F_2 + H^+ + e^-$	–3.03

Contd.

Standard oxidation potentials in alkaline solution

Couple	Equation	E^0_{296} volts
Li(0)–Li(I)	$Li \rightleftharpoons Li^+ + e^-$	+3.02
Cs(0)–Cs(I)	$Cs \rightleftharpoons Cs^+ + e^-$	3.02
Ca(0)–Ca(II)	$Ca + 2OH^- \rightleftharpoons Ca(OH)_2 + 2e^-$	3.02
Sr(0)–Sr(II)	$Sr + 2OH^- + 8H_2O \rightleftharpoons Sr(OH)_2{\cdot}8H_2O + 2e^-$	2.99
Rb(0)–Rb(I)	$Rb \rightleftharpoons Rb^+ + e^-$	2.99
Ba(0)–Ba(II)	$Ba + 2OH^- + 8H_2O \rightleftharpoons Ba(OH)_2{\cdot}8H_2O + 2e^-$	2.97
K(0)–K(I)	$K \rightleftharpoons K^+ + e^-$	2.922
La(0)–La(III)	$La + 3OH^- \rightleftharpoons La(OH)_3 + 3e^-$	2.76
Na(0)–Na(I)	$Na \rightleftharpoons Na^+ + e^-$	2.712
Mg(0)–Mg(II)	$Mg + 2OH^- \rightleftharpoons Mg(OH)_2 + 2e^-$	2.67
Th(0)–Th(IV)	$Th + 4OH^- \rightleftharpoons ThO_2 + 2H_2O + 4e^-$	2.64
Hf(0)–Hf(IV)	$Hf + 4OH^- \rightleftharpoons HfO(OH)_2 + H_2O + 4e^-$	2.60
B(0)–B(III)	$B + 4OH^- \rightleftharpoons H_2BO_3^- + H_2O + 3e^-$	2.5
Al(0)–Al(III)	$Al + 4OH^- \rightleftharpoons H_2AlO_3^- + H_2O + 3e^-$	2.35
Zr(0)–Zr(IV)	$Zr + 4OH^- \rightleftharpoons ZrO(OH)_2 + H_2O + 4e^-$	2.32
Be(0)–Be(II)	$2Be + 6OH^- \rightleftharpoons Be_2O_3^{-2} + 3H_2O + 4e^-$	2.28
P(0)–P(I)	$P + 2OH^- \rightleftharpoons H_2PO_2^- + e^-$	1.82
Si(0)–Si(IV)	$Si + 6OH^- \rightleftharpoons SiO_3^{-2} + 3H_2O + 4e^-$	1.73
P(0)–P(III)	$P + 5OH^- \rightleftharpoons HPO_3^{-2} + 2H_2O + 3e^-$	1.71
P(I)–P(III)	$H_2PO_2^- + 3OH^- \rightleftharpoons HPO_3^{-2} + 2H_2O + 2e^-$	1.65
Mn(0)–Mn(II)	$Mn + 2OH^- \rightleftharpoons Mn(OH)_2$	1.47
Cr(0)–Cr(III)	$Cr + 3OH^- \rightleftharpoons Cr(OH)_3 + 3e^-$	1.3
Ga(0)–Ga(III)	$Ga + 4OH^- \rightleftharpoons H_2GaO_3^- + H_2O + 3e^-$	1.22
Zn(0)–Zn(II)	$Zn + 4OH^- \rightleftharpoons ZnO_2^{-2} + 2H_2O + 2e^-$	1.216
Cr(0)–Cr(III)	$Cr + 4OH^- \rightleftharpoons CrO_2^- + 2H_2O + 3e^-$	1.2
P(III)–P(V)	$HPO_3^{-2} + 3OH^- \rightleftharpoons PO_4^{-3} + 2H_2O + 2e^-$	1.05
Sn(II)–Sn(IV)	$HSnO_2^- + 3OH^- + H_2O \rightleftharpoons Sn(OH)_6^{-2} + 2e^-$	0.96
Te(–II)–Te(0)	$Te^{-2} \rightleftharpoons Te + 2e^-$	0.92
S(IV)–S(VI)	$SO_3^{-2} + 2OH^- \rightleftharpoons SO_4^{-2} + H_2O + 2e^-$	0.90
Fe(0)–Fe(II)	$Fe + 2OH^- \rightleftharpoons Fe(OH)_2 + 2e^-$	0.877
P(–III)–P(0)	$PH_3(g) + 3OH^- \rightleftharpoons P + 3H_2O + 3e^-$	0.87
N(IV)–N(V)	$N_2O_4 + 4OH^- \rightleftharpoons 2NO_3^- + 2H_2O + 2e^-$	0.85
H(0)–H(I)	$H_2 + 2OH^- \rightleftharpoons 2H_2O + 2e^-$	0.828
Cd(0)–Cd(II)	$Cd + 2OH^- \rightleftharpoons Cd(OH)_2 + 2e^-$	0.815
Re(0)–Re(VII)	$Re + 8OH^- \rightleftharpoons ReO_4^- + 4H_2O + 7e^-$	0.81
Sn(0)–Sn(II)	$Sn + 3OH^- \rightleftharpoons HSnO_2 + H_2O + 2e^-$	0.79
Se(–II)–Se(0)	$Se^{-2} \rightleftharpoons Se + 2e^-$	0.78
Co(0)–Co(II)	$Co + 2OH^- \rightleftharpoons Co(OH)_2 + 2e^-$	0.73
As(III)–As(V)	$AsO_2^- + 4OH^- \rightleftharpoons AsO_4^{-3} + 2H_2O + 2e^-$	0.71
As(0)–As(III)	$As + 4OH^- \rightleftharpoons AsO_2^- + 2H_2O + 3e^-$	0.68
Sb(0)–Sb(III)	$Sb + 4OH^- \rightleftharpoons SbO_2^- + 2H_2O + 3e^-$	0.66

Contd.

Couple	Equation	E^0_{296} volts
Ni(0)–Ni(II)	$Ni + 2OH^- \rightleftharpoons Ni(OH)_2 + 2e^-$	0.66
Pb(0)–Pb(II)	$Pb + 2OH^- \rightleftharpoons PbO + H_2O + 2e^-$	0.578
Fe(II)–Fe(III)	$Fe(OH)_2 + OH^- \rightleftharpoons Fe(OH)_3 + e^-$	0.56
Pb(0)–Pb(II)	$Pb + 3OH^- \rightleftharpoons HPbO_2^- + H_2O + 2e^-$	0.54
S(–II)–S(0)	$S^{-2} \rightleftharpoons S + 2e^-$	0.508
S(–II)–S(0)	$HS^- + OH^- \rightleftharpoons S + H_2O + 2e^-$	0.478
N(II)–N(III)	$NO + 2OH^- \rightleftharpoons NO_2^- + H_2O + e^-$	0.46
Bi(0)–Bi(III)	$Bi + 3OH^- \rightleftharpoons BiOOH + H_2O + 3e^-$	0.46
Mn(II)–Mn(III)	$Mn(OH)_2 + OH^- \rightleftharpoons Mn(OH)_3 + e^-$	0.40
Cu(0)–Cu(I)	$2Cu + 2OH^- \rightleftharpoons Cu_2O + H_2O + 2e^-$	0.361
Se(0)–Se(IV)	$Se + 6OH^- \rightleftharpoons SeO_3^{-2} + 3H_2O + 4e^-$	0.35
Tl(0)–Tl(I)	$Tl + OH^- \rightleftharpoons TlOH + e^-$	0.3445
Cu(0)–Cu(II)	$Cu + 2OH^- \rightleftharpoons Cu(OH)_3 + 2e^-$	0.224
N(I)–N(III)	$N_2O_2^{-2} + 4OH^- \rightleftharpoons 2NO_2^- + 2H_2O + 4e^-$	0.18
Cr(III)–Cr(VI)	$Cr(OH)_3 + 5OH^- \rightleftharpoons CrO_4^{-2} + 4H_2O + 3e^-$	0.12
Cu(I)–Cu(II)	$Cu_2O + 2OH^- + H_2O \rightleftharpoons 2Cu(OH)_2 + 2e^-$	0.09
O(–I)–O(0)	$HO_2^- + OH^- \rightleftharpoons O_2 + H_2O + 2e^-$	0.076
Tl(I)–Tl(III)	$TlOH + 2OH^- \rightleftharpoons Tl(OH)_3 + 2e^-$	0.05
Te(0)–Te(IV)	$Te + 6OH^- \rightleftharpoons TeO_3^{-2} + 3H_2O + 4e^-$	0.02
N(III)–N(V)	$NO_2^- + 2OH^- \rightleftharpoons NO_3^- + H_2O + 2e^-$	–0.01
Se(IV)–Se(VI)	$SeO_3^{-2} + 2OH^- \rightleftharpoons SeO_4^{-2} + H_2O + 2e^-$	–0.03
Hg(0)–Hg(II)	$Hg + 2OH^- \rightleftharpoons HgO(r) + H_2O + 2e^-$	–0.0984
Pd(0)–Pd(II)	$Pd + 2OH^- \rightleftharpoons Pd(OH)_2 + 2e^-$	–0.1
Ir(0)–Ir(III)	$2Ir + 6OH^- \rightleftharpoons Ir_2O_3 + 3H_2O + 6e^-$	–0.1
N(I)–N(II)	$N_2O_2^{-2} \rightleftharpoons 2NO + 2e^-$	–0.10
Hg(0)–Hg(I)	$2Hg + 2OH^- \rightleftharpoons Hg_2O + H_2O + 2e^-$	–0.123
N(I)–N(III)	$N_2O + 6OH^- \rightleftharpoons 2NO_2^- + 3H_2O + 4e^-$	–0.15
Pt(0)–Pt(II)	$Pt + 2OH^- \rightleftharpoons Pt(OH)_2 + 2e^-$	–0.16
Cl(V)–Cl(VII)	$ClO_3^- + 2OH^- \rightleftharpoons ClO_4^- + H_2O + 2e^-$	–0.17
Co(II)–Co(III)	$Co(OH)_2 + OH \rightleftharpoons Co(OH)_3 + e^-$	–0.2
I(–I)–I(V)	$I^- + 6OH^- \rightleftharpoons IO_3^- + 3H_2O + 6e^-$	–0.26
Ag(0)–Ag(I)	$2Ag + 2OH \rightleftharpoons Ag_2O + H_2O + 2e^-$	–0.344
Cl(III)–Cl(V)	$ClO_2^- + 2OH^- \rightleftharpoons ClO_3^- + H_2O + 2e^-$	–0.35
O(–II)–O(0)	$4OH^- \rightleftharpoons O_2 + 2H_2O + 4e^-$	–0.401
Ni(II)–Ni(IV)	$Ni(OH)_2 + 2OH^- \rightleftharpoons NiO_2 + 2H_2O + 2e^-$	–0.49
I(–I)–I(I)	$I^- + 2OH^- \rightleftharpoons IO^- + H_2O + 2e^-$	–0.49
Mn(VI)–Mn(VII)	$MnO_4^{-2} \rightleftharpoons MnO_4^- + e^-$	–0.54
I(I)–I(V)	$IO^- + 4OH^- \rightleftharpoons IO_3^- + 2H_2O + 4e^-$	–0.56
Mn(IV)–Mn(VII)	$MnO_2 + 4OH^- \rightleftharpoons MnO_4^- + 2H_2O + 3e^-$	–0.57
Ag(I)–Ag(II)	$Ag_2O + 2OH^- \rightleftharpoons 2AgO + H_2O + 2e^-$	–0.57
Mn(IV)–Mn(VI)	$MnO_2 + 4OH^- \rightleftharpoons MnO_4^{-2} + 2H_2O + 2e^-$	–0.58
Cl(I)–Cl(III)	$ClO^- + 2OH^- \rightleftharpoons ClO_2^- + H_2O + 2e^-$	–0.59

Contd.

Couple	Equation	E^0_{296} volts
Br(–I)–Br(V)	$Br^- + 6OH^- \rightleftharpoons BrO_3^- + 3H_2O + 6e^-$	–0.61
Cl(–I)–Cl(V)	$Cl^- + 6OH^- \rightleftharpoons ClO_3^- + 3H_2O + 6e^-$	–0.62
I(V)–I(VII)	$IO_3^- + 3OH^- \rightleftharpoons H_3IO_6^{-2} + 2e^-$	ca. –0.70
Ag(II)–Ag(III)	$2AgO + 2OH^- \rightleftharpoons Ag_2O_3 + H_2O + 2e^-$	–0.74
Cl(–I)–Cl(III)	$Cl^- + 4OH^- \rightleftharpoons ClO_2^- + 2H_2O + 4e^-$	–0.76
N(I)–N(II)	$N_2O + 2OH^- \rightleftharpoons 2NO + H_2O + 2e^-$	–0.76
Br(–I)–Br(I)	$Br^- + 2OH^- \rightleftharpoons BrO^- + H_2O + 2e^-$	–0.76
O(–II)–O(–I)	$3OH^- \rightleftharpoons HO_2^- + H_2O + 2e^-$	–0.87
N(III)–N(IV)	$2NO_2^- \rightleftharpoons N_2O_4 + 2e^-$	–0.88
Cl(–I)–Cl(I)	$Cl^- + 2OH^- \rightleftharpoons ClO^- + H_2O + 2e^-$	–0.94
Cl(III)–Cl(IV)	$ClO_2^- \rightleftharpoons ClO_2 + e^-$	–1.15
	$O_2 + 2OH^- \rightleftharpoons O_3 + H_2 + 2e^-$	–1.24
O(–II)–O(–I)	$OH^- \rightleftharpoons OH + e^-$	–1.4

9.02 FORMAL POTENTIAL

The standard potential of a redox system is the potential at standard condition where the ions present are considered to be in simple form with full regard to activity effects. During actual analysis these ideal conditions are never met with. The activities of various species differ appreciably from respective concentrations. But in most cases we prepare solutions and conduct experiments in terms of concentrations, not activities. So the standard potentials of different cells and the inference drawn on this basis may often appear anomalous. This becomes more complicated when other electrolytes are present in the reactions and/or interactions like acid-base dissociation, etc. Hence instead of standard potential, formal potential is used which is the potential observed experimentally in a solution containing equal formal concentrations of the oxidised and reduced substances together with other specified substances at specified concentrations.

The activity (a) is an effective or thermodynamic concentrations and is related to molal concentration with some correction factor γ, the activity coefficient, such as $a = \gamma.m$ where a is the activity, γ is the coefficient of activity and m is the molar concentration. For dilute solution, potential can be obtained using Nernst equation $E = E_0 + \frac{0.0591}{n}\log\frac{[a_{Ox}]}{[a_{Red}]}$ at 25°C where E is the potential of the cell, E_0 is the standard potential, and a_{Ox} and a_{Red} are the corresponding activities, so we may write

$$E = E_0 + \frac{0.0591}{n}\log\frac{[\gamma_{Ox}.m_{Ox}]}{[\gamma_{Red}.m_{Red}]}$$

$$= \left\{E_0 + \frac{0.0591}{n}\log\frac{\gamma_{Ox}}{\gamma_{Red}}\right\} + \frac{0.0591}{n}\log\frac{m_{Ox}}{m_{Red}}$$

$$= E_0' + \frac{0.0591}{n}\log\frac{m_{Ox}}{m_{Red}}$$

Here E_0' is the formal potential, i.e., actual potential under experimental conditions. The measurement of formal potential of some species in different acid solutions can be mentioned. If there is no great difference in complexation of the oxidant or its conjugate reductant in various acids, the formal potentials are close to the standard potential and give us indications regarding their behavior. Thus, for Fe^{3+} / Fe^{2+} system, $Cr_2O_7^{2-}$ / Cr^{3+} system and Ce^{4+} / Ce^{3+} system in different acids are as follows:

Table 9.2 : Potentials of the systems in different acid medium

Acid	$Cr_2O_7^{2-}$ / Cr^{3+}	Fe^{3+}/Fe^{2+}
0.1M HCl	0.03	0.73
0.5M HCl	0.07	0.72
1M HCl	1.00	0.70
2M HCl	1.05	0.69
3M HCl	1.08	0.68
4M HCl	1.10	0.66
0.1M H_2SO_4	0.92	0.68
0.5M H_2SO_4	1.08	0.68
1M H_2SO_4	1.03	0.69
2M H_2SO_4	1.11	0.69
4M H_2SO_4	1.15	0.69
8M H_2SO_4	1.35	0.66
0.1M $HClO_4$	0.84	0.74
1.0M $HClO_4$	1.03	0.74
0.5M H_3PO_4+1M H_2SO_4	—	0.61

From these data complexation of Fe^{3+}/Fe^{2+} system, is least in $HClO_4$ and maximum in H_3PO_4. So Fe^{3+} is a powerful oxidising agent in $HClO_4$ and less in H_3PO_4 + H_2SO_4 mixture. $Cr_2O_7^{2-}$ /Cr^{3+} is least oxidant in HCl and maximum in H_2SO_4. For Ce^{4+} / Ce^{3+} system the table shows the ceric perchlorate in perchloric acid is least complexed and highest oxidising agent than that in either H_2SO_4 or HNO_3.

Acid (N)	H_2SO_4	HNO_3	$HClO_4$
1	1.44	1.61	1.70
2	1.44	1.62	1.71
4	1.43	1.61	1.75
6	—	—	1.82
8	1.42	1.56	1.87

The consideration of standard potential data leads us to the erroneous conclusion, which can be illustrated with the following example. The values of E_0 of the following reactions are:

$[Fe(CN)_6]^{3-} + e^- \rightleftharpoons [Fe(CN)_6]^{4-}$, $E_0 = +0.36$ volt

$I_2 + 2e^- \rightleftharpoons 2I^-$, $E_0 = 0.5453$ volt

The oxidant of the higher potential system, I_2 will oxidise the reductant of lower potential system, i.e., $[Fe(CN)_6]^{4-}$ to $[Fe(CN)_6]^{3-}$ and the reaction should proceed as follows:

$I_2 + 2[Fe(CN)_6]^{4-} \rightleftharpoons 2[Fe(CN)_6]^{3-} + 2I^-$

In practice, the reverse reaction takes place. The ferricyanide and ferrocyanide ions react with H^+ at low pH to form hydro-ferrocyanic acid, $H[Fe(CN)_6]^{3-}$, $H_2[Fe(CN)_6]^{2-}$, etc, and hydroferricyanic acid, $H[Fe(CN)_6]^{2-}$, $H_2[Fe(CN)_6]^-$ ions which are very weak acids. Among those, hydroferrocyanic acids are weaker and so, the activity of $[Fe(CN)_6]^{4-}$ decreases to a greater extent, thereby potential of the $[Fe(CN)_6]^{3-}/[Fe(CN)_6]^{4-}$ is much increased. It has been found that E_0' of $[Fe(CN)_6]^{3-}/[Fe(CN)_6]^{4-}$ in 1M HCl, H_2SO_4, or $HClO_4$ is 0.71 volt which goes high enough to I_2/I^- system and hence the reverse reaction takes place.

Thus, formal potential is very useful in analytical chemistry over the standard potential values of the system.

9.03 APPLICATION OF STANDARD POTENTIAL DATA

The redox potential data is very helpful in analytical process. A few applications are given below:

(A) Reaction prediction

In general, the oxidised form of a system will oxidise the reduced form of a system of higher positive potential value, e.g.,

$H_2S \rightleftharpoons 2H^+ + S + 2e^-$, $\quad E_0 = -0.141$ volt

$Cr_2O_7^{2-} + 14H^+ + 6e^- \rightarrow 2Cr^{3+} + 7H_2O$, $\quad E_0 = 1.36$ volts

So, $Cr_2O_7^{2-}$ will oxidise H_2S in acidic solution of 1-2M strength. The complete reaction will be obtained by electron balance

$$Cr_2O_7^{2-} + 8H^+ + 3H_2S = 2Cr^{3+} + 3S + 7H_2O$$

But the standard potential data are more theoretical than practical in character. These data only indicate the energy difference of the systems but bear no relation to reaction kinetics or reaction mechanism. Thus, direct reduction of Cu^{2+} ($E_0 = -0.34$volt) by hydrogen gas may be expected from potential data but not observed due to unfavourable kinetics. Again $Fe \rightarrow Fe^{3+}$, $E_0 = 0.04$ volt is energetically possible by H^+ but the process yields always Fe^{2+} not Fe^{3+} because of unfavourable mechanism.

(B) Combination of half-reactions

The potential value for a reaction can be obtained by combination of half-reactions involved.

I. If the two half-reactions give a complete reaction, then the reduction potentials of the half-reactions are additive algebraically.

II. If the combinations of two half-reactions give a third half-reaction, then the free energy changes characterising half-reactions are additive.

(I) *When two half-reactions give a complete reaction*: In this case the e.m.f. of the cell will be the algebraic sum of the potentials of the half-reactions. Say, two half-reactions are combined, then

(a) $Cu^{2+} + 2e^- \rightleftharpoons Cu$, $E_0 = 0.34$ volt and

$Al^{3+} + 3e^- \rightleftharpoons Al$, $\quad E_0 = -1.67$ volts.

Adding the two half-reactions

$2Al + 3Cu^{2+} \rightleftharpoons 3Cu + 2Al^{3+}$ and the cell will be $Al \mid Al^{3+} \parallel Cu^{2+} \mid Cu$, then

$$E = E_{Ox}\text{ (left)} - E_{Ox}\text{ (right)} = E_{Red}\text{ (right)} - E_{Red}\text{(left)} = 0.34 - (-1.67) = 2.01 \text{ volts.}$$

In writing the reaction, electron balance is necessary and in measuring potential value no such adjustment is necessary. Potential values measure the ease of loss of certain electrons per atom or ion and is independent of total number of atoms or ions of that species.

(b) Let us calculate the cell e.m.f. of the reaction

$$Cu + 2Ag^+ \rightleftharpoons Cu^{2+} + 2Ag$$

$E_0\ (Cu^{2+} \mid Cu) = 0.34$ volt and $E_0\ (Ag^+ \mid Ag) = 0.80$ volt.

The cell can be written as $Cu \mid Cu^{2+} \parallel Ag^+ \mid Ag$

E_0 (cell) = E_0 (red of right) – E_0 (red of left)

$$= E_{0\,(Ag^+ \mid Ag)} - E_{0\,(Cu^{2+} \mid Cu)} = 0.34 - 0.80 = -0.46 \text{ volt.}$$

(c) Calculate the cell e.m.f. of the cell for the reaction:

(i) $Co + Ni^{2+}(1M) \rightleftharpoons Co^{2+}(0.1M) + Ni$, $E_0\,(Co^{2+}/Co) = -0.28$ volt

(ii) $Co + Ni^{2+}(0.01M) \rightleftharpoons Co^{2+}(1M) + Ni$, $E_0\,(Ni^{2+}/Ni) = -0.25$ volt

(i) The cell will be $Co / Co^{2+}\ (0.1M) \parallel Ni^{2+}\ (1M) \mid Ni$

$$E\,(\text{Cell}) = E_{\text{red}(Ni^{2+}/Ni)} - E_{\text{red}(Co^{2+}/Co)}$$

$$E = E_{0\ Ni^{2+}/Ni} + \frac{0.0591}{2}\log\left[Ni^{2+}\right] - \left\{E_{0\ Co^{2+}/Co} + \frac{0.0591}{2}\log\left[Co^{2+}\right]\right\}$$

$$= -0.25 + 0.28 + \frac{0.0591}{2}\log\frac{\left[Ni^{+2}\right]}{\left[Co^{2+}\right]}$$

$$= 0.03 + \frac{0.0591}{2}\log\frac{1}{0.1}$$

$$= 0.03 + 0.029 = 0.059 \text{ volt.}$$

(ii) $$E = -0.25 + 0.28 + \frac{0.0591}{2}\log\frac{\left[Ni^{2+}\right]}{\left[Co^{2+}\right]}$$

$$= 0.03 + \frac{0.0591}{2}\log\frac{0.01}{1}$$

$$= 0.03 + \frac{0.0591}{2}\times -2$$

$$= 0.03 - 0.059 = -0.039 \text{ volt.}$$

The negative sign indicates that the cell reaction is not spontaneous and the cell reaction becomes spontaneous if concentration ratio of the two ions is changed.

(II) When the two half-reactions are combined to give a third half-reaction, then not the potential but the free energy changes characterising half-reaction are additive. The potential of the third half-reaction is then calculated from the energy change by the relation: $\Delta G^0_{298} = -nE^0_{298}F$. Here, n is the number of electron change in the half-reaction and F is the Faraday (F = 23066 calories per volt).

Let us calculate the potential of Mn^{3+}/Mn^{4+} system in the reaction

$Mn^{3+} + 2H_2O \rightleftharpoons MnO_2 + 4H^+ + e^-$

From the given reaction

(a) $Mn^{2+} + 2H_2O \rightleftharpoons MnO_2 + 4H^+ + 2e^-$, $E_0 = -1.28$ volt and

(b) $Mn^{2+} \rightleftharpoons Mn^{3+} + e^-$, $E_0 = -1.51$ volt.

The corresponding free energy change for the reaction (a) and (b) are:

(a) Mn^{2+}/Mn^{4+} $\Delta G^0_{298} = -2\ (-1.28)\ F = 2.56\ F$

(b) Mn^{2+}/Mn^{3+} $\Delta G^0_{298} = -1(-1.51)\ F = 1.51\ F$

Subtracting half-reaction (b) from (a) we get the half-reaction for Mn^{3+}/Mn^{4+} system and free energy change is (2.56 F – 1.51 F) = 1.05 F.

So potential for this is 1.05 F /1 F = 1.05 volt.

Another example may be stated here. We know the potential for the half-reactions as given

(a) $Tl \rightleftharpoons Tl^+ + e^-$, $E_0 = 0.3363$ volt

(b) $Tl^+ \rightleftharpoons Tl^{3+} + 2e^-$, $E_0 = -1.250$ volt

From these data we can calculate the potential for the system $Tl^0 \rightleftharpoons Tl^{3+} + 3e^-$ which is obtained by adding (a) and (b).

The free energy changes for (a) and (b) are obtained as

$\Delta G^0_{298} = -1\times (0.3363)\ F = -0.3363\ F$ and $\Delta F^0_{298} = -2\ (-1.250)\ F = 2.50\ F$

Adding these values we get (–0.3363 F + 2.50 F) = 2.1637 F

So, the e.m.f. for Tl/Tl^{3+} will be 2.1637 F/3 F = 0.7212 volt.

(C) Determination of equilibrium constant of a reaction

A redox reaction is reversible in nature and so the equilibrium is established after some time. The value of this equilibrium constant can be calculated from the reduction potential of the system and thus we can ascertain whether such reaction can find applications in quantitative analysis.

Let us take the reaction between Fe^{3+} ion and Sn^{2+} ion. The reaction can be written as

$2\ Fe^{3+} + Sn^{2+} = 2\ Fe^{2+} + Sn^{4+}$ and

$$K_{eq} = \frac{[Sn^{4+}][Fe^{2+}]^2}{[Sn^{2+}][Fe^{3+}]^2}$$

The equations for the potential of the systems are

$$E_{Fe^{3+}/Fe^{2+}} = 0.77 + \frac{0.0591}{1}\log\frac{[Fe^{3+}]}{[Fe^{2+}]} \quad \text{and}$$

$$E_{Sn^{4+}/Sn^{2+}} = 0.15 + \frac{0.0591}{2}\log\frac{[Sn^{4+}]}{[Sn^{2+}]}$$

Initially, the solution contains only Fe^{3+} and so potential of Fe^{3+}/Fe^{2+} is maximum and that for Sn^{4+}/Sn^{2+} is minimum. As the reaction proceeds $[Fe^{3+}]$ and $[Sn^{2+}]$ decrease whereas $[Fe^{2+}]$ and $[Sn^{4+}]$ increase resulting in the decrease in potential of Fe^{3+}/Fe^{2+} system and increase in the potential of Sn^{4+}/Sn^{2+} system. When the equilibrium is established, the two potentials will be equal and no transfer of electron will take place. So we can write

$$E_{Fe^{3+}/Fe^{2+}} = 0.77 + \frac{0.0591}{1}\log\frac{[Fe^{3+}]}{[Fe^{2+}]} = E_{Sn^{4+}/Sn^{2+}} = 0.15 + \frac{0.0591}{2}\log\frac{[Sn^{4+}]}{[Sn^{2+}]}$$

$$\text{or, } (0.77 - 0.15) = \frac{0.0591}{2}\log\frac{[Sn^{4+}]}{[Sn^{2+}]} - \frac{0.0591}{1}\log\frac{[Fe^{3+}]}{[Fe^{2+}]}$$

$$= \frac{0.0591}{2}\log\frac{[Sn^{4+}]}{[Sn^{2+}]} - \frac{0.0591}{2}\log\frac{[Fe^{3+}]^2}{[Fe^{2+}]^2}$$

$$= \frac{0.0591}{2} \log \frac{[Sn^{4+}][Fe^{3+}]^2}{[Sn^{2+}][Fe^{2+}]^2}$$

$$= \frac{0.0591}{2} \log K_{eq}$$

$\log K_{eq} = 2(0.77 - 0.15)/0.059 = 1.24/0.059 \approx 21$ and $K_{eq} = 10^{21}$.

So, at equilibrium, the concentration of Sn^{2+} and Fe^{3+} are very negligible, i.e., the reaction goes almost to completion. The general form of the equilibrium constant K is $\log K = n\,(E_0' - E_0'')/0.059$. This indicates that greater the difference between the standard potential of the system, greater will be the equilibrium constant and more complete will be the reaction. This statements can be justified with the following example: At the equivalence point of the reaction between Sn^{2+} and Fe^{3+}, the $\frac{1}{2}[Fe^{3+}] = [Sn^{2+}]$ or $[Fe^{3+}] = 2[Sn^{2+}]$ and similarly, $2[Sn^{4+}] = [Fe^{2+}]$, and hence we can write

$$\frac{[Fe^{2+}]}{[Fe^{3+}]} = \frac{[Sn^{4+}]}{[Sn^{2+}]}$$

$$\text{Since, } K = \frac{[Fe^{2+}]^2[Sn^{4+}]}{[Fe^{2+}][Sn^{2+}]} = \frac{[Fe^{2+}]^3}{[Fe^{3+}]^3} = 10^{21}$$

$\because \frac{[Fe^{2+}]}{[Fe^{3+}]} = 10^7$, i.e., $[Fe^{3+}]$ is very negligible compared to $[Fe^{2+}]$ and so, the reaction is a complete one.

When the potential difference between the systems is not large, the reaction is not complete, but these reactions can be used in analytical purpose by either changing concentrations or reaction medium such that the reaction proceeds towards completion as far as practicable, e.g.,

$$AsO_4^{3-} + 2I^- + 2H^+ \rightleftharpoons AsO_3^{3-} + I_2 + H_2O$$

The potential of the system is $E_{0\,AsO_4^{3-}/AsO_3^{3-}} = 0.57$ and E_0 of $I_2/I^- = 0.54$ and this reaction is carried out in large excess of I^- and high $[H^+]$.

(D) Stabilization of oxidation states of metals

Relative stability of different oxidation states of metal ions can be understood from its half-cell potential value, e.g., the half-cell reactions of Au^+ are

$$Au^+ \rightleftharpoons Au^{3+} + 2e^-, \quad E_0 = -1.41 \text{ volt}$$

$$Au \rightleftharpoons Au^+ + e^-, \quad E_0 = -1.68 \text{ volt}$$

When the difference in potential value of two half-reactions is low, the metal ion may undergo disproportionation reaction as $3Au^+ \rightleftharpoons 2Au + Au^{3+}$.

This may occur if the ions are in simple form. Solvent molecules always surround these ions. The number of solvent molecules that surround the ion, their type of attachment to the ion, and the distance of closest approach to the ion depend on the size, charge, and electronic configuration of the ion. Thus, by changing the environment, the potential can be changed and hence stability for Au^+,

$$Au + H_2O \rightleftharpoons Au(aq)^+ + e^-, \quad E_0 = -1.68 \text{ volt}$$

$$Au + 2Br^- \rightleftharpoons [AuBr_2]^- + e^-, \quad E_0 = -0.96 \text{ volt}$$

$Au + 2CNS^- \rightleftharpoons [Au(CNS)_2]^- + e^-$, $E_0 = -0.69$ volt

$Au + 2CN^- \rightleftharpoons [Au\,(CN)_2]^- + e^-$, $E_0 = 0.60$ volt

This shows that Au^+ is stabilised with CN^- forming complex ion and will not undergo dispropotionation. Similarly, the potential data in the Table 9.3 indicate that OH^-, CN^-, F^-, PO_4^{3-} can best stabilise Fe^{3+} and organic complexing agents can stabilise Fe^{2+}.

9.04 FACTORS CONTROLLING THE REDOX POTENTIAL

The standard potential of a half-reaction expressed by Nernst Equation includes unit activities of the constituents, but in most cases the activities are other than unity and hence standard potentials change. This can be illustrated with examples:

(a) Effect of concentration

Suppose, in Fe^{3+} / Fe^{2+}system, the potential is

$E = E_0 + \frac{0.0591}{1}\log\frac{[Fe^{3+}]}{[Fe^{2+}]}$ when $[Fe^{3+}] = [Fe^{2+}]$, then $E = E_0$. If, say $[Fe^{3+}] = 1$ gm ion per litre and $[Fe^{2+}] = 1\times10^{-4}$ gm ion per litre, then $E = 0.77 + \frac{0.0591}{1}\log\frac{1}{10^{-4}} = 1.002$ volt. Let us explain what difficulties we face during titration of Fe^{2+} with $Cr_2O_7^{2-}$ in 1-2N H_2SO_4 medium. Say, we titrate 10 cc of 0.1M Fe^{2+} solution in 2(N)H_2SO_4 using, say 0.1 M $K_2Cr_2O_7$ solution with barium diphenylamine sulphonate indicator ($E_0 = 0.79$) total volume = 100 cc and gradual addition of $K_2Cr_2O_7$ changes the potential as given below:

$K_2Cr_2O_7$	Excess in ml Fe^{2+}	$Cr_2O_7^{2-}$	$[Fe^{3+}]$ / $[Fe^{2+}]$	$[Cr_2O_7^{2-}]$ / $[Cr^{3+}]$	Potential
50	50	—	50:50	—	0.77
91	9	—	91:9 ≈ 10	—	0.828
99	1	—	99:1 ≈ 100	—	0.886
99.9	0.1	—	99.9:0.1 ≈ 1000	—	0.994
100	—	—	—	—	1.387
100.1	—	0.1	—	0.1:100 = 0.001	1.475
101	—	1	—	100:1 ≈ 0.01	1.487

Thus, with addition of $K_2Cr_2O_7$, more and more Fe^{2+} is oxidised to Fe^{3+}, the cell e.m.f. gradually increases, before equivalence point; the indicator gets oxidised and the colour changes to bluish violet. So to get exact equivalence point H_3PO_4 or NaF is used to form colourless complex $[Fe(HPO_4)]^+$ or $[FeF_6]^{-3}$ such that effective $[Fe^{3+}]$ is not allowed to increase.

(b) Effect of pH of the medium

As stated earlier, the half-reactions where H^+ or OH^- is not used up, their standard potential values remain constant throughout the entire pH range. But if H^+ or OH^- is involved in the half-reaction, their concentrations appear in the Nernst equation and so the potential is influenced by pH of the medium, e.g., in $[MnO_4^-]$/ $[Mn^{2+}]$ system, the reaction is

$MnO_4^- + 8H^+ + 5e^- \rightarrow Mn^{2+} + 4H_2O$

$$E = E_0 + \frac{0.0591}{5}\log\frac{[MnO_4^-][H^+]^8}{[Mn^{2+}]}$$

$$= E_0 + \frac{0.0591}{5}\log\frac{[MnO_4^-]}{[Mn^{2+}]} + \frac{0.0591}{5}\log[H^+]^8$$

$$= E_0 + \frac{0.0591}{5}\log\frac{[MnO_4^-]}{[Mn^{2+}]} + \frac{0.0591}{5}\times 8\log[H^+]$$

$$= E_0 + \frac{0.0591}{5}\log\frac{[MnO_4^-]}{[Mn^{2+}]} - \frac{0.0591}{5}\times 8\text{pH}$$

So, maximum oxidation power of MnO_4^- is obtained at minimum pH of the solution, i.e., higher acid concentration. In permanganometric or dichromatometric reactions, when $[H^+]$ is raised to higher power, maximum oxidising power is obtained using acids of 1-2N strength with N/10, N/100 and still lower acid concentrations; $E < E_0$ (1.51), e.g., at pH = 6, E = 0.95.

MnO_4^- (E=1.51volts) can easily oxidise $Cl^- \rightarrow Cl_2$, $Br^- \rightarrow Br_2$ and $I^- \rightarrow I_2$ at low pH. At pH 6, MnO_4^- cannot oxidise Br^-or Cl^- whose potential is higher than that of MnO_4^-/Mn^{2+} system at this pH.

Another example is the reaction between AsO_4^{3-}/AsO_3^{3-} and I_2/I^- system. The half-reactions are

$AsO_4^{3-} + 2H^+ + 2e^- = AsO_3^{3-} + H_2O$, $E_0 = 0.56$ volt

$I_2 + 2e^- = 2I^-$, $E_0 = 0.54$ volt

$$\text{At } 25°\text{C}, \quad E_{AsO_4^{3-}/AsO_3^{3-}} = 0.56 + \frac{0.0591}{2}\log\frac{[AsO_4^{3-}][H^+]^2}{[AsO_3^{3-}]}$$

$$= 0.56 - 0.0591\,\text{pH} + \frac{0.0591}{2}\log\frac{[AsO_4^{3-}]}{[AsO_3^{3-}]}$$

At higher acid concentration, say in 1-2 M acid strength [pH $\approx$ 0] and then only AsO_4^{3-} can oxidize $I^- \rightarrow I_2$. (This technique is followed in the identification of AsO_4^{3-} in presence of PO_4^{3-} or AsO_3^{3-}.) But at high pH, say at pH = 8,

$$E = 0.56 - 0.0591\times 8 + \frac{0.0591}{2}\log\frac{[AsO_4^{3-}]}{[AsO_3^{3-}]}$$

$$= 0.088 + \frac{0.0591}{2}\log\frac{[AsO_4^{3-}]}{[AsO_3^{3-}]}$$

This value is much lower than that of I_2/I^- system. Hence, I_2 will then oxidize AsO_3^{3-} to AsO_4^{3-}, i.e., the reverse reaction occurs. In practice, $NaHCO_3$ is used that maintains the pH to 8. If stronger alkalis like NaOH, Na_2CO_3 are used, then pH > 9 and I_2 will react with strong alkali giving IO^-. This can be generalised as, when H^+ is used up in the half-reaction, the reaction is to be carried out in strong acid medium, and when H^+ is produced in the half-reaction, then H^+ produced must be neutralized by $NaHCO_3$, etc.

Another example is the identification of NO_2^-.

In neutral or alkaline medium, $E_{NO_2^-/NO}$ is $NO_2^- + H_2O + e^- = NO + OH^-$, $E = 0.46$ volt.

This value is lower than that of E_{I_2/I^-} system. So, if KI is added to KNO_2 solution (pH = 7), no noticeable change takes place. If few drops of HCl or H_2SO_4 are added, vigorous reaction takes place with evolution of NO and precipitation of dark grey iodine.

$$2NO_2^- + 2I^- + 4H^+ \rightleftharpoons I_2\downarrow + 2NO\uparrow + 2H_2O$$

In acid medium, the standard potential for NO_2^- / NO

$HNO_2 + H^+ + e^- = NO + H_2O$, $E_0 = 0.99$ volt; it is higher than that of I_2/I^- system.

(c) Change of standard potential with precipitation

This effect is also due to change of concentration of the oxidised or reduced form from the reaction medium, e.g., iodometric estimation of copper $E_{Cu^{2+}/Cu^+} = 0.15$ volt and for $E_{I_2/I^-} = 0.54$, $S_{CuI} = 10^{-12}$.

The reaction that takes place in this case, $2Cu^{2+} + 4I^- \rightarrow 2CuI + I_2$

From the solubility product value of CuI,

$$S_{CuI} = [Cu^+][I^-] = 10^{-12}, \quad \therefore [Cu^+] = \frac{10^{-12}}{[I^-]}$$

$$\text{So, } E_{Cu^{2+}/Cu^+} = E_0 + \frac{0.0591}{1}\log\frac{[Cu^{2+}]}{[Cu^+]}$$

$$= 0.15 + \frac{0.0591}{1}\log\frac{[Cu^{2+}]}{\frac{10^{-12}}{[I^-]}} = 0.15 + \frac{0.0591}{1}\log\frac{[Cu^{2+}][I^-]}{10^{-12}}$$

$$= 0.15 + \frac{0.0591}{1}\times 12 + \frac{0.0591}{1}\log[Cu^{2+}][I^-]$$

$$= 0.86 + \frac{0.0591}{1}\log[Cu^{2+}][I^-]$$

The formal potential of Cu^{2+} / Cu^+ is now high enough than that of I_2/I^- system, and Cu^{2+} will quantitatively liberate, I_2 that can be titrated by $S_2O_3^{2-}$ and hence Cu^{2+} can be estimated.

Let us see now whether Cu^{2+} is estimated quantitatively. Say, 25 ml of 0.2 M Cu^{2+} is mixed with 25 ml of 1M I^-. Now, the volume is doubled, so strength is halved. $[Cu^{2+}]$=0.1M and $[I^-]$ = 0.5M. After reaction, say, x M Cu^{2+} is left at the equilibrium.

$$\log K = \frac{1(0.86 - 0.54)}{0.0591} = \frac{0.32}{0.0591} = 5.3 \quad \therefore K = 1.6 \times 10^5$$

At equilibrium, $[I^-]$ is the difference of its concentration initially less I^- consumed by Cu^{2+}.

$$[I^-] = \{0.5 - 2(.1 - x)\}\text{M} = (0.5 - 0.2 + 2x)\text{M} = (0.3 + 2x)\text{M} \approx 0.3\text{M}$$

$$K = \frac{[I^-][CuI]}{[Cu^{2+}][I^-]} \quad \text{or,} \quad 1.6 \times 10^5 = \frac{1}{[Cu^{2+}]\,[I^-]}$$

$$\text{or, } [Cu^{2+}] = x = \frac{1}{(0.3)^3 \times 1.6 \times 10^5} = 6.9 \times 10^{-5} \text{ gm ion/lit.}$$

So, the reaction is complete and estimation is quantitative.

Another example can be discussed. The E_0 for $[Fe\,(CN)_6]^{3-}$ /$[Fe\,(CN)_6]^{4-}$ system is 0.36 volt which is much below I_2/I^-. So ordinarily, $Fe\,(CN)_6]^{3-}$ will not oxidize I^- to I_2.

We have, $$E_{0[Fe(CN)_6]^{3-}/[Fe(CN)_6]^{4-}} = 0.36 + \frac{0.0591}{1}\log\frac{[Fe(CN)_6]^{3-}}{[Fe(CN)_6]^{4-}}$$

In neutral or slightly acid medium, if $ZnSO_4$ solution is added, the ferrocyanide ion reacts with Zn^{2+} as follows:

$2[Fe(CN)_6]^{4-} + 2K^+ + 3Zn^{2+} = K_2Zn_3\,[Fe(CN)_6]_2$

So, as long as Zn^{2+} is present in solution, $K_2Zn_3\,[Fe(CN)_6]_2$ is precipitated, thus, removing ferrocyanide ion from solution, potential of ferricyanide will increase and will oxidise I^- ion. The liberated iodine is estimated by $Na_2S_2O_3$ solution and hence Zn^{2+} can be estimated.

1ml N $Na_2S_2O_3$ = 0.3293 gm $K_3[Fe(CN)_6]$ = 0.0982 gm Zn

In practice, $ZnSO_4$ is taken in a conical flask, starch solution is added as indicator, KI solution is added, then 2 cc of $K_3[Fe(CN)_6]$ is added at a time and the liberated iodine is estimated by standard $Na_2S_2O_3$ solution when no more I_2 is liberated, all Zn^{2+} has been precipitated and that will be the end point of the titration.

(d) Change of potential due to complex formation

Sometimes, addition of some complexing agent may form complex ion of any of the oxidised or reduced form decreasing the effective concentration or activity of any one species and hence may change the standard potential of the system. The titration of $Fe^{2+}$with $Cr_2O_7^{2-}$ in 1N H_2SO_4 using barium diphenylamine sulphonate indicator may be discussed here. As stated before (in the Section 9.04 change of E_0 with concentration), addition of NaF removes Fe^{3+} as $[Fe\,F_6]^{3-}$ or $[FeF]^{2+}$, $[FeF_2]^+$, etc., For the simple reaction

$$Fe^{3+} + F^- \rightleftharpoons [FeF]^{2+}, \qquad K_{\text{instablity}} = 10^5 = \frac{[FeF]^{2+}}{[Fe^{3+}][F^-]}$$

$$\text{or, } [Fe^{3+}] = \frac{[FeF]^{2+}}{10^5 \times [F^-]}$$

$[FeF_6]^{3-}$ is less stable and paramagnetic due to half-filled high spin *d*-shell but ionisable as it is outer orbital complex.

The potential for Fe^{3+}/Fe^{2+} system now becomes

$$E = E_0 + \frac{0.0591}{1}\log\frac{[Fe^{3+}]}{[Fe^{2+}]}$$

$$= 0.77 + \frac{0.0591}{1} \log \frac{\{[FeF]^{2+}\}}{10^5 \times [F^-][Fe^{2+}]}$$

$$= 0.77 + \frac{0.0591}{1} \log 10^{-5} + \frac{0.0591}{1} \log \frac{\{[FeF]^{2+}\}}{[F^-][Fe^{2+}]}$$

$$= 0.472 + \frac{0.0591}{1} \log \frac{\{[FeF_2]^{2+}\}}{[F^-][Fe^{2+}]}$$

This potential goes below that of I_2/I^- system and hence from a mixture of Cu^{2+} and Fe^{2+}/Fe^{3+} direct estimation of Cu^{2+} can be done using NaF or NH_4HF_2 in excess. This also masks the colour of Fe^{3+} and so clear end point can be obtained.

Another example is the displacement of zinc from zinc salt solution by copper. Their reduction potentials are $Zn^{2+}/Zn = 0.76$ and $Cu^+/Cu = 0.52$. Ordinarily, Zn displaces Cu from copper salt solution. When excess KCN is added to this solution, both Cu^{2+} and Zn^{2+} form complexes with CN^-, but $[Cu(CN)_4]^{3-}$ is most stable, where Cu is in +1 state, and decreasing the concentration of Cu^+ and E_{Cu^+/Cu^0} then falls below that of Zn^{2+}/Zn system and the reaction occurs.

$Zn + 4CN^- \rightleftharpoons [Zn(CN)_4]^{2-} + 2e^-$, $E_0 = 1.26$ $K_{instability}$ of $[Cu(CN)_4]^{3-} = 5\times10^{-28}$

$Cu + 4CN^- \rightleftharpoons [Cu(CN)_4]^{3-} + e^-$, $E_0 = 0.43$ $K_{instability}$ of $[Zn(CN)_4]^{2-} = 2\times10^{-17}$

Another example can be cited with Co^{3+}/Co^{2+} system.

$Co^{3+}(aq) + e \rightleftharpoons Co^{2+}(aq)$, $E_0 = 1.842$ volt

So, oxidation of $Co^{2+} \rightarrow Co^{3+}$ is very difficult. When complexing agent like NH_3 or CN^- is added to Co^{2+} salt solution, the process takes place immediately even in presence of atmospheric oxygen and potential falls to 0.1.volt.

$[Co(NH_3)_6]^{3+} + e^- \rightleftharpoons [Co(NH_3)_6]^{2+}$, $E_0 = 0.1$ volt

NH_3 is a strong ligand in the spectrochemical series. In its presence, quenching of *d* orbital of cobalt takes place, t_{2g} is more established than the e_g state, and the Co^{2+} has $t_{2g}{}^6 e_g{}^1$.

Now, NH_3 being a strong ligand, forces the $e_g{}^1$ electron to remove to higher energy 4*d* state and is removed. So, instability constant for $Co(NH_3)_6^{3+}$ complex is very low compared to that of $[Co(NH_3)_6]^{2+}$ and $[Co^{3+}] << [Co^{2+}]$. Hence, $E_{0\,Co^{3+}/Co^{2+}}$ decreases.

The oxidation potential data of some complex ions are given in the table.

$K_{instability}$ of $[Co(NH_3)_6]^{2+} = 1.25 \times 10^{-5}$

$K_{instability}$ of $[Co(NH_3)_6]^{3+} = 6 \times 10^{-36}$

Table 9.3 : Standard reduction potentials of some metal ions showing the effect of complex formation (aqueous medium, 25 °C)

Couple	E_o volt
$Fe(H_2O)_6^{3+}/Fe(H_2O)_6^{2+}$	0·77
$Fe(phen)_3^{+3}/Fe(phen)_3^{2+}$	1·14
$Fe(bipy)_3^{+3}/Fe(bipy)_3^{2+}$	~ 1·1

Contd.

Couple	E_0 volt
$FeF_6^{3-}/Fe(H_2O)_6^{2+} + 6F^-$	> 0·4
$Fe(CN)_6^{3-}/Fe(CN)_6^{4-}$	0·36
Fe (edta)$^-$/Fe (edta)$^{2-}$	0·12 (pH, 4-6)
$Fe(OH)_3/Fe(OH)_2 + OH^-$	– 0·56
$Co(H_2O)_6^{3+}/Co(H_2O)_6^{2+}$	1·84
Co (edta)$^-$/Co (edta)$^{2-}$	0·60
$Co(ox)_3^{3-}/Co(ox)_3^{4-}$	0·57
$Co(OH)_3/Co(OH)_2 + OH^-$	0·17
$Co(NH_3)_6^{3+}/Co(NH_3)_6^{2+}$	0·10
$Co(en)_3^{3+}/Co(en)_3^{2+}$	–0·26
$Co(CN)_6^{3-}/Co(CN)_5^{3-} + CN^-$	–0·83
$Cu(H_2O)_6^{2+}/Cu^+$ (aq)	0·15
$Cu(NH_3)_4^{2+}/Cu(NH_3)_2^+ + 2NH_3$	–0·01
Cu^{2+}(aq) + $Cl^-/CuCl$	0·54
Cu^{2+} (aq) + I^-/CuI	0·86
Cu^{2+} (aq) + $2CN^-/Cu(CN)_2^-$	1·10
Ag^+/Ag	0.80
$Ag(NH_3)_2^+/Ag + 2NH_3$	0·37
$Ag(S_2O_3)_2^{3-}/Ag + 2S_2O_3^{2-}$	0·02
$Ag(CN)_2^-/Ag + 2CN^-$	–0.31

9.05 DISPROPORTIONATION REACTION AND COMPROPORTIONATION REACTION

The disproportionation reaction is a self oxidation-reduction reaction in which an element or an ion in its intermediate oxidation state spontaneously changes to its stable higher and/or lower oxidation states. So, a part of the atom or an ion is oxidised and other part is reduced, both taking place to an equivalent extent. The examples are:

(a) In neutral aqueous medium

(i) $3Au^+ \rightleftharpoons Au^{3+} + 2Au^0$

(ii) $\overset{+}{Cu_2}SO_4 \rightleftharpoons \overset{0}{Cu} + \overset{+2}{Cu}SO_4$

(b) In acid medium

$$3\overset{+6}{Mn}O_4^{2-} + 4H^+ \rightleftharpoons 2\overset{+7}{Mn}O_4^- + \overset{+4}{Mn}O_2 + 2H_2O$$

(c) In cold dilute alkali medium

$$\overset{0}{Cl_2} + 2OH^- \rightleftharpoons Cl^- + (O\overset{+}{Cl})^- + 2Cl^-$$

$$3(O\overset{+}{Cl})^- \rightleftharpoons \overset{+5}{Cl}O_3^- + 2Cl^-$$

(d) In strong alkaline medium

$$\overset{0}{P_4} + 3NaOH + 3H_2O \rightleftharpoons \overset{-3}{P}H_3 + 3NaH_2\overset{+}{P}O_2$$

(e) Heating in solid state

(i) $4K\overset{+5}{Cl}O_3 \rightleftharpoons 3K\overset{+7}{Cl}O_4 + K\overset{-1}{Cl}$

(ii) $4NaH_2\overset{+}{P}O_2 \rightleftharpoons Na_4\overset{+5}{P_2}O_7 + 2\overset{-3}{P}H_3 + H_2O$

(f) Heating or catalytic decomposition

$2H_2\overset{-}{O_2} \rightleftharpoons 2H_2O^{-2} + \overset{0}{O_2}$

All the above reactions can be balanced considering change in their oxidation numbers. As for example, the reaction c of the above is shown below:

c(I) (i) $\overset{0}{Cl_2} - 2e^- \rightleftharpoons 2Cl^+$ or, $\overset{0}{Cl_2} + 4OH^- - 2e^- \rightleftharpoons 2O\overset{+1}{Cl^-} + 2H_2O$

(ii) $\overset{0}{Cl_2} + 2e^- \rightleftharpoons 2Cl^-$ i.e., $Cl_2 + 2e^- \rightleftharpoons 2Cl^-$ on adding (i) + (ii) gives,

$2\overset{0}{Cl_2} + 4OH^- \rightleftharpoons 2O\overset{+}{Cl^-} + 2Cl^- + 2H_2O$ or, $\overset{0}{Cl_2} + 2OH^- \rightleftharpoons O\overset{+}{Cl} + Cl^- + H_2O$

c(II) $Cl^+ \rightarrow Cl^{5+} + 4e^-$

$2(Cl^+ + 2e^- \rightarrow Cl^-)$ on adding these two $3\,Cl^+ \rightleftharpoons Cl^{5+} + 2\,Cl^-$ so the ionic equation will be

$3OCl^- \rightleftharpoons ClO_3^- + 2Cl^-$

The equilibrium constants of these reactions are high. These reactions can be stopped or checked by changing the concentrations of the reactants or products by neutralization, complexation or precipitation. It has been found that if the difference of E_0 values of two successive oxidation states of an element or an ion is 0.3-0.4 volt, they may undergo disproportionation reaction. e.g.,

Cu^0 —— 0.5 —— Cu^+ —— 0.15 —— Cu^{2+}

(Cu^0 to Cu^{2+}: 0.35)

E_0 for Cu / Cu^+(0.5) and Cu^+ / Cu^{2+}(0.15), the difference is 0.35volt and so, disproportionation reaction occurs. If Cl^- is present in the solution, Cu_2Cl_2 ($S_{Cu_2Cl_2} = 10^{-7}$) is precipitated, $[Cu^+]$ is lowered and disproportionation reaction is checked. But in presence of CN^-, Cu^+ forms stable $[Cu(CN)_4]^{-3}$ complex [$K_{inst}\,Cu(CN)_4^{3-} = 10^{-30}$] and $[Cu^+]$ is so much reduced and disproportionation reaction stops.

The reduction potential diagram (Latimer diagram) of manganese system can be given as

MnO_4^- —— 0.56 —— MnO_4^{2-} —— 2.26 —— MnO_2 —— 0.95 —— Mn^{+3} —— 1.51 —— Mn^{+2} —— 1.18 —— Mn

(MnO_4^- to MnO_2: 1.70; MnO_2 to Mn^{+2}: 1.23; MnO_4^- to Mn^{+2}: 1.51)

From this diagram it can be said that both Mn^{3+} and MnO_4^{2-} undergo disproportionation reaction.

$Mn^{3+} - e \rightarrow MnO_2^-$, $E_{Ox} = -0.95$ V, $\Delta G_0^- = 0.95$ F

$Mn^{3+} + e \rightarrow Mn^{2+}$, $E_{Red} = 1.51$ V, $\Delta G_0^- = -1.51$ F

The difference in free energy is –0.56 F; so changes take place. For MnO_4^- systems

$2MnO_4^{2-} - 2e = 2MnO_4^-$, $E_{Ox} = -0.56$ V, $\Delta G_0^- = 1.12$ F

$MnO_4^{2-} + 2e = MnO_2$, $E_{Red} = 2.26$ V, $\Delta G_0^- = -4.52$ F

Adding we get, change in free energy is –3.4 F and so disproportionation reaction occurs. Comproportionation reaction is reverse reaction of the disproportionation reaction.

Here, two different oxidation states of an ion or an element in two different compounds react to give one compound having one oxidation state, e.g.,

$$\overset{+2}{Cu}Cl_2 + \overset{0}{Cu} \rightleftharpoons \overset{+}{Cu_2}Cl_2\downarrow \qquad \overset{-3}{N}H_4\overset{+3}{N}O_2 = \overset{0}{N_2} + 2H_2O$$

$$\overset{+4}{S}O_2 + 2H_2\overset{-2}{S} \rightleftharpoons 3\overset{0}{S}\downarrow + 2H_2O \qquad \overset{-3}{N}H_4\overset{+5}{N}O_3 = \overset{+1}{N_2}O + 2H_2O$$

By passing H_2S gas slowly through cold saturated solution of SO_2 at 0°C (Wakenroder's solution) dithionic acid, $H_2S_2O_6$ is formed.

$$H_2\overset{-2}{S} + 3H_2\overset{+4}{S}O_3 \rightleftharpoons H_2\overset{+5}{S_2}O_6 + 3H_2O$$

9.06 REDOX INDICATOR

It has been discussed in the acid-base titration section that the indicators undergo sharp colour change at the pH near the equivalence point of an acid-base titration. Likewise, a redox indicator indicates by its sudden colour change in the redox potential of a redox titration near the equivalence point. E_0 value of the indicators will be intermediate between the E_0 value of the solution to be titrated and that of the titrant.

Thus, redox indicators are mostly organic compounds capable of undergoing reversible oxidation or reduction in the titration medium, having distinct and sharp colour change between the oxidised or reduced form with slight variation of the potential. Indicators are of two types: (a) internal indicator and (b) external indicator. Again, some redox titration can be done without adding indicator, e.g., titration with $KMnO_4$ and I_2. In permanganometric titration, as long as reducing agent is present in the titrating solution, $KMnO_4$ will not impart any colour, but when all reducing agents are removed, addition of single drop imparts purple colour. So, $KMnO_4$ may be called self- or auto-indicator.

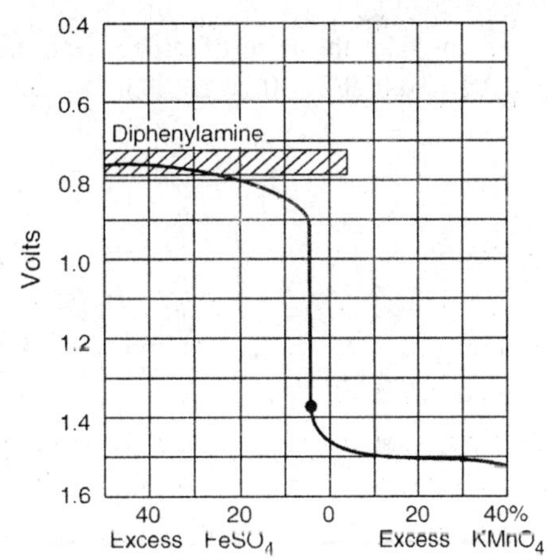

Fig. 9.2 Curve for titration of $FeSO_4$ solution with permanganate (at $[H^+] = 1$).

Similarly, reducing agents can be titrated with I_2 solution when dark brown iodine solution becomes colourless due to the change, $I_2 + 2e^- \rightleftharpoons 2I^-$. Towards the end point, dark brown colour becomes fade and so starch solution is used to form a deep blue adsorption complex with iodine. Titration is complete when this deep blue colour disappears with single drop. (Care must be taken at this stage, starch is to be added when concentration of iodine is low in solution, otherwise drifted end point is obtained.)

Internal indicators are added in the solution to be titrated. As for example, diphenylamine, barium diphenylamine sulphonate, etc., are used during the titration of Fe^{2+} with $K_2Cr_2O_7$ solution in 1.5 N acid medium. Here, solution changes from light greenish to bluish violet colour with slight variation of potential.

External redox indicators are not added in the solution to be titrated, e.g., titration of Fe^{2+} with $K_2Cr_2O_7$ as stated above. Freshly prepared potassium ferricyanide solution, taken as an external indicator, is taken on the spot plate. Towards the end point of titration, one drop of the solution is added to ferricyanide solution on the spot plate when blue colour does not appear, titration is complete. External indicators now are not in use as satisfactory internal indicators are available.

Choice of indicator:

The indicators will behave as a reversible redox system indicated by the reaction,

$In_{Ox} + ne^- \rightleftharpoons In_{Red}$

The Nernst equation at potential E can be written as,

$E = E_{0\,In} + \frac{0.0591}{n}\log\frac{[In_{Ox}]}{[In_{Red}]}$ where $E_{0\,In}$ is the standard potential of the indicator.

The colour of the solution depends upon the ratio $[In_{Ox}]/[In_{Red}]$. It has been found that if the oxidised form is coloured, then intensity of colour of the solution depends when $[In_{Ox}] \approx 10[In_{Red}]$. So, the equation of $E = E_{0\,In} + \frac{0.0591}{n}\log 10 = E_{0\,In} + \frac{0.0591}{n}$ (colour of In_{Ox}) when the reduced form is coloured, the E will be $E = E_{0\,In} + \frac{0.0591}{n}\log\frac{1}{10} = E_{0\,In} - \frac{0.0591}{n}$ (colour of In_{Red}). Consequently, the range of potential over which sharp change in colour will be obtained is $E = E_{0\,In} \pm 0.0591/n$.

If both the oxidised and reduced forms are coloured, and their intensities differ considerably, the intermediate colour of the solution will be obtained when $E = E_0 \pm 0.06$ volt. For sharp change in colour $E = E_0 \pm 0.15$ volt, i.e., 0.15 volt from the formal potential of the other systems present in the reaction.

Let us take the case of diphenylamine indicator (E_0= 0.76). So, the potential range will be $E = 0.76 \pm 0.059/2 = 0.76 \pm 0.03$.

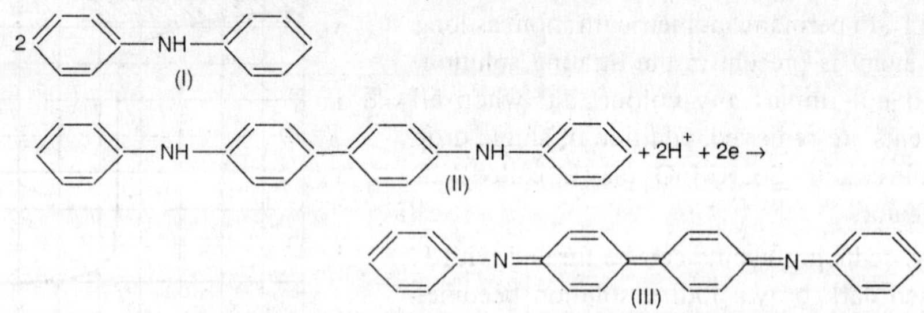

Fig. 9.3 (I) Diphenylamine, (II) Diphenyl benzidine (colourless), (III) Diphenyl benzidine (violet).

Therefore, at $E = 0.73$ the reduced form, i.e., colourless form predominates and at $E = 0.79$, i.e., the oxidised form (bluish-violet form) predominates. In the section—change of potential with concentration—it has been shown that when 91cc of the Fe^{2+} has been oxidised then potential of the redox solution becomes 0.828volt, i.e., before theoretical end point, indicator gives colour and that will give erroneous result. To avoid this, NaF or H_3PO_4 is added to the solution. This F^- or PO_4^{3-} will form complex with the Fe^{3+} and reduces the active concentration of Fe^{3+} in solution. So, potential will not increase and colour of the reduced form of the indicator predominates. When all the Fe^{2+} are consumed by $K_2Cr_2O_7$, then addition of one drop of $K_2Cr_2O_7$ increases the potential at 1.475 volt, the oxidised form of the indicator predominates and colour becomes bluish-violet. The titration curve of Fe^{3+}/Fe^{2+} system with $K_2Cr_2O_7$ is given above. A sharp break in the curve indicates the end point.

Table 9.4 : Some Oxidation-Reduction indicators

Indicator	Colour Change		Reduction Potential E_0 (volts) at pH = 0
	Oxidised Form	Reduced Form	
5-Nitro-1,10-phenanthroline ferrous sulphate (nitroferroin)	Pale blue	Red	1.25
1,10-Phenanthroline ferrous sulphate (ferroin)	Pale blue	Red	1.06
5-Methyl-1,10-phenanthroline ferrous sulphate (methylferroin)	Pale blue	Red	1.02
N-Phenylanthranilic acid	Purple red	Colourless	1.08 (?)
2,2′-Dipyridyl ferrous sulphate	Faint blue	Red	0.97
5,6-Dimethylferroin	Pale blue	Red	0.97
4,7-Dimethyl-1,10-phenanthroline ferrous sulphate (4,7-dimethylferroin)	Pale blue	Red	0.88
Diphenylaminesulphonic acid	Red-violet	Colourless	0.85
3,3′-Dimethylnaphthidine-disulphonic acid	Reddish-purple	Pale green	0.80
Diphenylbenzidine	Violet	Colourless	0.76
Diphenylamine	Violet	Colourless	0.76
3,3′-Dimethylnaphthidine	Purplish-red	Colourless	0.71
Starch-I_3^-, KI	Blue	Colourless	0.53
Methylene blue	Blue	Colourless	0.52

CHAPTER

10

ACIDS-BASES

10.01 INTRODUCTION

The acid and base are two distinct classes of chemical compounds having totally different chemical properties. Acids are substances, which possess one or more H-atoms, which can be replaced by metals or group of atoms acting like metals. They are sour in taste and turn blue litmus red. The acid reacts with metallic carbonates producing carbon dioxide. They react with bases forming salt and water. Some examples of acids are HCl, H_2SO_4, HNO_3, etc.

$$Zn + HCl \rightarrow ZnCl_2 + H_2$$

But the same metal reacts with sodium hydroxide with the liberation of hydrogen. So, can sodium hydroxide be called acid? The answer is negative, because sodium hydroxide though contains hydrogen atom does not possess all the other properties of the acid.

$$Zn + NaOH \rightarrow Na_2ZnO_2 + H_2$$

Sodium hydroxide is slippery in touch and turns red litmus blue. So, metallic oxides and hydroxides constitute a separate class known as the bases. If the bases are soluble in water like NaOH, KOH, etc., they are called alkalis.

10.02 ARRHENIUS' THEORY

The first modern approach to the acid-base concept was by Arrhenius in his theory of electrolytic dissociation. According to this theory, an acid is a hydrogen-containing compound, which gives H^+ ion in water solution and a base is a compound that gives OH^- ion in water medium.

$$HA \xrightarrow{H_2O} H^+ + A^- \quad \text{Acid}$$

$$MOH \xrightarrow{H_2O} M^+ + OH^- \quad \text{Base}$$

The bare hydrogen ion or proton actually does not exist in aqueous solution because of its high hydration energy (256 kcal per mole). It gets bound to one water molecule by co-ordination with a lone pair of electron on the oxygen atom of water, producing the hydroxonium, oxonium or hydronium ion.

$H_2O + H^+ \rightarrow H_3O^+$ (hydronium ion)

Therefore, we can write

$HA + H_2O \rightarrow H_3O^+ + A^-$

This theory was successful in explaining acid-base behavior only in aqueous medium. According to this concept, neutralization reaction of acids and bases is basically a reaction between H^+ and OH^- ions in solution.

H^+ (aq.) + OH^- (aq.) $\rightarrow H_2O$ (liq.)

Thus, the constant heat of neutralization of a strong acid by a strong base is justified. Even the strength of an acid or base could be expressed quantitatively in terms of an ionic equilibrium of the type,

$HA \rightleftharpoons H^+ + A^-$; $K = C_H^+ . C_A^- / C_{HA}$

LIMITATIONS

Though it has wide applicability in aqueous solution, it has some limitations also. The acidic or basic property of a substance is not an inherent property but requires the presence of water to exhibit the properties. e.g., HCl is regarded as an acid in water medium only and not in some other solvents or in gaseous state.

(i) It cannot explain the acidic or basic character of some substances in non-aqueous solvents, e.g., NH_4NO_3 acts as an acid in liquid NH_3 though it cannot produce H^+ ion. Similarly, aniline, pyridine, etc., show basic character though they have no OH^- ions.

(ii) It cannot account for the acidic character of certain salts like BF_3, $AlCl_3$, etc.

(iii) The neutralization process is limited to aqueous solution only, although salt formation can take place in any other solvents or in absence of solvent.

10.03 BRONSTED-LOWRY THEORY—THE PROTON DONOR/ACCEPTOR SYSTEM

In 1923, Bronsted and Lowry independently proposed new definitions for acids and bases.

According to this theory, an acid is a substance that is capable of releasing a proton, i.e., an acid is a proton donor, and a base is a substance that is capable of combining with a proton, i.e., a base is a proton acceptor. This definition leads to the fundamental relationship:

Acid (A) $\rightleftharpoons$ base (B) + proton (H^+)

These definitions are more general because even ions can behave as acids or bases. Moreover, these definitions are not restricted to reactions taking place in aqueous solutions only. Bronsted acids may be a neutral molecule, cationic or anionic species while Bronsted bases may also be neutral molecule, cationic or anionic species.

Bronsted acids Molecule: $HCl \rightarrow H^+ + Cl^-$

Cationic: $[Al(H_2O)_6]^{3+} \rightarrow H^+ + [Al(H_2O)_5OH]^{2+}$

Anionic: $HCO_3^- \rightarrow H^+ + CO_3^{2-}$

Bronsted bases Molecule: C_5H_5N (py) + $H^+ \rightarrow C_5H_5NH^+$ (pyH)$^+$ py = pyridine

Cationic: $[Al(H_2O)_5OH]^{2+} + H^+ \rightarrow [Al(H_2O)_6]^{3+}$

Anionic: $CO_3^{2-} + H^+ \rightarrow HCO_3^-$

From the above examples, it is obvious that acid-base reactions according to Bronsted-Lowry concept involve transfer of proton from an acid to a base. A substance can act as an acid only if another substance capable of accepting a proton is present. It is interesting to note here that both Arrhenius as well as Bronsted concepts require acid to be a source of proton. Hence, all Arrhenius acids are also Bronsted acids. But there is a difference in the definition of bases. According to Arrhenius, a base requires to be source of OH^- ions in aqueous solution but Bronsted theory requires base to be a proton acceptor. Hence, Arrhenius bases may not be Bronsted bases. For example, NaOH is base according to Arrhenius concept because it gives OH^- ions in aqueous medium but does not accept a proton as such.

CONJUGATE ACID-BASE PAIR

Consider the equilibrium reaction

$$\underset{\text{acid}}{HCl} + \underset{\text{base}}{H_2O} \rightleftharpoons \underset{\text{acid}}{H_3O^+} + \underset{\text{base}}{Cl^-}$$

In this forward reaction, HCl donates its proton to H_2O and is, therefore, an acid. Water, on the other hand, accepts a proton from HCl, and is, therefore, a base by the protonic definition. In the backward reaction, the H_3O^+ ion is an acid while Cl^- ion may now be looked upon as a base because it accepts a proton from H_3O^+ ion. The equilibrium thus involves two acid-base systems functioning simultaneously.

$$\text{acid}_1 \rightleftharpoons \text{base}_1 + H^+$$

$$\text{base}_2 + H^+ \rightleftharpoons \text{acid}_2$$

$$\text{adding,} \quad \text{acid}_1 + \text{base}_2 \rightleftharpoons \text{acid}_2 + \text{base}_1$$

The acid-base pairs HCl-Cl^- or H_3O^+-H_2O may be called conjugate acid-base pairs. Chemical species, which can be formed from each other mutually by the gain or loss of proton, is called *conjugate acid-base pairs.* Thus, an acid reacts with a base to form another acid and another base, and the reaction proceeds towards the weaker acid and base.

In this theory, neutralization in terms of salt formation has no room and shows that the reaction is independent of solvent. But though the reaction is independent of solvent yet it has some influence over the acid-base character of the substance. Solvent has also acid-base character. Hence solvents can be classified as

(i) Protogenic (acidic) has tendency to release proton, e.g., H_2SO_4, CH_3COOH, HF, etc.

(ii) Protophillic (basic) has tendency to accept proton, e.g., NH_3, N_2H_4, NH_2OH, etc.

(iii) Amphiprotic (amphoteric) has tendency to release and accept proton, e.g., H_2O, alcohol, etc.

(iv) Aprotic (neutral) has no tendency to release or accept proton, e.g., C_6H_6, CCl_4, etc.

The substances which are acids, can behave as a base in presence of a strong acid, e.g., HNO_3. In water nitric acid behaves as an acid as it liberates proton, $HNO_3 + H_2O \rightarrow H_3O^+ + NO_3^-$

But in presence of non-aqueous solvents like HF or H_2SO_4, $HNO_3 + H_2SO_4 \rightarrow HSO_4^- + H_2NO_3^+ \rightarrow HSO_4^- + (H_2O + NO_2^+)$.

Let us again consider the above reaction between HCl and H_2O.

$$\underset{\text{acid}_1}{HCl} + \underset{\text{base}_2}{H_2O} \rightleftharpoons \underset{\text{acid}_2}{H_3O^+} + \underset{\text{base}_1}{Cl^-}$$

In the equilibrium mixture two acids, HCl and H_3O^+ ion are competing to donate protons to a base. Since HCl wins, it is the stronger acid compared to H_3O^+. Similarly, there is also competition between the two bases, H_2O and Cl^- ion. Here, H_2O is the stronger base. Thus we can find that the stronger acid HCl forms the weaker conjugate base Cl^- while the stronger base H_2O furnishes the weaker conjugate acid H_3O^+ ion. Thus stronger acid and stronger base will exchange proton in such a way that equilibrium is shifted to produce another weaker acid and weaker base. So, it may be concluded that conjugate bases of strong acids must be weak in nature and vice versa. The stronger acid and weaker base form one conjugate acid-base pair and weaker acid and stronger base form another conjugate acid-base pair. A few conjugate acid-base pairs are listed in Table 10.1.

Table 10.1 : Some typical conjugate acid-base systems

$Acid_1$	+	$Base_2$	→	$Acid_2$	+	$Base_1$
HNO_3	+	H_2O	→	H_3O^+	+	NO_3^-
H_2SO_4	+	H_2O	→	H_3O^+	+	HSO_4^-
HSO_4^-	+	H_2O	→	H_3O^+	+	SO_4^{2-}
$[Fe(H_2O)_6]^{3+}$	+	H_2O	→	H_3O^+	+	$[Fe(H_2O)_5\,OH]^{2+}$
NH_4^+	+	H_2O	→	H_3O^+	+	NH_3
H_2O	+	NH_3	→	NH_4^+	+	OH^-
H_2O	+	CO_3^{2+}	→	HCO_3^-	+	OH^-
H_2O	+	HS^-	→	H_2S	+	OH^-
HCl	+	CH_3COOH	→	$CH_3COOH_2^+$	+	Cl^-
$HClO_4$	+	H_2SO_4	→	$H_3SO_4^+$	+	ClO_4^-
$HClO_4$	+	HF	→	H_2F^+	+	ClO_4^-
HF	+	HNO_3	→	$H_2NO_3^+$	+	F^-
$CO(NH_2)_2$	+	NH_3	→	NH_4^+	+	NH_2CONH^-
$HCOOH$	+	$CO(NH_2)_2$	→	$H_2N.CO.NH_2^+$	+	$HCOO^-$
H_2	+	OD^-	→	DOH	+	H^-
$C_6H_5CH_3$	+	CH_3Na	→	CH_4	+	$C_6H_5CH_2Na^+$

Some more conjugate acid-base pairs have been given in the following equations:

$Acid_1$		$Base_2$		$Acid_2$		$Base_1$	
CH_3COOH	+	H_2O	→	H_3O^+	+	CH_3COO^-	(i)
H_2O	+	CO_3^{2-}	→	HCO_3^-	+	$OH^{\cdot}$	(ii)
HCO_3^-	+	NH_3	→	NH_4^+	+	CO_3^{2-}	(iii)
HCl	+	HCO_3^-	→	H_2CO_3	+	Cl^-	(iv)

It may be noticed that in equation (i) H_2O is behaving as a base whereas in equation (ii) it is behaving as an acid. Similarly, HCO_3^- ion in equation (iii) acts as an acid and in equation (iv) it acts as a base. Such substances which can act as acids as well as bases are called *amphoteric* or *amphiprotic substances*.

10.04 THE SOLVENT SYSTEM

The protonic definition of acids and bases given by Bronsted can be extended to the reactions occurring in non-aqueous solvents containing hydrogen such as NH_3, N_2H_4, HF, CH_3COOH, HCN, H_2SO_4 and alcohols.

Studies on certain non-aqueous systems, particularly in liquid ammonia, reveal that reactions in this solvent may be explained by assuming auto-ionisation of the solvent similar to water.

$H_2O + H_2O \rightleftharpoons H_3O^+ + OH^-$ and

$NH_3 + NH_3 \rightleftharpoons NH_4^+ + NH_2^-$

Thus, all those compounds which can give H_3O^+ ions in water will act as acids and all the compounds which can give OH^- ions in water will behave as bases. Similarly, substances furnishing

NH_4^+ ions in liquid ammonia may also be expected to behave as an acid in that medium and substances furnishing NH_2^- ions in liquid ammonia are formally similar to the compounds giving OH^- ions in water, i.e., bases.

Acids in liquid ammonia: NH_4Cl, NH_4NO_3, $(NH_4)_2SO_4$, etc.

Bases in liquid ammonia: KNH_2, $NaNH_2$, etc.

Similarly, in N_2O_4 as solvent, substances such as NOCl which yields NO^+ ions are acids while substances such as $NaNO_3$ which yields NO_3^-, ions are bases.

$$\underset{\text{Acid}}{N_2O_4} + \underset{\text{Base}}{N_2O_4} \rightleftharpoons \underset{\text{Acid}}{2NO^+} + \underset{\text{Base}}{2NO_3^-}$$

The auto-ionisation of some non-protonic solvents is shown below:

$$SO_2 + SO_2 \rightleftharpoons SO^{2+} + SO_3^{2-}$$
$$BrF_3 + BrF_3 \rightleftharpoons BrF_2^+ + BrF_4^-$$
$$COCl_2 \rightleftharpoons COCl^+ + Cl^-$$

These observations may be generalized for both protonic as well as non-protonic solvents, and acids and bases may be defined in general as follows:

An acid is a substance that, either by direct dissociation or by reaction with the solvent, gives the cation characteristic of the solvent and a base is a solute that, either by direct dissociation or by reaction with the solvent, gives the anion characteristic of the solvent.

Just as with Arrhenius definition, neutralization here is a reaction between an acid and a base to produce salt and the solvent. Neutralization reactions in non-aqueous solvents are given below:

Solvent	Acid		Base		Salt		Solvent
In liquid ammonia	NH_4Cl	+	$NaNH_2$	$\rightleftharpoons$	NaCl	+	$2NH_3$
In N_2O_4	NOCl	+	$NaNO_3$	$\rightleftharpoons$	NaCl	+	N_2O_4
In CH_3COOH	HCl	+	CH_3COONa	$\rightleftharpoons$	NaCl	+	CH_3COOH
In SO_2	$SOCl_2$	+	$[N(CH_3)_4]_2\ SO_3$	$\rightleftharpoons$	$2[N(CH_3)_4]Cl$	+	$2SO_2$

It may be seen from the following reactions that there is a complete analogy between the solvolytic and amphoteric behaviours in aqueous solvents.

Solvolytic behaviour

In liq. NH_3, $AlCl_3 + NH_3 \rightarrow [Al(NH_2)]^{2+} + H^+ + 3Cl^-$

In H_2O, $AlCl_3 + H_2O \rightarrow [Al(OH)]^{2+} + H^+ + 3Cl^-$

Amphoteric behaviour

In liq. NH_3, $Zn(NH_2)_2 + 2NH_2^- \rightarrow [Zn(NH_2)_4]^{2-}$

In H_2O, $Zn(OH)_2 + 2OH^- \rightarrow [Zn(OH)_4]^{2-}$

UTILITY OF THE CONCEPT

This concept of solvent system is able to include many non-aqueous as well as non-protonic acid-base system. Each solvent may be looked upon as giving rise to an acid-base system of its own. Separately, these may be treated along the same line of approach as that of Arrhenius for aqueous systems.

LIMITATIONS

1. It limits acid-base phenomenon to solvent systems only and fails to consider the inherent acid or base character of a substance independent of a solvent.

2. This theory does not consider a number of acid-base reactions included in the protonic definition.
3. It can not explain the neutralization reactions occurring without the presence of ions.

Thus, this theory can simply be said to be an extension of Arrhenius water-ion system.

LUX-FLOOD DEFINITION

The Lux-Flood theory describes the acid-base behaviour in terms of the oxide ion. This acid-base concept is applicable to non-protonic systems which are not included in the Bronsted-Lowry definition. For example, the reaction between calcium oxide and silicon dioxide (silica) at high temperature takes place as

$$\underset{\text{base}}{CaO} + \underset{\text{acid}}{SiO_2} \rightarrow CaSiO_3$$

Here the base, i.e., CaO is an oxide donor and the acid, i.e., SiO_2 is an oxide acceptor. This definition of acid-base is limited to systems such as molten oxides.

10.05 USANOVICH CONCEPT

The most comprehensive of all the acid-base theories is the Usanovich (1939) concept according to which an acid is any chemical species capable of giving cations, combining with anions or electrons or neutralizing a base to give a salt. In other words, an acid is a cation, a cation donor, an electron pair acceptor.

A base, on the other hand, is defined as any chemical species capable of giving up anions or electrons, combining with cations or neutralizing an acid to give a salt.

The scope of the definition is thus very wide, covering all the previous acid-base definitions and also includes many oxidation-reduction reactions as a special case of acid-base reactions. Some additional examples of acid-base reactions according to this concept are:

Acid		Base		Salt	
SO_2	+	Na_2O	=	Na_2SO_3	SO_2 accepts the anion O^{2-}
Cl_2	+	2Na	=	2NaCl	Chlorine accepts an electron
$Fe(CN)_2$	+	4KCN	=	$K_4[Fe(CN)_6]$	$Fe(CN)_2$ adds further anion CN^-

Usanovich here considered

(i) the general trend of acidic and basic character in the periodic table and

(ii) the aspect of "coordination unsaturation".

(iii) In this context, his main observations may be summarized as follows:

(a) The acidic character of the oxides of the elements increases from left to right in the periodic table. Thus, the oxides of the alkali and alkaline earth metals are essentially basic whereas those of P, S, Cl are acidic. Acidic character of the oxides of elements decreases from the top to bottom, e.g., As_2O_3 is more acidic than Sb_2O_3.

(b) The higher valence of an element is associated with greater acidic character, e.g., $Sn^{IV}S_2$ is more acidic than $Sn^{II}S$.

(c) High charge of cations favours acidic property (Na_2O vs. Al_2O_3), while high charge of anions promote basic property (Na_2O vs. NaCl).

(d) The term "coordination unsaturation" was used to mean the tendency of an atom to increase its covalence. A coordinately unsaturated cation can exhibit acid function (by adding an anion), while a similarly unsaturated anion can display basic property (by adding a cation). The S-atom in SO_2 molecule is coordinately unsaturated; as such SO_2 can accept an anion like O^{2-} and act as an acid.

UTILITY

The definition is the most general of the different acid—base definitions. At the same time, it states that the donation or acceptance of electrons need not take place as shared pairs. It also includes some oxidation-reduction reactions.

LIMITATIONS

It is extremely general—its scope is so large that almost all—chemical reactions may be considered as acid-base reactions.

10.06 LEWIS THEORY: THE ELECTRON DONOR-ACCEPTOR SYSTEM

This theory explains the acid-base phenomenon in terms of electronic structure of the acid and the base. According to Lewis (1923), an acid is a substance (molecule, radical or ion) that acts as an electron pair acceptor while a base is a substance that acts as an electron pair donor. We denote a Lewis acid by *A* and a Lewis base by *B*, without considering the presence of more than one lone pair. The fundamental reaction of Lewis acids and bases is the formation of a compound *A-B*, in which *A* and : *B* get bonded together by sharing the electron pair supplied by the base. Here neutralization means formation of a coordinate compound followed by association, dissociation or rearrangement as the case may be. The proton (H^+) is a Lewis acid because it can take an electron pair. It may be noted that all Bronsted bases are also Lewis bases but all Bronsted acids are not Lewis acids. Lewis bases generally need one or more lone pairs of electrons and, therefore, they can also accept a proton (Bronsted base). Thus, all Lewis bases are also Bronsted bases. On the other hand, Bronsted acids are those which are proton donors, for example, HCl, H_2SO_4, etc. But they may not be capable of accepting a pair of electrons. Hence, all Bronsted acids are not Lewis acids.

Lewis acids should, therefore, possess an empty orbital, which may accommodate a pair of electrons. So, such substances have the following possibilities:

(i) A molecule containing a central atom with an incomplete octet can complete its octet by accepting an electron pair. As for example, BF_3 is a Lewis acid. Other electron-deficient molecules such alkyls and halides of Be, B and Al falls in this class.

(ii) A metal cation can bond to an electron pair supplied by the base. e.g., hydration of Cu^{2+} ion where the water molecule binds the central cation through the lone pair on O-atom of water which acts as the Lewis base. The central cation Cu^{2+} is the Lewis acid as they have stable, low-lying orbital, which is empty or partly filled.

(iii) A molecule or ion with a complete octet or having a multiple bond between atoms of dissimilar electronegativity can rearrange its valence electrons and accept an additional electron pair. As for example, CO_2 acts as a Lewis acid when it forms HCO_3^- by accepting an electron pair from an O-atom in an OH^- ion.

(iv) A molecule or ion may be able to expand its octet to accept an electron pair if the central atom has vacant *d*-orbitals, e.g., the formation of the complex $(SiF_6)^{2-}$ when two F^- ions (Lewis base) bound to SiF_4 (the Lewis acid).

Lewis bases are capable of donating a pair of electrons to form a coordinate bond, e.g., ether, alcohol, amines, pyridines, etc. Like acids they can be similarly classified as:

(i) Molecules having at least one pair of unshared electrons or so, called *lone pair*, e.g., H_2O and NH_3 are Lewis bases. The basic character is found to be inversely proportional to the number of lone pair present in the molecule. NH_3 with one lone pair is more basic in character than H_2O with two lone pairs.

(ii) Generally, all negative ions, e.g., F^-, Cl^- and I^- are the typical examples.

$$SiF_4 + 2F^- \rightleftharpoons [SiF_6]^{2-}$$
$$BF_3 + F^- \rightleftharpoons BF_4^-$$
$$I_2 + I^- \rightleftharpoons I_3^-$$
$$AlCl_3 + Cl^- \rightleftharpoons AlCl_4^-$$

(iii) Compounds containing carbon-carbon double bonds:

The loosely bound π-electron cloud of a carbon-carbon π-bond can be donated to Lewis acids like metal ions, electron-deficient compounds like BF_3, $AlBr_3$, etc., forming the so-called π-complexes. This type of compounds is generally formed by transition metals. In this type of organometallic compounds, the normal σ-bond is formed through the π cloud of electrons of the $> C = C <$. The most well-known example is the Zeise's salt, $K[PtCl_3\text{-}\eta\text{-}(C_2H_4)]$.

UTILITY OF LEWIS CONCEPT

(i) It is more general and also includes compounds having no protons.

(ii) It can explain the basic properties of metallic oxides and acidic properties of non-metallic oxides.

(iii) It includes all types of neutralization processes along with that taking place in aqueous media.

(iv) It introduces a parallelism between acid-base character and oxidising and reducing power of a substance. A good electron donor (strong base) will be a strong reducing agent. Similarly, a strong acid (electron acceptor) will be a good oxidising agent; though some limitations are still there in such generalizations.

LIMITATIONS OF LEWIS CONCEPT

(i) It is too general and includes all the coordination compounds and coordination reactions.

(ii) It does not provide any idea about the relative strengths of acids and bases.

(iii) Normally, formation of coordination compounds is slow. Therefore, acid-base reactions should also be slow, but in actual practice, acid-base reactions are extremely fact.

10.07 HARD AND SOFT ACIDS AND BASES

The Lewis definition considered acids and bases in terms of their ability to accept or donate electron pairs. As such, the strength of any acid or base is determined by the nature of reaction involved in any particular electron transfer process. A qualitative correlation between the various Lewis acids and bases is done by classifying the acids and bases into two classes: hard and soft. The criterion of hardness (or softness) is ascribed to the "hardness" of the electron cloud associated with any species. A firmly held electron cloud with low polarisability makes a species "hard", while an easily polarisable electron cloud makes the species "soft". A third category with intermediate character appears in the borderline.

Experimental method to determine the hardness or softness of a substance is based on the following standard reaction: $BH^+ + CH_3Hg^+ \rightleftharpoons CH_3HgB^+ + H^+$, H^+ is the hardest acid and CH_3Hg^+ is the typical soft acid and convenient as reference in studying acid-base equilibrium.

A hard base will push the equilibrium to the left and a soft base will drive it to the right. In the reaction, $OH^- + CH_3HgSO_3^- \rightleftharpoons CH_3HgOH + SO_3^{2-}$, hard base OH^- displaces borderline base SO_3^{2-} from the soft acid CH_3Hg^+.

Soft and Hard acids

Soft acids are readily polarised. The acceptor atoms generally

(i) have a large size,

(ii) have zero or low positive oxidation state, and

(iii) do not have a noble gas electronic configuration or several easily excitable valence electrons.

A hard acid is difficult to polarise. The acceptor atoms generally have

(i) a small size,

(ii) a high positive oxidation state, and

(iii) a noble gas electronic configuration, i.e., absence of any outer electrons, which are easily excited to higher states.

Soft and Hard bases

Soft Lewis bases are those in which the donor atoms are marked by

(i) low electronegativity,

(ii) high polarisability, and

(iii) partly filled orbitals; empty orbitals are low lying.

Hard Lewis bases are those in which the donor atoms have the following characteristics:

(i) high electronegativity,

(ii) low polarisability, and

(iii) presence of filled orbitals; empty orbitals may exist at high energy level.

Table 10.2 shows some hard/soft acids and hard/soft bases respectively with the intermediate ones in between the hard and soft.

Table 10.2 : Classification of some common Lewis acids and bases

Acids	Bases
(i) Hard: H^+, Li^+, Na^+, K^+, Be^{2+}, Mg^{2+}, Ca^{2+}, Sr^{2+}, Mn^{2+}, Al^{3+}, Ga^{3+}, Sc^{3+}, In^{3+}, La^{3+}, Ce^{3+}, Gd^{3+}, Lu^{3+}, Ce^{4+}, Cr^{3+}, Co^{3+}, Fe^{3+}, As^{3+}, Si^{4+}, Ti^{4+}, Zr^{4+}, Hf^{4+}, Th^{4+}, U^{4+}, Cr^{6+}, Mn^{7+}, Cl^{3+}, Cl^{7+} Pu^{4+}, VO^{2+}, MoO^{3+}, VO_2^{2+}, I^{5+}, I^{7+}, WO^{4+}, SO_3, BF_3, $(CH_3)_2Sn^{2+}$, etc.	(i) Hard: H_2O, NH_3, RNH_2, N_2H_4, OH^-, O^{2-}, ROH, RO^-, R_2O, F^-, Cl^-, NO^{3-}, ClO_4^-, Ch_3COO^-, CO_3^2, 2-SO_4^2, 2-PO_4^{3-}, etc.
(ii) Borderline: Fe^{2+}, Co^{2+}, Ni^{2+}, Cu^{2+}, Ru^{2+}, Zn^{2+}, Pb^{2+}, Sn^{2+}, Sb^{3+}, Bi^{3+}, Rh^{3+}, Ir^{3+}, Ru^{3+}, Os^{2+}, NO^+, SO_2, R_3C^+, $B(CH_3)_3$, BBr_3, etc.	(ii) Borderline: $C_6H_5NH_2$, C_5H_5N, N_2, N_3^-, Br^-, NO_2^-, SCN^-, SO_3^{2-}, etc.
(iii) Soft: Cu^+, Ag^+, Au^+, Tl^+, Pd^{2+}, Pt^{2+}, Pt^{4+}, Hg_2^{2+}, Cd^{2+}, CH_3Hg^+, Hg^{2+}, $Co(CN)_5^{3-}$, BH_3, etc.	(iii) Soft: R_2S, RSH, R_3P, $(RO)_3P$, CO, R_3As, RNC, C_2H_4, C_6H_6, RS^-, R^-, SCN^-, CN^-, H^-, I^-, $S_2O_3^{2-}$, etc.

SHAB PRINCIPLE AND STABILITY OF THE COMPLEX A:B

Ralph G. Pearson (1963) proposed the principle of soft and hard acids and bases, which is abbreviated as SHAB principle. According to this principle, the complex *A*:*B* is most stable when *A* and *B* are either both soft or both hard. The complex is least stable when one of the reactants (namely A and B) is very hard the other is very soft. For the typical Lewis acid-base reaction,

A	+	B	$\rightarrow$	A:B
Lewis acid (acceptor)		Lewis base (donor)		adduct or complex

The principle states that if there is a choice of reaction between an acid and two bases, or between a base and two acids, a hard acid will prefer to combine a hard base and a soft acid will prefer to combine with a soft base and thus a more stable product will be obtained.

UTILITY OF THE SHAB PRINCIPLE

In the hard-soft terminology, a large number of chemical reactions may be understood. The guiding principle is that hard-hard and soft-soft combinations of acids and bases are always more stable than hard-soft combinations. For this, let us consider some typical examples:

(i) AgI_2^- is stable, but AgF_2^- does not exist. It will be seen from Table 10.2 that Ag^+ is a soft acid, F^- is a hard base and I^- is a soft base. Hence, AgI_2^- (soft acid + soft base) is a stable complex and AgF_2^- (soft acid + hard base) does not exist. Similarly, it can be shown that CoF_6^{3-} (hard acid + hard base) is more stable than CoI_6^{3-} (hard acid + soft base).

(ii) The existence of certain metal ores can also be rationalized by applying the SHAB principle. Thus, hard acids such as Mg^{2+}, Ca^{2+} and Al^{3+} occur in nature as $MgCO_3$, $CaCO_3$ and Al_2O_3, and not as sulphides (MgS, CaS and Al_2S_3), since the anions CO_3^{2-} and O^{2-} are hard bases and S^{2-} is a soft base.

(iii) Ammonia, amines, water and fluoride ions prefer to combine with Be^{2+}, Ti^{4+}, Co^{3+}, etc., giving stable complexes. On the other hand, phosphines (R_3P), thioethers (R_2S) and other species with P and S as donor atoms prefer to combine with Pt^{2+}, Pd^{2+}, Hg^{2+}, etc.

(iv) BF_3, a hard acid, combines readily with a further fluoride ion; but BH_3 (present as a dimer B_2H_6), a soft acid, prefer the softer base hydride ion.

$BF_3 + F^- \rightarrow BF_4^-$, $\quad B_2H_6 + 2H^- \rightarrow 2BH_4^-$

From this it is easy to understand why the following reactions proceed as:

$BF_3H^- + BH_3F^- \rightarrow BF_4^- + BH_4^-$

$CF_3H + CH_3F \rightarrow CF_4 + CH_4$

(v) Hard-hard and soft-soft combinations similarly determine the course of many other reactions, for example,

$LiI + CsF \rightleftharpoons LiF + CsI$

$HgI_2 + BeI_2 \rightleftharpoons BeF_2 + HgI_2$

(vi) Numerous chemical aspects have been interpreted by the hard soft concept. The catalytic power of metals may be understood from the fact that the soft metal atoms adsorb soft bases on their surface. The tendency of hard solvents to dissolve hard solutes and soft solvents to dissolve soft solutes can be justified from the solubility point of view. Even speeds of electrophilic and nucleophilic substitution reactions may be related to hardness and softness of the species involved.

LIMITATIONS

Though hard-soft classification is a useful concept, but it does not lead directly to a scale of acid-base strength. Inherent acid-base strengths are also not taken into consideration. The very common reaction between ethanol and acetic acid producing ester and water may be interpreted in either of two ways:

1. The acid produces H^+ ion while the alcohol produces OH^- ion.
2. The alcohol produces H^+ ion while the acid produces OH^- ion.

Both the interpretations are justifiable by the hard-hard combination of H^+ with OH^-. The actual involvement of the second mechanism has to be understood by other means.

Sometimes the hard-soft principle fails to keep parity with inherent acid-base strengths. The reaction,

$$CH_3^+(g) + H_2\,(g) \rightarrow CH_4\,(g) + H^+\,(g),$$

should be favourable in view of the soft-soft combination of CH_3^+ with H^-. But actually the combination is endothermic by about 360 kJ per mole. (This has to be explained by the greater acidity of H^+ relative to CH_3^+.) Similarly, hard-soft combinations takes place in many cases like $SO_3^{2-} + HF \rightarrow HSO_3^- + F^-$.

Here, the soft base SO_3^{2-} replaces the hard base F^- and combines with the hard acid H^+. The soft SO_3^{2-} is thus stronger than the hard F^-. In fact, the hard-soft idea germinated from a study of the behaviour of metals towards different non-metal donor atoms (Ahrland Chat and Davies 1958). Certain metals prefer to form stable complexes when the first member of a given group of non-metals in the periodic table acts, as donor atoms, e.g., N, O, F. These metals are marked as class-*a*. Alkali and alkaline earth metals, and the first row transition metals in high oxidation states (Na^+, K^+, Ca^{2+}, Fe^{3+}, Co^{3+}, etc.) belong to this class. The other class (class-*b*) of metal ions of lighter transition metals of lower oxidation state and some heavier metals like Cu^{+1}, Hg^{+1}, Pt^{2+}, Pd^{2+} can form stable complexes with heavier non-metals like P, S, Cl.

The metals which form most stable complexes with heavier members of a group of non-metals (like P, S, Cl) are marked as class-*b*. Lighter transition metals in the low oxidation state and heavier transition metals belong to this class (Cu^+, Ag^+, Hg^+, Pt^{2+}, Pd^{2+}, etc.). Thus, class-*a* metal ions bind most strongly to the least polarisable atom of a family, whereas class-*b* metal ions bind most strongly to a highly polarisable atom of the family.

The features which bring about class-*a* behaviour are small size, high positive oxidation state and absence of easily distorted outer electrons, i.e., low polarisability and hence class-*a* metals are called *hard acids*. Similarly, class-*b* behaviour is associated with a low or zero oxidation state, large size and presence of easily distorted outer electrons in the metals and hence class-*b* metals are called *soft acids*.

10.08 STRENGTHS OF ACIDS AND BASES

The strength of acids and bases are measured by the ease with which the proton is released (by an acid) or captured (by a base). A strong acid has a strong tendency to donate a proton and a strong base has a strong tendency to accept a proton. Such strengths may be expressed by the dissociation constant as developed by the protonic definition and this yields the same result as the Arrhenius definition for aqueous solutions. In this part we shall try to correlate strengths of acids and bases in terms of their dissociation constants. General trends of acid-base strengths in different compounds will also be considered.

(A) DISSOCIATION CONSTANTS OF ACIDS AND BASES

According to the Arrhenius definition, the strength of an acid HA may be expressed in terms of the equilibrium constant for its ionisation in aqueous solution liberating a proton:

$$HA \rightleftharpoons H^+ + A^-;\; K = (C_{H^+} \cdot C_{A^-})/C_{HA}$$

Similar expressions may be obtained from the protonic definition taking water as a reference base (for acids).

$$\underset{\text{acid}_1}{HA} + \underset{\text{base}_2}{H_2O} \rightleftharpoons \underset{\text{acid}_2}{H_3O^+} + \underset{\text{base}_1}{A^-}; \qquad K = \left(C_{H_3O^+.A^-}\right)/C_{HA}.C_{H_2O}$$

$$K' = KC_{H_2O} = \left(C_{H_3O^+.A^-}\right)/C_{HA}$$

The strength of a base is similarly expressed by the equilibrium constant for ionisation to produce a hydroxyl ion (OH^-) in aqueous solution.

$$BOH \rightleftharpoons B^+ + OH^- \qquad K = C_{B^+}.C_{OH^-}/C_{BOH}$$

or, $B + H_2O \rightleftharpoons BH^+ + OH^-$ $\quad K = C_{BH^+}.C_{OH^-}/C_{H_2O}.C_B$ $\quad$ or, $K' = KC_{H_2O} = C_{BH^+}.C_{OH^-}/C_B$

It is often convenient to express the dissociation constant of a weak acid by K_a and that of a weak base by K_b. The dissociation constants of some weak acids and bases are given in Table 10.3.

Table 10.3 : Dissociation constants of weak acids and bases in aqueous solution at 298 K

Acids / Bases	Equilibrium	K_a or K_b
Phosphoric acid	$H_3PO_4 + H_2O \rightleftharpoons H_3O^+ + H_2PO_4^-$	$7{\cdot}5 \times 10^{-3}$ (K_1)
	$H_2PO_4^- + H_2O \rightleftharpoons H_3O^+ + HPO_4^{2-}$	$6{\cdot}2 \times 10^{-8}$ (K_2)
	$HPO_4^- + H_2O \rightleftharpoons H_3O^+ + HPO_4^{3-}$	$1{\cdot}0 \times 10^{-12}$ (K_3)
Monochloroacetic acid	$ClCH_2COOH + H_2O \rightleftharpoons H_3O^+ + ClCH_2COO^-$	$1{\cdot}4 \times 10^{-3}$
Hydrogen fluoride	$HF + H_2O \rightleftharpoons H_3O^+ + F^-$	$6{\cdot}8 \times 10^{-4}$
Nitrous acid	$HNO_2 + H_2O \rightleftharpoons H_3O^+ + NO_2^-$	$4{\cdot}5 \times 10^{-4}$
Acetic acid	$CH_3COOH + H_2O \rightleftharpoons H_3O^+ + CH_3COO^-$	$1{\cdot}8 \times 10^{-5}$
Carbonic acid	$H_2CO_3 + H_2O \rightleftharpoons H_3O^+ + HCO_3^-$	$4{\cdot}2 \times 10^{-7}$ (K_1)
	$HCO_3^- H_2O \rightleftharpoons H_3O^+ + CO_3^{2-}$	$4{\cdot}8 \times 10^{-11}$ (K_2)
Hydrogen sulphide	$H_2S + H_2O \rightleftharpoons H_3O^+ + HS^-$	$1{\cdot}1 \times 10^{-7}$ (K_1)
	$HS^- + H_2O \rightleftharpoons H_3O^+ + S^{2-}$	$1{\cdot}0 \times 10^{-14}$ (K_2)
Hydrogen cyanide	$HCN + H_2O \rightleftharpoons H_3O^+ + CN^-$	$4{\cdot}8 \times 10^{-10}$
Phenol	$C_6H_5OH + H_2O \rightleftharpoons H_3O^+ + C_6H_5O^-$	$1{\cdot}3 \times 10^{-10}$
Trimethylamine	$(CH_3)_3N + H_2O \rightleftharpoons (CH_3)_3NH^+ + OH^-$	$7{\cdot}4 \times 10^{-3}$
Dimethylamine	$(CH_3)_2NH + H_2O \rightleftharpoons (CH_3)_2NH_2^+ + OH^-$	$7{\cdot}4 \times 10^{-4}$
Methylamine	$CH_3NH_2 + H_2O \rightleftharpoons CH_3NH_3^+ + OH^-$	$5{\cdot}0 \times 10^{-4}$
Aniline	$C_6H_5NH_2 + H_2O \rightleftharpoons C_6H_5NH_3^+ + OH^-$	$4{\cdot}0 \times 10^{-10}$

Polybasic acids and polyacidic bases have more than one dissociation constant, corresponding to each successive stage of dissociation. The successive dissociation constants of a polybasic acid become gradually smaller, as may be observed from the data for H_2S or H_3PO_4 ($K_1 > K_2 > K_3$). It is due to the fact that release of another proton becomes more and more difficult as the negative charge on the anion increases. The K_a and K_b values generally obtained are used for the comparison of relative acidity or basicity of given acids or bases. If the acidities of two acids, say, CH_3COOH and HCN is compared in terms of their proton donating tendencies towards a common base (generally water), the experimental observation will make it clear that CH_3COOH is a stronger acid than HCN, and CN^- is a stronger base than CH_3COO^- ion.

$$CH_3COOH + H_2O \rightleftharpoons H_3O^+ + CH_3COO^- \qquad (K_a = 1.8 \times 10^{-5})$$

$$HCN + H_2O \rightleftharpoons H_3O^+ + CN^- \qquad (K_a = 4.0 \times 10^{-10})$$

[The law of mass action as applied above to obtain K_a and K_b is valid only for weak electrolytes, which undergo reversible dissociation in water. For strong electrolytes, dissociation is almost complete and there is a little or no equilibrium between the ions the undissociated molecules in solution.]

(B) PERIODIC VARIATIONS OF ACIDIC AND BASIC PROPERTIES

We have critically discussed the different concepts of acids and bases. Now we are going to study the general trend in acid or base strength among the compounds considered to be as acids in the protonic concept.

The protonic acid may be broadly classified as:

(i) Hydracids—the acidic hydrogen atom is directly linked to some second element, i.e., binary hydrides like HF, H_2S, HCl, HCN.

(ii) Oxyacids—the acidic hydrogen is linked to some central atom via an oxygen atom, e.g., H_2SO_4, $HClO_4$.

The strength of acid will be discussed in two parts for the two separate classes.

(a) Hydracids of the elements of the same period

The hydracids of the elements of the second period of the periodic table are CH_4, NH_3, H_2O and HF. Thus, CH_4 has negligible acidic properties, but NH_3 donates a proton (H^+) to strong bases to form NH_2^- ion, H_2O loses a proton even more readily and HF is a fairly strong acid. The increase in the acidic properties of these hydracids is due to the fact that as we move from CH_4 to HF, the stability of their conjugate bases, viz., CH_3^-, NH_2^-, OH^- and F^- increases in the order $CH_3^- < NH_2^- < OH^- < F^-$.

The increase in acidic properties is supported by the successive increase in the dissociation constant values of these hydracids: $CH_4\ (= 10^{-58}) < NH_3\ (= 10^{-35}) < H_2O\ (= 10^{-14}) < HF\ (= 10^{-4})$.

The variation of acid character along a period may also be explained from the increasing electronegativity of the elements linked to hydrogen. Here, size of the atom does not change appreciably from one element to the next; greater electronegativity of an element makes the release of proton easier due to greater shift of the bonding electron cloud towards it. So, acid property increases.

(b) Hydracids of the elements of the same group

The electro-negativity alone cannot explain the increase of acidity down a group (rather this leads to opposite expectations). The size of the atom increases as we down the group and upsets the electro-negativity factor. As the atoms become larger, bonding with the small hydrogen atom becomes weaker (poorer overlap of atomic orbital); the increase of acid character parallels this decrease in the bond strength.

(i) *Hydracids of group VA elements* (NH_3, PH_3, AsH_3, SbH_3 and BiH_3)

All these hydracids show basic character. With increase in size and decrease in electronegativity from N to Bi, there is a decrease in electron density in sp^3 hybrid orbital and thus electron donor capacity (i.e., basic character) decreases.

(ii) *Hydracids of group VIA elements* (H_2O, H_2S, H_2Se and H_2Te)

Aqueous solutions of the hydracids of this group elements behave as weak diprotic acids and ionise as: $H_2R \rightleftharpoons H^+ + HR^-$, $\quad HR^- \rightleftharpoons H^+ + R^{2-}$

The strength of the hydracids as acids increases in the order: $H_2O < H_2S < H_2Se < H_2Te$. This order is supported by the successive increase of their dissociation constants given as follows: $H_2O\ (K_a = 1.3\times10^{-16}) < H_2S\ (K_a = 1.1\times10^{-7}) < H_2Se\ (K_a = 2\times10^{-4}) < H_2Te\ (K_a = 2.3\times10^{-3})$.

The increasing acid character reflects decreasing trend in the electron donor ability of OH^-, HS^-, HSe^-, or HTe^- ions. The increasing acid character is explained by saying that as the charge density on the conjugate base decreases from OH^- to HTe^-, the proton is less tightly held in higher members and, therefore, acidic character increases.

(iii) *Hydracids of group VIIA elements* (HF, HCl, HBr, and HI)

The aqueous solutions of these hydracids show acidic character which increases in the order HF < HCl < HBr < HI. This order is explained by saying that as we pass from HF to HI, there is a gradual decrease in the bond energies of H–X bonds (where X = F, Cl, Br, I)

[H–F = 135 kcal/mole, H–Cl = 103 kcal/mole, H–Br = 88 kcal/mole and H–I = 71 kcal/mole]. This decreasing order of bond energies increases the tendency of HX molecule to split up.

(c) Oxy acid

(i) The acidic character of oxy acid of the same element which is in different oxidation states increases with the increase of its oxidation state. The following series illustrate this rule (called oxidation number or oxidation state rule).

$$\overset{+1}{\text{HOCl}} < \overset{+3}{\text{HClO}_2} < \overset{+5}{\text{HClO}_3} < \overset{+7}{\text{HClO}_4}$$

$$\overset{+4}{\text{H}_2\text{SO}_3} < \overset{+6}{\text{H}_2\text{SO}_4}; \quad \overset{+3}{\text{HNO}_2} < \overset{+5}{\text{HNO}_3}$$

The order of acidity can be explained with the oxidation rule with reference to the oxy acids of halogens. It is well known that the stronger is acid weaker will be its conjugate base and vice versa. Now the conjugate bases of the acid are: ClO^-, ClO_2^-, ClO_3^- and ClO_4^- respectively.

$$\underset{\text{acid}_1}{\text{HA}} + H_2O \rightleftharpoons \underset{\text{conjugate base}_1}{H_3O^+ + A^-}$$

Now greater is the stability of A^-, lesser will be its basic strength or greater will be the tendency of HA to release H^+, in other words, stronger will be the acid HA. Among the conjugate bases of the oxy acids, the stability increases with increase in the number of oxygen atoms in the anion. It is because, more is the number of oxygen atom in the ion, greater will be the dispersal of the negative charge and thus, more will be the stability.

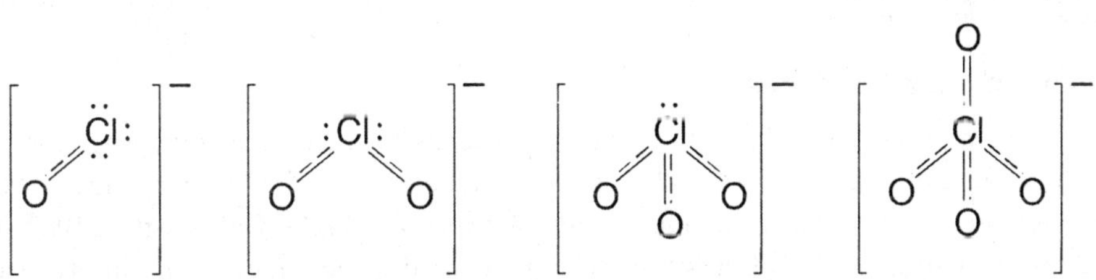

+1 +3 +5 +7

Oxidation state Charge delocalisation increases →

Similarly, H_2SO_4 is stronger acid than H_2SO_3, since the S-atom in sulphuric acid has more coordinated oxygen atom.

When the oxidation state rule as given above is applied to the oxides of phosphorus, viz., H_3PO_2, H_3PO_3 and H_3PO_4, it is predicted that the acidic character of these acids should be in the order $H_3PO_2 < H_3PO_3 < H_3PO_4$, but the experimental observation shows the reverse order, i.e., $H_3PO_2 > H_3PO_3 > H_3PO_4$.

The experimental order can be explained when the structures of these acids are considered. There are two types of oxygen atoms:

(i) protonated oxygen atom attached with a proton,

(ii) non-protonated oxygen atom having no proton.

The proton attached to any oxygen atom has a far greater chance of dissociation than that linked directly with P-atom (which is the central atom). The inductive effect of the single oxygen atom acts on the single O–H bond in H_3PO_2, but it is distributed over two and three O–H bonds in H_3PO_3 and H_3PO_4 respectively. Thus in this series, since the number of protonated oxygen atoms and consequently the number of dissociable protons increase from one in H_3PO_2 to three in H_3PO_4, the acidity of these acids decreases in the order: $H_3PO_2 \geq H_3PO_3 > H_3PO_4$.

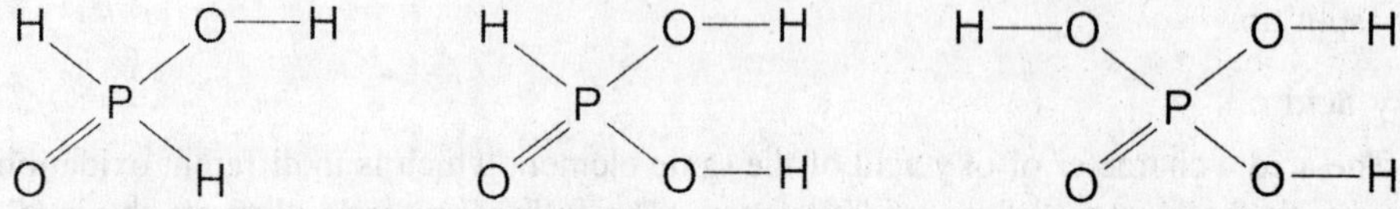

H_3PO_2 monobasic $\quad$ H_3PO_3 dibasic $\quad$ H_3PO_4 tribasic

The trends in acidity of oxy acids H_aXO_b can be generalized from an empirical relation given by J. Ricci (1948).

pK_a or $-\log K_a = 8 - 9m - 4n$

where K_a = ionisation constant of the acid

m = formal charge on the central atom

$n = b - a$ = number of non-protonated oxygen atom

(ii) The acid character of oxy acids of different elements which are in the same oxidation state decreases with increase in the atomic number of the central atom. Examples are:

$HOCl > HOBr > HOI$; $\quad HClO_4 > HBrO_4 > HIO_4$; $\quad H_2SO_3 > H_2SeO_3 > H_2TeO_3$

As the atomic number of the central atom increases, its electro-negativity decreases and its size increases. As a result of this, the tendency of the acid to lose a proton to water decreases. This makes the acid a weaker one.

Basicity of metallic oxides

In a given periodic group, the basicity of oxides tends to increase as one progresses down the periodic table. For example, in Gr. IIA, BeO is amphoteric but the heavier oxides (MgO, CaO, SrO, BaO) are distinctly basic and the basicity increases as MgO < CaO < SrO < BaO. This trend parallels the increase in size of the cation (charge remaining the same) and is closely related to the Fajans' polarising ability of the metal ion. The same effect is seen in Gr. III, oxides B_2O_3 is acidic, Al_2O_3 is amphoteric and Sc_2O_3 is quite basic.

10.09 ACIDITY CAUSED BY METAL ION ON HYDRATION AND HYDROLYSIS REACTION

When metallic salts are dissolved in water, the metal ion gets surrounded by the water molecule, which is generally called *hydration*. It is not a reaction beyond co-ordination of water molecule to the cation. $\quad Na^+ + nH_2O \rightarrow [Na(H_2O)_n]^+$

Hydration energy increases with the increase in the charge/size ratio of the cation.

Under favourable conditions, one or more protons may dissociate from the co-ordinated aqua complex like $[Al(H_2O)\ ]^{3+} + H_2O \rightarrow H_3O^+ + [Al(H_2O)_5OH]^{2+}$.

This is a case of hydrolysis where rupture of an O–H bond takes place and the solution becomes acidic. Cations that hydrolyze are those which are either small (e.g., Be^{2+}) and /or are highly charged (Fe^{3+}, Sn^{4+}), i.e., which have a high charge/size density (ionic potential).

10.10 THE LEVELLING EFFECT

Any acid stronger than H_3O^+ in water donates a proton to H_2O and forms H_3O^+. Consequently, no acid stronger than H_3O^+ can survive in water. So, water is said to be responsible for levelling effect, which brings all stronger acids down to the acidity of H_3O^+. Thus, water acts as an appreciably good base towards all the common 'strong' acids like HCl, HNO_3 and H_2SO_4. All these appear to be equally strong in water medium as they are almost completely dissociated. An almost similar effect can be found for base in water. Any base strong enough to react completely with water to produce the OH^- ion will be levelled. So, water acts as an acid for all common 'strong' bases like NaOH, KOH, etc., and they are comparably strong in water medium. This tendency of water to equalize the strengths of many acids (or bases) is termed as levelling effect.

PROBLEMS

PROBLEMS ON ATOMIC STRUCTURE

1. Explain the curve of the ground state hydrogen orbital. Find out the difference between Bohr atom and wave mechanical atom.

2. What is the difference between orbit and orbital?

3. Calculate the wavelength of an electron that has been accelerated through a potential difference of 100 volts.

4. Calculate the wavelength of an electron moving with a velocity nearly 1/10th of that of light.

5. Calculate the wavelength of a rifle bullet of 2.2 gm moving with a velocity of 3×10^4 cm/s and interpret the result.

6. Calculate the frequency of radiation which gives a spectral line of $\lambda = 4500.0$ Å

$$\nu = \frac{c}{\lambda} = \frac{3\times10^8}{4500\times10^{-10}} = 6.6\times10^{14}\,s^{-1}$$

7. Calculate the energy in kJ per mole associated with a radiation of $\lambda = 300$ nm (1 nm = 10^{-9} m)

$\nu = \frac{c}{\lambda} = 3 \times 10^8 /300 \times 10^{-9}\,s^{-1} = 1 \times 10^{15}\,s^{-1}$

$\Delta E = h\nu = 6.626 \times 10^{-34} \times 1 \times 10^{15}$ J / atom

$= 6.626 \times 10^{-34} \times 1 \times 10^{15} \times 6.02 \times 10^{23} \times 1 / 1000$ kJ / mole

$= 398.9$ kJ / mole.

8. The energy of the electron in first Bohr's orbit of hydrogen atom is 2.178×10^{-18} J. Calculate the velocity of the electron in this orbit.

$E = + 2.178 \times 10^{-18} = 2\pi^2 z^2 e^4 m/h^2$, but $v = 2\pi z e^2/h$

$2.178 \times 10^{-18} = \pi e^2 mv/h$

$\therefore v = 2.178 \times 10^{-18} \times 6.626 \times 10^{-34} /3.14 \times (1.9 \times 10^{-19})^2 \times 9.16 \times 10^{-31}$

$= 2.18 \times 10^6$ metres / second.

9. Find the wavelength of radiation emitted when an electron falls from $n = 4$ to $n = 1$ in a hydrogen atom. The ionisation energy of hydrogen is 2.179×10^{-18} J.

$\Delta E = E_4 - E_1 = 2.179 \times 10^{-18}\ (1 - 1/16) = \frac{15}{16} \times 2.179 \times 10^{-18}$

$= 2.043 \times 10^{-18}$ J

Now, $\Delta E = h\nu$ or, $\nu = \Delta E/h = 2.043 \times 10^{-18}\,J/6.6 \times 10^{-34}$

$= 3.07 \times 10^{15}\,s^{-1}$

$\lambda = c / \nu = 3 \times 10^8 /3.07 \times 10^{15} = 9.97 \times 10^{-8}$m = 97.7 nm.

10. The threshold frequency of Cu is 1.1×10^{15}/ s. When UV light of frequency 1.5×10^{15} / s is shined on a Cu surface, what is the maximum energy of the photoelectron?

Incident energy = work function ($h\nu_0$) + K.E.

$h\,(1.5 \times 10^{15}) = h\,(1.1 \times 10^{15})$ + K.E.

$\therefore$ K.E. $= h\,(1.5 \times 10^{15} - 1.1 \times 10^{15})$

$= 6.63 \times 10^{-34}\ (0.4 \times 10^{15})$ J

$= 2.7 \times 10^{-19}$ J

Now, 1 eV = 1.6×10^{-19} J, $\therefore$ K.E. = 1.7 eV.

11. What is the lowest frequency of light to cause photoemission from the metal surface that requires 1.9 eV to eject an electron? [Ans. 4.1×10^{14} sec^{-1}]

12. Photoelectrons are emitted with maximum speed 7×10^5 m/sec from a surface when light of frequency 8×10^{14} cycles per second is shined on it. What is the threshold frequency for this surface?

13. Light of wavelength 5×10^{-7} m fall on potassium surface that requires 2 eV to eject an electron from the surface. What is the kinetic energy in eV of the ejected electron?

14. Calculate the de Broglie wavelength of a projectile of mass 10 mg and velocity 1 m/s.

$\lambda = h / mv = 6.626 \times 10^{-34}/10 \times 10^{-6} \times 1 = 6.626 \times 10^{-28}$ m.

15. Calculate the uncertainty in velocity Δv of an electron when uncertainty in its position is 10^{-10} m.

$\Delta x \times \Delta p = h / 4\pi.$

$\therefore \Delta v = h / 4\pi m \times \Delta x = 6.626 \times 10^{-34} / 4 \times 3.14 \times 9 \times 10^{-31} \times 10^{-10}$

$= 5.79 \times 10^5$ m/s.

16. The radius of Bohr's 3rd orbit is 476 pm for hydrogen. Calculate the radius of 4th orbit.

We know $r \propto n^2$, $\therefore r_4 / r_3 = 4^2 / 3^2 = 16 / 9$

$r_4 = 16/9 \times 476 = 846.2$ pm (1 pm = 10^{-12} cm).

17. Calculate the longest wavelength of line in the Balmer series for hydrogen spectrum.

The longest wavelength will have lowest energy, i.e., transition from $n = 3$ to $n = 2$.

$1/\lambda = x\bar{v} = R_H (1/2^2 - 1/3^2) = 109760 (1/4 - 1/9) = 109760 \times 5/36$

$\therefore \lambda = 6.56 \times 10^{-5}$ cm = 656 nm.

18. The energy of electron in hydrogen atom is -21.7×10^{-19} J. Calculate the energy required to remove the electron from 2nd orbit of hydrogen. What is the wavelength of light to be used to cause this transition?

The energy of the electron in 2nd quantum state will be

$E_2 = 21.7 \times 10^{-19} / Z^2 = 5.425 \times 10^{-19}$ J.

The wavelength of the light corresponding to this energy is

$$\Delta E = hv = hc / \lambda, \quad \therefore \lambda = hc / \Delta E$$

$\therefore \lambda = 6.626 \times 10^{-34} \times 3 \times 10^8 / 5.425 \times 10^{-19} = 3.66 \times 10^{-7}$ m = 366 nm.

19. According to Bohr's theory, the energy of an electron in the nth quantum level of hydrogen atom is $E_n = -2.176 \times 10^{-18}/n^2$ J Calculate the largest wavelength of light that can remove an electron from 3rd Bohr orbit of He^{+1} ion.

For He^+ ion the energy of electron will be

$E_n = -2.176 \times 10^{-18} Z^2 / n^2$ J $= -2.176 \times 10^{-18} \times Z^2 / n^2$ J

So, $E_3 = -2.176 \times 10^{-18} \times 2^2 / 3^2$ J $= -9.67 \times 10^{-19}$ J

The wavelength of light corresponding to this energy is

$E = hv = hc / \lambda$ i.e., $\lambda = hc/E$

$\lambda = 6.626 \times 10^{-34} \times 3 \times 10^8 / 9.67 \times 10^{-19} = 2.055 \times 10^{-7}$ m = 205.5 nm.

20. Considering the Rydberg constant to be $R_H = 109670\ cm^{-1}$, calculate the value of the longest wavelength of line in the Lyman series of hydrogen spectrum.

$\nu = R_H\ (1/n_1^2 - 1/n_2^2)$ The largest wavelength, i.e., the shortest or the smallest wave number will be the transition $2 \to 1$.

$\bar{\nu} = 109670 \times [\ 1/1^2 - 1/2^2\ ] = 3/4 \times 109670\ cm^{-1}$

$\lambda = \frac{1}{\nu} = 4/3 \times 109670\ cm = 1.215 \times 10^{-5}\ cm = 1215\ Å$.

21. Estimate the difference in energy between 1st and 2nd Bohr orbit of hydrogen atom. At what minimum atomic number, transition from 2 to 1 energy level would result in the emission of X-ray with $\nu = 3 \times 10^{-8}$ m? Which hydrogen like species does this atomic number correspond to?

In the hydrogen atom the energy difference,

$\Delta E = E_2 - E_1 = 2.18 \times 10^{-18}\ [1/1^2 - 1/2^2\ ]\ Z^2\ J = 1.635 \times 10^{-18}\ J \quad [Z = 1]$

The energy of the X-ray corresponds to $\lambda = 3 \times 10^{-8}$ m is $h\nu = hc/\lambda$

$= 6.626 \times 10^{-34} \times 3 \times 10^8 / 3 \times 10^{-8}\ J = 6.626 \times 10^{-18}\ J$

This energy will be equivalent to the energy $2.18 \times 10^{-18}\ (1/1^2 - 1/2^2\ )\ Z^2$.

$\because\ 2.18 \times 10^{-18} \times 3 \times Z^2/4 = 6.626 \times 10^{-18}$

$\therefore\ Z^2 = 6.626 \times 10^{-18} \times 4 / 2.18 \times 10^{-18} \times 3 = 4, \quad \because\ Z = 2.$

So, the corresponding species will be He^+.

22. What transition in the hydrogen spectrum will have the same wavelength as the Balmer transition $n = 4$ to $n = 2$ of He^+ spectrum?

Energy of transition for He^+ from 4 to 2 is

$\Delta E = 2.18 \times 10^{-18}\ (1/2^2 - 1/4^2)\ Z^2\ J = 2.18 \times 10^{-18} \times 3 \times 4 / 16\ J$

$= 2.18 \times 10^{-18} \times 3 / 4\ J.$

Since, for hydrogen atom $Z = 1$,

$\therefore\ 2.18 \times 10^{-18}\ (1/n_1^2 - 1/n_2^2)\ J = 2.18 \times 10^{-18} \times 3/4J,\ \because\ (1/n_1^2 - 1/n_2^2) = 3/4$ and this is possible when $n_1 = 1$ and $n_2 = 2$.

Thus, for hydrogen transition 2 to 1 will have the same energy difference for He^+ for the transition 4 to 2.

23. How much energy is required to ionise hydrogen atom in which the electron is in the 5th quantum level? [*Ans.* 8.68×10^{-13} ergs / atom]

24. Compute the energy required to remove the electron from Li^{+2} (g). The ionisation energy of ground state hydrogen atom is 13.6 eV. [*Ans.* 122.4 eV]

25. What is the energy in ergs required to shift the electron of hydrogen atom from 1st to 5th orbit and what is the wavelength of line emitted when it returns to the ground state? The ground state energy of electron is -2.17×10^{-11} erg. [*Ans.* Energy = 2.08×10^{-11} erg, $\lambda = 956$ Å]

26. Calculate the de Broglie wavelength of a proton moving with 25% of the velocity of light. The mass of proton = 1.675×10^{-24} gm. [*Ans.* 5.27×10^{-13} cm]

27. Given wavelength of the first line in the Balmer series of the hydrogen spectrum to be 6564 Å, compute the wavelength of 2nd and 3rd lines. [*Ans.* 4862 and 4341 Å]

28. Hydrogen atom in its ground state excited with monochromatic radiation of wavelength 975 Å. How many different lines are possible in the spectrum? Calculate the longest wavelength among them. Assume ionisation energy of hydrogen is 13.6 eV.

Energy given to the hydrogen atom, $E = hc / \lambda$

$E = 6.63 \times 10^{-34} \times 3 \times 10^8 / 975 \times 10^{-10} = 6.63 \times 3 \times 10^{-16} / 975\ J$

$= 6.63 \times 3 \times 10^{-16} / 975 \times 1.6 \times 10^{-19}\ eV = 12.75\ eV.$

Difference = 13.6 – 12.75 = 0.85 eV.

Energy level at which the electron promoted is n^2 = 13.6 / 0.85 = 16

$\therefore$ $n = 4$.

So, total six (6) transitions are possible.

The longest wavelength will be transition 4 to 3.

$\Delta E = 2.71 \times 10^{-18} (1/3^2 - 1/4^2) = 2.71 \times 10^{-18} \times 7/144$.

$\because hc / \lambda = \Delta E,$

$\therefore \lambda = hc / \Delta E = 6.6 \times 10^{-34} \times 3 \times 10^8 \times 144 / 2.71 \times 10^{-18} \times 7$

$= 18.83 \times 10^{-7}$ m.

29. The ionisation energy of hydrogen-like Bohr atom is $4R$ (Rydberg). (i) What is the wavelength of the radiation when the electron jumps from first excited state to the ground state and (ii) what is the radius of the first Bohr orbit for this atom?

Ionisation energy, $E = -RZ^2 = 4R$, $\quad Z^2 = 4$ or, $Z = 2$.

Energy radiated from 1st excited state; i.e., 2 to 1.

$h\nu = RZ^2 (1/1^2 - 1/2^2) = Rz^2 \times 3/4 = 4R \times 3/4$

$= 3R = 3 \times 2.2 \times 10^{-18} = 6.6 \times 10^{-18}$ J

$\therefore \lambda = hc / E = 6.6 \times 10^{-34} \times 3 \times 10^{-8} / 6.6 \times 10^{-18} = 3 \times 10^{-8}$ cm.

r = Bohr radius for hydrogen / Z = 0.532 Å / 2 = 0.266 Å.

30. At what speed must the electron move round the nucleus of hydrogen atom so that it must not be pulled into the nucleus? (radius = 0.5×10^{-10} m, mass = 9.1×10^{-31} kg, charge = 1.6×10^{-19} coulomb).

[*Ans.* 2.25×10^6 m/s]

31. Find the energy and momentum of an X-ray photon whose wavelength is 2×10^{-11} m.

32. Find the de Broglie wavelength of a ten-ton truck moving with a speed of 60 mph.

33. An electron is confined in a box of 10^{-9} m in length. What is the uncertainty in its velocity? If a proton is confined in the same box, what will be its uncertainty in velocity?

34. Calculate the energy in kcal/mole of the photon of an electromagnetic radiation of wavelength 3800 Å.

λ = 3800 Å = 3800×10^{-8} cm $\quad c = 3 \times 10^{10}$ cm/sec $\quad \nu = 3 \times 10^{10} / 3800 \times 10^{-8} = 7.9 \times 10^{14}$

$E = h\nu = 6.625 \times 10^{-27} \times 7.9 \times 10^{14} = 5.23\ 10^{-12}$ erg/quantum

E for one mole = $5.23 \times 10^{-12} \times 6.023 \times 10^{23}$ erg/mole

$= 5.23 \times 10^{-12} \times 6.023 \times 10^{23} / 4.2 \times 10^7$ cal/mole

= 75.26 kcal / mole.

35. Find the de Broglie wavelength of an electron whose speed is 1×10^8 m/sec and that of a proton whose kinetic energy is 1 meV.

36. Find out the number of waves made by a Bohr electron in one complete revolution in its 3rd orbit.

We know, $mvr = nh / 2\pi$ and $\lambda = h / mv$. So, $hr / \lambda = 3h / 2\pi$ or, $3\lambda = 2\pi r$

i.e., circumference of the 3rd orbit is three times the wavelength of the electron.

37. Calculate the uncertainty in the position of a particle when the uncertainty in momentum is (a) 1×10^{-7} gm cm s^{-1} and (b) zero.

(a) We know $\Delta x . \Delta p = h / 4\pi$

$\therefore \Delta x = 6.62 \times 10^{-34} / 4 \times 3.14 \times 10^{-7}$J sec m = 0.527×10^{-27} m.

(b) $\Delta p = 0$, so Δx is infinity.

38. Calculate the uncertainty in the velocity of a bullet of 10 gm whose position is known with ± 0.01 mm.

Now, $m = 10$ gm $= 1 \times 10^{-2}$ kg, $\Delta x = 0.01$ mm $= 1 \times 10^{-5}$ m

$\Delta x.\ \Delta p = \Delta x \times m\ \ \Delta v = h/4\pi$

$\Delta v = 6.62 \times 10^{-34}/4 \times 3.14 \times 10^{-2} \times 10^{-5} = 0.527 \times 10^{-27}$ m/sec.

39. In Balmer series of hydrogen spectra, there is a line corresponding to wavelength 4344 Å. Calculate the number of higher orbit from which the electron drops to generate other line.

$\nu = c / \lambda = 3 \times 10^{8}$ m / sec^{-1}/ 4344 $\times 10^{-10}$ m $= 6.907 \times 10^{14}$ sec^{-1}

Let the number of higher level from where electrons drops is $n_2 = Rc\ [1 / n_1^2 - 1 / n_2^2]$

or, 6.907×10^{14} sec^{-1} $= 3.289 \times 10^{15}\ [1/4 - 1/n^2]$

or, $n^2 - 4 /4\ n^2 = 6.907/3.289 \times 10 = 0.21$

or, $n^2 - 4 = 0.21 \times 4n^2 = 0.84n^2$

or, $n^2 - 0.84n^2 = 4$

or, $0.16n^2 = 4$

or, $n^2 = 4/0.16 = 25, \quad n = 5.$

Hence, the electron drops from the 5th quantum state.

PROBLEMS ON RADIOACTIVITY

1. Uranium mineral (pitchblende) contains 15 gm Pb / 100 gm of Uranium. What is the age of the mineral? Given, $t_{1/2}$ of $^{238}U_{92} = 4.5 \times 10^9$ years.
2. An archaeological specimen containing $^{14}C_6$ gives 40 counts in 5 minutes / gm-atom of Carbon. A freshly cut wood gives 20.3 counts / minute / gm of Carbon. The counter used recorded at 5 counts / minute as background count without $^{14}C_6$ sample. What is the age of the sample? ($t_{1/2}$ of $^{14}C_6 = 5668$ years)
3. Calculate the binding energy per mole of Oxygen atom $^{16}O_8$, given mass of proton = 1.007277 amu, mass of neutron = 1.008665 amu, mass of electron = 0.0005486 amu, actual mass of $^{16}O_8$ = 15.99491 amu, Δm = 0.13701amu.

   $E = \Delta mc^2 = 0.13701 \times 1.6605 \times 10^{-27} \times (3 \times 10^8)^2$ J / atom

   $= 0.13701 \times 1.6605 \times 10^{-27} \times (3 \times 10^8)^2 \times 6.023 \times 10^{23}$ J / mole

   $= 1.23 \times 10^{13}$ J / mole.
4. A radio element has $t_{1/2}$ = 35 sec. Calculate its decay constant and time required for the activity to reduce to 1% of original value.

   $\lambda = 0.693 / 35 = 1.98 \times 10^{-2}$ sec^{-1} and

   time required = $t = (2.303 / \lambda) \log (N_0 / N)$

   or, $t = (2.303 / 1.98 \times 10^{-2}) \log (100 / 1) = 232.6$ sec.
5. A hospital buys 4 gm Co-60 for therapeutic use. How much Co-60 will be left after 20 years, if $t_{1/2}$ of Co-60 = 5 years?

   20 years = 5 × 4, i.e., 4 half-lives

   ∴ amount left = $(1/2)^4 \times 4 = 0.25$ gm.
6. The amount of ^{14}C in a wood sample is found to be 1/10th of the amount present in a fresh wood. What is the age of the wood, if $t_{1/2}$ of ^{14}C is 5770 years?

   $\lambda = 0.693 / 5770 = 1.2 \times 10^{-4}$ / year

   ∴ Age of the wood = $t = (2.303/1.2 \times 10^{-4}) \log (1/0.1)$

   = 19192 years.
7. An Uranium mineral contains U-238 and Pb-206 in the ratio 4:1. Calculate the age of the mineral ($t_{1/2}$ of U-238 = 4.5×10^9 years).

   1 gm-atom of Pb-206 ≅ 1 gm-atom U-238

   206 gm-atom of Pb-206 ≅ 238 / 206 gm atom U-238 = 1.155 gm

   So, $N_0 = 4 + 1.155 = 5.155$ and $N = 4$

   $\lambda = 0.693 / 4.5 \times 10^9 = 1.54 \times 10^{-10}$ / year

   ∴ age, $t = (2.303 / 1.54 \times 10^{-10}) \log (5.155 / 4) = 1.65 \times 10^9$ years.
8. Radioactive decay is first order process–C-14 in wood decays with $t_{1/2}$ = 5770 years. What is the decay constant? What fraction would remain after 11540 years?

   $\lambda = 0.693 / 5770 = 1.2 \times 10^{-4}$/year.

   11540 year = 2 times $t_{1/2}$.

   So, amount of C-14 remaining will be $(1/2)^2 = 0.25$.

9. An experiment required minimum β activity produced at the rate 346 β per minute.$t_{1/2}$ of ${}_{42}Mo^{99}$ which is a β emitter is 66.6 hours. Find out the minimum amount of ${}_{42}Mo^{99}$ required for carrying out the experiment in 6.909 hours.

λ for ${}_{42}Mo^{99} = 0.693/66.6 \times 60 = 1.734 \times 10^{-4}$ / min.

Minimum activity at the end of 6.909 hours = 346

Let at the beginning the activity = x ; So by formula,

$\lambda = 2.303/\ 6.909 \times 60 \log x/346 = 1.734 \times 10^{-4}$

$\log x/346 = 1.734 \times 10^{-4} \times 6.909 \times 60/2.303 = 312.1 \times 10^{-4}$

$x/346 = 1.0745$ or, $x = 372$.

So, the initial disintegration will be 372 min^{-1}

Now, 372 disintegration will be obtained from $372/1.734 \times 10^{-4}$,

that will weigh $(99/6.023 \times 10^{23}) \times (372/1.734 \times 10^{-4}) = 3.53 \times 10^{-16}$ gm.

10. A radioactive element has $t_{1/2}$ = 30 days. What will be its average life?

$\lambda = 0.693/30 = 0.0231$/day ; so $= 1/\ 0.0231 = 43.29$ days.

11. Radium emiits 3.7×10^{10} β-particles/gram /second. Again it is found that Radium produces He gas at the rate of 1.07% 10^{-4} cc (STP)/gm/day. Calculate Avogadro's no. from this data.

β-particle emitted/gm/day will be $3.7 \times 10^{10} \times 24 \times 60 \times 60$ and this β particles are converted to He gas.

So, 1.07×10^{-4} ml should contain $3.7 \times 10^{10} \times 24 \times 60 \times 60$ He atoms.

$\therefore$ 22.4 litres will have $3.7 \times 10^{10} \times 24 \times 60 \times 60 \times 22400/\ 1.07 \times 10^{-4} = 6.6 \times 10^{23}$

12. What mass of ${}_{27}Co^{60}$ has activity of 1 curie? Given $t_{1/2}$ = 5.25 years.

$dn/dt = \lambda N = 3.7 \times 10^{10}$ /sec. $\therefore N = 3.7 \times 10^{10}/\lambda$

$\lambda = 0.693/5.25 \times 365 \times 24 \times 60 \times 60 = 4.18 \times 10^{-9}$

$\therefore N = 3.7 \times 10^{10}/\ 4.18 \times 10^{-9} = 8.85 \times 10^{18}$

6.023×10^{23} atom weighs 60 gm

$\therefore 8.85 \times 10^{18}$ atom weighs $60 \times 8.85 \times 10^{18} /6.023 \times 10^{23}$

$= 8.8 \times 10^{-4}$ gm $= 8.8 \times 10^{-7}$ kg.

13. A radioactive element emits one β-particle for each decay process. Measurement shows that an average of 8.4 β-particles is emitted each second by 2.5 mg of the substance. The atomic weight is 230. Calculate the $t_{1/2}$ of the substance.

Now, 2.5 mg of the substance contains atoms $= 0.0025 \times 6.02 \times 10^{23} /230 = 6.54 \times 10^{18}$

Now, $dN/dt = \lambda N = 8.4$ $\qquad \lambda = 8.4/6.54 \times 10^{18} = 1.28 \times 10^{-18}$ /sec

$t_{1/2} = 0.693 / \lambda = 0.693 /1.28 \times 10^{-18} = 5.41 \times 10^{17}$ sec

$= 5.41 \times 10^{17} /365 \times 24 \times 60 \times 60 = 1.7 \times 10^{10}$ years.

14. Express mass of an electron and 1amu in energy unit.

$E_m = 9.11 \times 10^{31}$ kg

$E = m_0c^2 = 9.11 \times 10^{-31} \times 3^2 \times (10^8)^2$ J

$= 9.11 \times 10^{-31} \times 3^2 \times (10^8)^2 /1.602 \times 10^{-13}$ MeV $= 0.511$ MeV.

1 amu $= 1.66 \times 10^{-27}$ kg

$E = mc^2 = 1.66 \times 10^{-27} \times 3^2 \times (10^8)^2$ J $= 1.66 \times 10^{-27} \times 3^2 \times (10^8)^2/1.602 \times 10^{-13} = 931$ MeV.

15. What is the power output of a $_{92}U^{235}$ reactor if it takes 30 days to use up 2 kg of fuel, if each fission gives 185 MeV usable energy?

The number of $_{92}U^{235}$ nuclei in 2 kg = $2 \times 1000 \times 6.023 \times 10^{23}/235 = 5.12 \times 10^{24}$.

So, fission rate = $5.12 \times 10^{24}/ 30 \times 24 \times 60 \times 60 = 1.975 \times 10^{18}$/sec.

Power = $1.975 \times 10^{18} \times 185 \times 10^{6} \times 1.602 \times 10^{-19} = 58.5$ MW.

16. $_{19}K^{40}$ isotope decays to $_{18}Ar^{40}$ with $t_{1/2} = 1.37 \times 10^{9}$ years. Moon rock shows the ratio of the $_{19}K^{40}$ to be 1/7. Estimate the age of the rock. [*Ans.* 4.1×10^{9} years]

17. Radium-C after disintegration reduces its activity to 1/128 of the original value in 14 sec. What is the $t_{1/2}$ of Ra-C?

$$t_{1/2} = \frac{2.303}{\lambda} \log \frac{N_0}{N_t}.$$

$$\lambda = 2.303/140 \log \frac{1}{\frac{1}{128}} = 2.303/140 \text{ sec} \log 2^7 = 2.303/140 \times 7 \times 0.3010$$

$t_{1/2} = 0.693 / \lambda = .693 \times 140/2 = 20$ min.

18. A small volume of a solution containing radioactive sodium isotope of activity 12000 dpm was injected in the blood of a patient. After 30 hours, activity of 1.0 cc of the blood was found to be 0.5 dpm. If $t_{1/2}$ of Na-isotope is 15 hours, estimate the volume of blood.

[*Ans.* 6000 cc]

19. One gram of Radium^{226} gives 11.6×10^{17} α-particle per year. Calculate the Avagadro's number ($t_{1/2}$ = 1590 yr).

As 1α is ejected from 1 atom, $-dN/dt = 11.6 \times 10^{17}$ $\lambda = 0.693/1590$

so, $-dN/dt = \lambda N$ or, $N = 11.6 \times 10^{17} \times 1590 / 0.693$

This number of atom present in 1 gm.

∴ Avogadro's number = $11.6 \times 10^{17} \times 1590 \times 226 / 0.693 = 6.0 \times 10^{23}$.

20. Hydrogen isotope $_1H^3$ is radioactive and emits an electron with $t_{1/2}$ = 12.5 years. What does $_1H^3$ become after decay? What percentage of an original sample of $_1H^3$ will remain after 25 years of its preparation?

21. In certain stars three particles join in a single reaction to form $_6C^{12}$ nucleus, whose mass is 12.0038 amu. Mass of β-particle is 4.003873 amu. How much energy is evolved in this reaction?

22. If in each fission of $_{92}U^{235}$ produces 200 MeV, how many fissions must occur per second to produce a power of 1 kilowatt?

PROBLEMS ON SOLID STATE

1. CsBr has CsCl structure (body-centered cubic type lattice). Its density is 4.49 gm cm^{-3}. Calculate the side of the unit cell.

   Molar volume of CsBr = 212.8 / 4.49
   $= 47.39$ cc

   Volume of each molecule $= 47.39 / (6.023 \times 10^{23})$ cc
   $= 7.868 \times 10^{-23}$ cc

   Volume of unit cell $= 7.868 \times 10^{-23}$ cc

   Considering that $\sqrt[3]{7.868\times10^{-23}}$ unit cell contains one molecule, side of unit cell will be
   $= 4.285 \times 10^{-8}$ cm
   $= 4.285$ Å.

2. Aluminium metal (at.wt.26.98) crystallizes in the cubic system with $a = 4.05$ Å. Its density is 2.7 gm/cc. Determine the unit cell type and radius of the aluminium atom.

   Molar volume of Al
   $= 6.023 \times 10^{23} \times (4.05 \times 10^{-8})^3 / n = 26.98 / 2.7$
   $\therefore n = 4.003 \approx 4$ [considering a unit cell contains n atoms]

   Thus, the cubic lattice is face-centered type.

   Diagonal of the face is $a\sqrt{2} = 4.05 \text{ Å} \times \sqrt{2}$
   $= 5.727 \text{ Å} = 4r$
   $\therefore r = 1.432$ Å.

3. Vanadium crystallises with a body centered cubic structure with a unit cell of each side 3.011 Å. Calculate the atomic radius and density. Atomic weight of vanadium is 50.94.

   Body diagonal of the cube $= \sqrt{3}a - 4r$

   $$\therefore 4r = \sqrt{3} \times 3.011 \text{ Å}$$

   $$\therefore r = \frac{\sqrt{3} \times 3.011 \text{Å}}{4}$$
   $= 1.304$ Å.

   Now, volume of unit cell $= (3.011 \times 10^{-8})^3$ and each cell contains two atoms.

   So, the volume of one atom $= (3.011 \times 10^{-8})^3 / 2$

   Volume per gram atom $= (3.011 \times 10^{-8})^3 / 2 \times 6.023 \times 10^{23} = 50.94 / \rho$

   $\therefore \rho = 50.94 \times 2 / (3.011 \times 10^{-8})^3 \times 6.023 \times 10^{23}$
   $= 6.199$ gm/cc.

4. NaBr has NaCl structure. Its density is 3.203 gm/cc. Calculate the unit cell side (at. wt. of Br 79.9).

   Molar volume $= 102.9 / 3.203$

   So volume of one molecule will be $102.9 / (3.023 \times 6.023 \times 10^{23})$ = volume of unit cell.

   $\because a^3 / 4 = 102.9 / (3.023 \times 6.023 \times 10^{23})$,

   $\therefore a = 5.976$ Å.

5. CsI has the lattice of two interpenetrating simple cube structures, I^- ion being at the centre of the lattice of Cs^+ ions. The unit cell side length is 4.562 Å. Calculate the density of CsI (Cs = 132.9, I = 127).

Unit cell contains one ion pair Cs^+I^-.

Volume of one ion pair is $(4.562 \times 10^{-8})^3 = a^3$

So, molar volume = $6.023 \times 10^{23} \times (4.562 \times 10^{-8})^3$

It is equal to (132.9+127) / ρ

$\therefore \rho = (132.9+126.9)/6.023 \times 10^{23} \times (4.562 \times 10^{-8})^3$

= 4.546 gm/cc.

6. Silver (at. wt. =107.88) crystallises with fcc lattice for which the side length of the unit cell is 4.0774 Å. Density of Ag is 10.53 gm/cc. Calculate the Avogadro's number.

Molar volume of Ag is $107.88 / 10.53 = N_o \times (4.0744 \times 10^{-8})^3 / 4$

$\therefore N_o = 107.88 \times 4/ 10.53 \times (4.0774 \times 10^{-8})^3$

$= 6.045 \times 10^{23}$.

7. An element of atomic mass 60 having fcc structure has a density of 6.023 gm/cc. What is the edge length of unit cell?

Now, mass of an atom = $60 / (6.023 \times 10^{23}) = 10^{-22}$ gm

As fcc has 4 atoms per unit cell, so mass of an unit cell is 4×10^{-22} gm.

So, volume of a unit cell = $4 \times 10^{-22} / 6.023 = 6.64 \times 10^{-23}$.

Therefore, edge length of a unit cell is $\sqrt[3]{6.64 \times 10^{-23}}$

$= 4.05 \times 10^{-8}$ cm

= 4.05 Å

= 405 pm.

PROBLEMS ON BORN-HABER CYCLE

1. Calculate the lattice energy of RbCl from the following data:
   ΔH_s of Rb = + 20.6 kcal mole^{-1}, ΔH_D of Cl_2 = 57.6 kcal mole^{-1}
   ΔH_i of Rb = + 95.9 kcal mole^{-1}, ΔH_A of Cl_2 = – 90.7 kcal mole^{-1}
   ΔH_f of RbCl = –105 kcal mole^{-1}
   We know, $\Delta H_f = \Delta H_s + \Delta H_i + 1/2\ \Delta H_D + \Delta H_A + \Delta H_L(U)$
   or, $-105 = 20.6 + 95.9 + 57.6/2 - 90.7 + \Delta H_L(U)$
   or, $\Delta H_L(U) = -159.6$ kcal mole^{-1}.
2. Calculate the lattice energy of MgS from the following data, energies are in kcal mole^{-1}. Given,
   ΔH_s of Mg = 36.5, ΔH_D of S_8 133.2 × 8, ΔH_i of Mg = 520.6, ΔH_A of S = –72.4
   ΔH_f of MgS = –82.2
   $\Delta H_f = \Delta H_s + \Delta H_i + 1/8\ \Delta H_D + \Delta H_A + \Delta H_L(U)$
   or, $-82.2 = 36.5 + 520.6 + 133.2 - 72.4 + \Delta H_L(U)$
   or, $\Delta H_L(U) = -700.1$ kcal mole^{-1}.
3. Calculate the electron affinity of Cl_2 from the following data:
   $\Delta H_L(U)$ of KCl = – 169.4 kcal mole^{-1}, ΔH_f of KCl = –104.2 kcal mole^{-1}
   ΔH_s of K = + 21.5 kcal mole^{-1}, ΔH_i of K = + 100 kcal mole^{-1}
   ΔH_D of Cl_2 = 57.8 kcal mole^{-1}
   $\Delta H_f = \Delta H_s + \Delta H_i + 1/2\ \Delta H_D + \Delta H_A + \Delta H_L(U)$
   $-104.2 = 21.5 + 100 + 28.9 + \Delta H_A - 169.4$
   or, $\Delta H_A = -85.2$ kcal mole^{-1}
   = –3.7 eV (1 eV = 23 kcal mole^{-1})
4. Calculate the lattice energy of KI from the following data, energies are in kcal mole^{-1}.
   ΔH_f of KI = –78.9 kcal/mole, ΔH_s of K = + 21.7 kcal/mole,
   ΔH_i of $K_{(g)}$ = 99.6 kcal/mole, ΔH_D of I_2 = 50.8 kcal/mole,
   ΔH_A of I_2 = –76.6 kcal/mole.
   $\Delta H_f = \Delta H_s + \Delta H_i + 1/2\Delta H_D + \Delta H_A + \Delta H_L$
   $-78.9 = 21.7 + 99.6 + 25.4 - 76.6 + \Delta H_L$
   $\Delta H_L = -78.9 - 21.7 - 99.6 - 25.4 + 76.6$
   = 149.6 kcal/mole.
5. Draw the Born-Haber Cycle for the formation of $MgCl_2$.

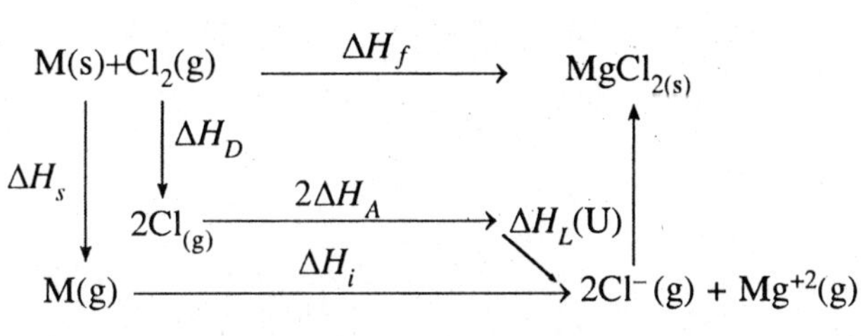

APPENDIX

Particle in a one-dimensional box

The simplest application of Schrödinger equation is found in the treatment of a particle confined to move within a box, rectangular in shape with dimensions *xyz* lying along *X*, *Y* and *Z* axes. The particle moves only inside the box and it has no existance outside the box, i.e., it has infinite potential energy at the side walls. The particle when comes in contact with the side wall, it is reflected without loosing energy rather than penetration. Thus the total energy is constant. The potential energy within the box is also constant and for convenience we assume it to be zero, i.e., $V = 0$.

Within this box the wave function of the particle can be represented by ψ_{xyz}. For simplicity we consider the box to the one-dimensional, the particle moves between $x = 0$ and $x = L$, so the wave function will be ψ_x along *x*-axis. For realistic particle this function will be well-behaved function, i.e., it will be continuous, finite and single-valued.

As the particle now is confinedd within one-dimensional square well of infinitely high wall, its bahaviour can be represented by time-independent Schrödinger equation in *x*-direction

$$\frac{d^2\psi(x)}{dx^2} + \frac{8\pi^2 m}{h^2}(E - V)\psi_{(x)} = 0$$

and when $V = 0$, it is $\dfrac{d^2\psi(x)}{dx^2} + \dfrac{8\pi^2 m}{h^2} E_x \psi_{(x)} = 0$

or, $\dfrac{d^2\psi_{(x)}}{dx^2} + k^2\psi_{(x)} = 0$, where $k^2 = \dfrac{8\pi^2 mE_x}{h^2}$

The general solution of this differential equation is

$$\psi_{(x)} = A\sin kx + B\cos kx$$

At the boundary conditions, i.e., at $x = 0$ and at $x = L$, $\psi_{(x)}$ must be zero.

$0 = A \sin(0) + B \cos(0)$

or, $0 = A.0. + B.1$ or $B = 0$

so $\psi_{(x)} = A \sin kx$

At the other wall where $x = L$

$0 = A \sin kL$

As $A \neq 0$, $\sin kL = 0$

or, $kL = n\pi$, where $n = 1, 2, 3, ...,$ etc.

or, $k = \dfrac{n\pi}{L}$, i.e., $\psi_{(x)} = A\sin\left(\dfrac{n\pi}{L}x\right)$

Equating k^2, we have

$$k^2 = \frac{8\pi^2 mE_x}{h^2} = \frac{n^2\pi^2}{L^2} \quad \text{or,} \quad En = \frac{h^2 n^2}{8mL^2}$$

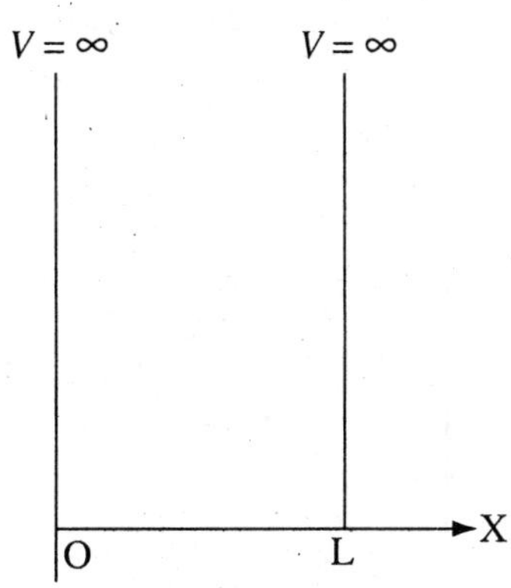

One-dimentional box

From the equation of the energy it is seen that

(a) $E_n \propto n$, i.e., for each value of n, E_n has definite value, so energy is quantised and also explains its discontinuous character.

(b) Other things remaining same $E_n \propto \frac{1}{L}$, i.e., in smaller box energy of the particle will be greater.

(c) Since $n \neq 0$, (as whole wave function disappear) the particle will have minimum energy $\frac{h^2}{8L^2m}$, even at absolute zero. This is known as zero point energy.

(d) As $E_n \propto n^2$, when n increases, the successive energy levels get closer and ultimately merge into a continuum.

$$\text{So, } \Psi_x = A\sin\frac{n\pi x}{L} = A\sin\left(\frac{\sqrt{8\pi^2 mE_x}}{h}\right)$$

This is a well-behaved function. For each value of n, ψ_x has a definite value, i.e., it is a single-valued function, ψ_x is contineous, and only the value of A is unknown; it can be known by applying the normalization condition. Since the particle is in the box, its probability is unity. Hence

$$\int_0^L A^2\sin^2\left(\frac{n\pi x}{L}\right)dx = 1 \text{ or, } A^2\int_0^L \frac{1}{2}\left(1-\cos\frac{2\pi nx}{L}\right)dx = 1$$

From the solution we get $A = \sqrt{\frac{2}{L}}$ and the wave function becomes $\psi_x = \sqrt{\frac{2}{L}}\sin\frac{n\pi x}{L}$.

The nature of the plot of ψ_x with different values of n which may be both (+)ve or (–)ve and corresponding probability densities which is (+)ve are shown in the diagram.

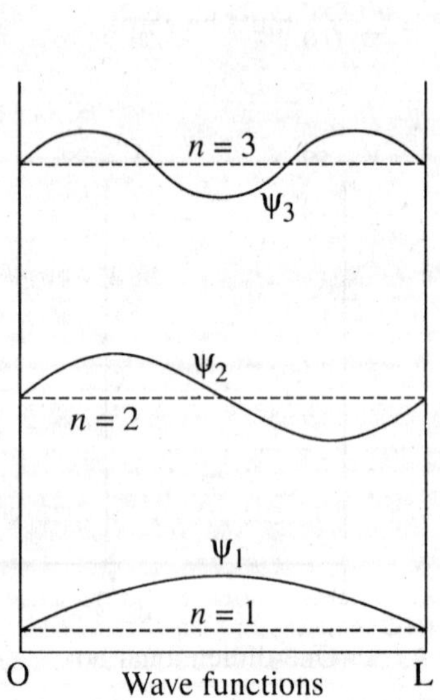

Wave functions in one-dimensional box

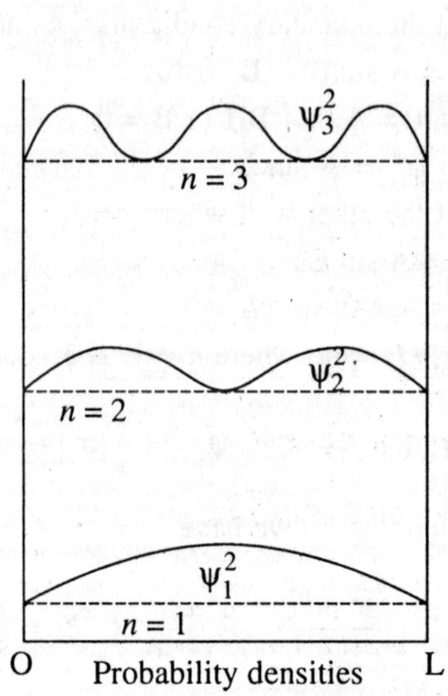

Probability densities in one-dimensional box

From the plot of probability densities, density is maximum at $\frac{L}{2}$ for $n = 1$, for $n = 2$, the node is at $\frac{L}{2}$ and maximum probability is at $\frac{L}{4}$ and $\frac{3L}{4}$, and for $n = 3$, nodes are at the distances $\frac{L}{3}$ and $\frac{2L}{3}$, and maximum probability are at $\frac{L}{6}, \frac{3L}{6}$ and $\frac{5L}{6}$. The total number of nodes for any value of n is $(n - 1)$.

Three-dimensional box

Like before the ψ function will be $\psi_{(xyz)} = X_{(x)}Y_{(y)}Z_{(z)}$ and the Schrödinger equation will take the form as

$$\left(\frac{d^2}{dx^2}+\frac{d^2}{dy^2}+\frac{d^2}{dz^2}\right)\psi_{(xyz)}+\frac{8\pi^2 m}{h^2}E\,\psi_{(xyz)}=0$$

$$\text{or, } \left(\frac{d^2}{dx^2}+\frac{d^2}{dy^2}+\frac{d^2}{dz^2}\right)X_{(x)}Y_{(y)}Z_{(z)}+\frac{8\pi^2 m}{h^2}E\,X_{(x)}Y_{(y)}Z_{(z)}=0$$

$$\text{or, } Y_{(y)}Z_{(z)}\frac{d^2X_{(x)}}{dx^2}+X_{(x)}Z_{(z)}\frac{d^2Y_{(y)}}{dy^2}+X_{(x)}Y_{(y)}\frac{d^2Z_{(z)}}{dz^2}+\frac{8\pi^2 m}{h^2}E\,X_{(x)}Y_{(y)}Z_{(z)}=0$$

$\frac{d^2}{dx^2}$ will have no effect on $Y_{(y)}Z_{(z)}$ and so on.

Or multiplying with $\frac{1}{X_{(x)}Y_{(y)}Z_{(z)}}$ we have

$$\frac{1}{X_{(x)}}\frac{d^2X_{(x)}}{dx^2}+\frac{1}{Y_{(y)}}\frac{d^2Y_{(y)}}{dy^2}+\frac{1}{Z_{(z)}}\frac{d^2Z_{(z)}}{dz^2}=-\frac{8\pi^2 mE}{h^2}$$

The sum of each term in the L.H.S. is a constent. So each term will be equal to a constant, say $-k^2$.

$$\therefore\ \frac{1}{X_{(x)}}\frac{d^2X_{(x)}}{dx^2}=-k_{(x)}^2,\ \frac{1}{Y_{(y)}}\frac{d^2Y_{(y)}}{dy^2}=-k_{(y)}^2 \text{ and } \frac{1}{Z_{(z)}}\frac{d^2Z_{(z)}}{dz^2}=-k_{(z)}^2$$

$$\text{and } k_{(x)}^2+k_{(y)}^2+k_{(z)}^2=\frac{8\pi^2 mE}{h^2}$$

Solution of each term will be like that in one-dimensional potential box.

So, $\frac{1}{X_{(x)}}\frac{d^2X_{(x)}}{dx^2}-k_x^2$ or, $\frac{d^2X_{(x)}}{dx^2}+k_x^2X_{(x)}=0$ and the normalized function will be

$$X_{(x)}=\sqrt{\frac{2}{L_x}}\sin\frac{n_x\pi x}{L_x}$$

so total wave function will be

$$\psi_{xyz} = \sqrt{\frac{8}{L_x L_y L_z}} \cdot \sin\frac{n_x \pi x}{L_x} \cdot \sin\frac{n_y \pi y}{L_y} \cdot \sin\frac{n_z \pi z}{L_z}$$

and $E = E_x + E_y + E_z = \frac{h^2}{8m}\left[\frac{n_x^2}{L_x^2} + \frac{n_y^2}{L_y^2} + \frac{n_z^2}{L_z^2}\right]$

so energy of the particle is also quantized.

For a cubical box, when $L_x = L_y = L_z = L$ (say)

$$E = \frac{h^2}{8mL}\left[n_x^2 + n_y^2 + n_z^2\right]$$

It may be easily shown that $E_{112} = E_{121} = E_{211}$. Thus states having different n_x, n_y and n_z values may have same energy. Such states having same energy but different n_x, n_y, n_z are called *degenerate states*.

BOOKS FOR SUPPLEMENTARY READING

1. Arnikar, H. J., *Essential of Nuclear Chemistry*
2. Basrnard, D. T., *Theoretical Basis of Inorganic Chemistry*
3. Burns, D. T., Carter, A. H. and Townshend, A., *Reactions of Elements and Their Compounds*
4. Beiser, A., *Concepts of Modern Physics*
5. Barlar, J. C. and Emeleus, H. J., *Comprehensive Inorganic Chemistry*
6. Cotton, F. A. and Wilkinson, G., *Basic Inorganic Chemistry*
7. Cartmell, E. and Fowles, G. W. A., *Valence and Molecular Structure*
8. Day, M. C. and Selbin, J., *Theoretical Inorganic Chemistry*
9. Drago, R. S., *Physical Methods in Inorganic Chemistry*
10. Dutta, P. K., *General and Inorganic Chemistry*
11. Dauglas, B., Mc Daniel, D. H. and Alexander, J. J., *Concepts and Models of Inorganic Chemistry*
12. Emeleus, H. J. and Sharpe, A. G., *Modern Concepts of Inorganic Chemistry*
13. Glasstone, S., *Textbook of Physical Chemistry*
14. Glasstone, S., *Source Book of Atomic Energy*
15. Huheey, J. E., *Inorganic Chemistry*
16. Lee, J. D., *Concise Inorganic Chemistry*
17. Maron, S. H. and Prutton, C. F., *Principles of Physical Chemistry*
18. Moeller, T., *Inorganic Chemistry*
19. Moore, W. J., *Physical Chemistry*
20. Mukherjee, S. N., *Physical Chemistry* (Vol. III)
21. Rajam, J. B., *Atomic Physics*
22. Rakshit, P. C., *Physical Chemistry*
23. Sanderson, R. T., *Chemical Periodicity*
24. Sharpe, A. G., *Inorganic Chemistry*

INDEX

A

B

C

D

University Question Papers

University Question Papers

2005

PART—I

SECOND PAPER

Full Marks—100

(New Syllabus)

Answer *one* question from each of Units **1** to **5**. Question No. **11** is Compulsory

UNIT—1

1. (a) (i) Discuss the origin and physical significance of magnetic quantum number.
   (ii) Identify the possible Bohr-Sommerfeld orbits for $n = 4$. 4+2
   (b) In a hydrogenic ion the energy required to excite the electron from its ground state to first excited state is 40.8 eV. Find out the energy required to release the electron from the ion. 3
   (c) State Hund's rules and hence find out the ground state term for carbon atom. 3
   (d) What is radial distribution function? Show diagrammatically the variation of radial distribution function with '*r*' for the orbitals, 3*s* and 3*p*, in a hydrogen atom. 3

2. (a) Explain the concept of Nuclear Binding Energy and hence the binding energy curve. How does the curve give information about nuclear fission and fusion as energy source? 2+1+2
   (b) What are half-life and mean-life of a radio-element? Using radioactive disintegration law, find mathematical expression for each of them. From their relation find which one is greater. 2+3+1
   (c) A small amount of a radioactive material of half-life 20 days got inadvertently spread in a laboratory making the level of radiation 40 times the permissible safety level. After how many days the laboratory would be safe to work? 3
   (d) Show that the de Broglie wavelength of the electron in the first Bohr orbit of hydrogen atom is $2\pi a_0$. [where a_0 = 1st Bohr radius]. 2

UNIT—2

3. (a) Define formal charge. Draw the Lewis structure of carbonate ion and calculate the formal charge on each atom. 4
   (b) Construct the M.O. energy level diagram for HF molecule and discuss on which atom the bonding and anti-bonding orbitals are concentrated. 3+1
   (d) Explain why the conductivity of Ge is enhanced manifolds when trace amount of As is added to it. 3

4. (a) Define lattice energy of an ionic solid. Frame the Born-Haber cycle for the formation of NaCl crystal starting from solid sodium metal and gaseous chlorine, and hence calculate the lattice energy of crystalline NaCl using the following data:

   Sublimation energy of metallic Na = 108 kJ/mole

Bond dissociation energy of Cl_2 = 240 kJ/mole

Ionisation energy of Na = 495 kJ/mole

Electron affinity energy of Cl atom = –359 kJ/mole

Heat of formation of solid NaCl = –410 kJ/mole

(b) What do you mean by radius ratio? What information can be obtained from it? Find out the limiting value of the radius ratio of cubic coordination. 2+2

(c) Using VSEPR theory predict the shapes of XeO_2F_2 and BrF_4^-. Indicate the state of hybridisation of the central atom in each case. 2+2

(d) Indicate the effects of the following ionisation processes on bond orders and bond lengths:

(i) $O_2 + e \rightarrow O_2^-$; (ii) $N_2 - e \rightarrow N_2^+$ 2

UNIT—3

6. (a) Define 'ionisation potential'. Discuss the factors controlling the ionisation potential values of elements. 5

(b) What is meant by ionic radii? Discuss Pauling's method for the determination of univalent radii. 1+4

(c) The atomic radii of Zr and Hf are almost indentical—Explain. 2

(d) State Slater's rules and using these calculate the effective nuclear charge for the following electrons: (i) 4*s* electron in Zn atom ($Z = 30$), (ii) 3*d* electron in Zn atom. 4

UNIT—5

9. (d) Arrange the following ions in increasing order of acidity with explanation: 2

$[Sc(H_2O)_6]^{3+}$, $[Mn(H_2O)_6]^{2+}$, $[Ni(H_2O)_6]^{2+}$

(e) What is HSAB principle? Explain with example characteristics of soft and hard acids and bases. 4

10. (a) What do you mean by standard electrode potential ($E°$) of redox couple? How does it differ from formal potential? 2+2

(b) Explain the observation: 3

Cu^{2+} ion readily liberates iodine from iodide but in presence of ethylenediamine it does not.

[Given: $E^o_{Cu^{2+}/Cu^+} = 0.15$ volt, $E^o_{Cu^{2+}/CuI^+} = 0.87$ volt and $E^o_{\frac{1}{2}I_2/I^-} = 0.54$ volt.]

(c) In I(N) H_2SO_4 medium, calculate the redox potential values at the following three stages of titration of 0.1(N) Fe^{2+} with 0.1(N) $KMnO_4$:

(i) 25 ml Fe^{2+} + 24.95 ml $KMnO_4$

(ii) 25 ml Fe^{2+} + 25 ml $KMnO_4$

(iii) 25 ml Fe^{+2} + 25.05 ml $KMnO_4$

[Given: $E^o_{Fe^{3+}/Fe^{2+}} = 0.77$ volt and $E^o_{MnO_4^-/Mn^{2+}} = 1.51$ volt.]

(d) An aqueous solution which is 0.3(M) with HCl contains Zn^{2+} and Cd^{2+} ions. The concentration of each cation in the solution is 0.02 gm ion/lit. The solution is saturated with H_2S. Show that only Cd^{+2} is precipitated as CdS.

[Given: K_{sp} of ZnS = 1.1×10^{-21} gm ion^2/lit^2
K_{sp} of CdS = 8×10^{-28} gm ion^2/lit^2
$K_{H_2S} = 1.3\times10^{-20}$ gm ion^2/lit^2
$[H_2S]$ in saturated solution of H_2S = 0.1 gm mole/lit.] 4

COMPULSORY QUESTION

11. Answer the following questions: 2×10

(a) Account for the decrease in first ionisation energy between nitrogen and oxygen.

(b) $H\hat{O}H$ bond angle in H_2O is larger than the $H\hat{S}H$ bond angle in H_2S—explain.

(c) Solubility of HgX_2 (X = Cl, Br, I) decreases from chloride to iodide—explain.

(e) CO and N_2 are isoelectronic but they differ greatly in donor properties—explain.

(f) Indicate with reasons the direction of the following reactions:

(i) $BF_3H^- + BH_3F_4^- = BF_4^- + BH_4$; (ii) $HgS + 2HCl = HgCl_2 + H_2S$.

(g) Explain the respective mode of decay of the radionuclei $^{14}_{6}C$ and $^{13}_{7}N$.

(h) Arrange the bases HS^-, F^- and I^- in order in increasing proton affinity with explanation.

(i) Complete the following reactions:

(i) $N_2O_5 + H_2O_2 \longrightarrow$; (ii) $I_2O_5 + CO \longrightarrow$

(j) Balance by ion-electron method:

$NaNO_3 + Zn + NaOH \longrightarrow NH_3 + Na_2ZnO_2 + H_2O$

2006

PART—I

SECOND PAPER

Full Marks—100

The figures in the margin indicate full marks

Candidates are required to give their answers in their own word as far as practicable

[New Syllabus (2003-2004) and Old Syllabus (1998-1999) Combined]

Answer *one* question from each Units **1** to **5**. Question No. **11** is Compulsory

UNIT—1

1. (a) Which quantum number solely determines the allowed energies for a hydrogenic atom? Write the expression for energy and explain the terms and sign involved therein. Which condition corresponds to zero energy? 1+2+2+1

(b) In an atom, the angular momentum of an electron is $\sqrt{6}\,.\,\frac{h}{2\pi}$. What is the minimum value of the principal quantum number of the electron? $1\frac{1}{2}$

(c) "The intrinsic spin angular momentum of an electron is defined by two quantum numbers"—explain the statement. 2

(d) Calculate the weight of 1.0 curie of U_{92}^{238}. Given $T_{\frac{1}{2}} = 4.5 \times 10^9$ years and 1 curie = 3.7×10^{10} dps. $3\frac{1}{2}$

(e) An electron in a hydrogen atom in its ground state absorbs 1.5 times as much energy as the minimum required for it to escape from the atom. What is the wavelength of the emitted electron? (Given: 13.6 eV energy is needed for ionisation and $m = 9.106 \times 10^{-28}$ gm = mass of electron) 3

2. (a) State Hund's rules and hence find out the Ground Term for gaseous Cr atom (Z = 24). 2+2

(b) Give the mass numbers and atomic numbers of the most long-lived radioactive representative and the end product in each of the natural radioactive disintegration series. 4

(c) Give the mathematical expression of the Radioactive Decay Law and hence give the physical significance of decay constants. 2

(d) The binding energy per nucleon for C_6^{12} is 7.68 MeV and that for C_6^{13} is 7.47 MeV. Calculate the energy required to remove a neutron from C_6^{13}. 3

(e) What do you understand by magic numbers? Explain. 3

UNIT—2

3. (a) Write down the Born-Lande equation and explain the terms involved in it. Mention about its application. 4

(b) On the basis of change in the value of lattice enthalpy, comment on the products of the reaction between Group I metals and dioxygen. 3

(c) On the basis of VSEPR theory write the possible structures of ClF_3 and predict the most favoured structure showing your arguments. 3

(d) Explain the effect of polarising power and polarisability on the properties of ionic compounds. 3

(e) State and explain Bent's rule and hence discuss the shape and bond angles in CH_2F_2 molecule. 3

4. (a) Mention the conditions for linear combination atomic orbitals relating to the formation of molecular orbitals. 3

(b) Construct the M.O. diagram of CO_2 molecule showing the arrangement of electrons. Find out the number of bonding and non-bonding electrons. 4

(c) Draw the resonating structures of thiocyanate and cyante ions showing the formal chrges wherever necessry and predict the most important structure in each case. 2+2

(d) What is meant by 'partial ionic character of a covalent bond''? What are its consequences? 3

(e) When Ge is doped with Ga it becomes a p-type semiconductor—Explin. 2

UNIT—3

6. (a) Discuss the Alred-Rochow's absolute electronegativity and show its relation with Pauling's relative electronegativity. 4

(b) A, B, C and D are the elements of the same period in the Periodic Table. The ionisation energies of the elements are listed below:

	I_1/eV	I_2/eV
A	5.39	75.6
B	21.5	41.0
C	6.11	11.87
D	17.4	35.0

Using the above letters identify (i) the alkali metal, (ii) the alkaline earth metal, (iii) the non-metal and (iv) the noble gas. 4

(c) Although the electron affinity of fluorine atom is lower than that of chlorine atom, F_2 is much more reactive than Cl_2. Account for the higher reactivity of F_2 with respect to the formation of solid halides MX or MX_2. 3

(d) What is meant by inert pair effect? Explain with a suitable example. 3

(e) Compare the oxidising ability of H_2SO_4 and H_2SeO_4. 2

UNIT—5

9. (a) What are superacids? How the acidity of such solutions is measured? Explain with an example. 3

10. (a) Apply solubility product principle and common ion effect in separations of Group II and Group IIIB metal sulphides in qualitative inorganic analysis. 6

(b) "Disproportionation and comproportionation reactions are actually redox cell reaction." Justify the statement with suitable examples. 2+2

(c) In an aqueous solution containing Fe^{3+} and Fe^{2+} ions, the redox potential is +0.70 volt at 25°C. Taking the E_0 value of Fe^{3+}/Fe^{2+} couple as +0.77 volt, calculate the Fe^{3+}/Fe^{2+} ratio in the solution. 4

(d) What happens when I_2 is added to excess aqueous acidic $HClO_3$? Write the half reactions. Predict which net reaction will occur.

Given E_0 values for I_2/I^- and ClO_3^-/ClO_4^- couples are 0.54 V and –1.20 V respectively. 2

COMPULSORY QUESTION

11. Answer the following questions: 2×10

(a) An electron and a proton have the same kinetic energy. Which will have greater de Broglie wavelength and why?

(b) Between H_2 and He_2^{2+} which one is more stable and why?

(c) Electron affinity of Mn^{3+} is greater than that of Fe^{3+}—explain.

(d) Give the Lewis dot structures and sketch the shapes of XeO_3 and IOF_5.

(e) Find the minimum energy required to split an O_8^{16} nucleus into an He_2^4 and C_6^{12} nucleus. The binding energies of O_8^{16}, C_6^{12} and He_2^4 are 127.6, 92.1 and 28.3 MeV respectively.

(j) Give the name and structure of a redox indicator. Show its mode of action with equation.

2007

PART—I

SECOND PAPER

Full Marks—100

Answer *one* question from each Units **1** to **5**. Question No. **11** is Compulsory

UNIT—1

1. (a) (i) State Heisenberg's uncertainty principle and explain its significance.

(ii) Show that de Broglie's hypothesis applied to an electron moving in a circular orbit leads to Bohr's postulate of quantisation of angular momentum. 2+2

(b) What is radial distribution function? Show diagrammatically the variation of radial distribution function with '*r*' for the 3*s*, 3*p* and 3*d* orbitals in a hydrogen atom. 1+3

(c) "Radionuclides with n : p ratio above the stability ratio range decay by β-emission and not by neutron emission"—explain. 3

(d) For the hydrogen spectrum show that Lyman series occurs between 912 Å and 1216 Å, and Balmer series occurs between 3647 Å and 6564 Å ($r = 1.0968 \times 10^7\ m^{-1}$). 3

(e) Deduce ground state term symbol for atom having atomic number 22. 2

2. (a) What is the significance of the term "orbital magnetic quantum number"? Draw the vector orientation of the m_l values corresponding to $l = 2$ in magnetic field. Terms are all of their usual significance. 1+2

(b) A small amount of solution containing ^{24}Na ($t_{1/2}$ = 15 hours) is injected into blood stream of a man. The initial activity of the injected sample is $2 \times 10^3\ s^{-1}$. The activity of 1 cm^3 of blood sample taken after 5 hours is 16 min^{-1}. Find the total volume of blood of the man. 3

(c) (i) What do you mean by "Fissile Nucleus"? Name two such nuclei with their atomic and mass numbers.

(ii) What do you mean by moderator nuclei? State the characteristics of a good moderator. 2+2

(d) Find the radius of the first orbit of Be^{3+} ion and velocity of the electron in this orbit. 1+3

(Radius of the first Bohr orbit of H atom = 5.29×10^{-11} m,
$\varepsilon_0 = 8.85 \times 10^{-12}\ J^{-1}\ C^2\ m^{-1}$, $e = 1.602 \times 10^{-19}$ C.)

(e) Which of the following arrangements would be more stable and why? 2

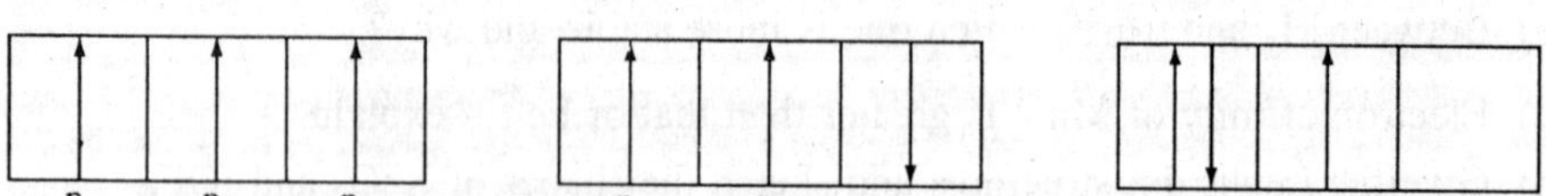

UNIT—2

3. (a) Draw the possible resonance structures of nitrate ion and indicate with reasons the greatest contributing structure. 3+1

(b) State the basis of 'radius ratio rule' for ionic compounds. Calculate the limiting radius ratio for tetrahedral lattice structure. 1+3

(c) Discuss the kind of defect observed in the crystal structure of ZnO when heated. What is the consequence of heating? 2

(d) Explain the Linear Symmetrical Structure of HF_2^- ion. 2

(e) Construct the M.O. energy level diagram of $BeF_2(g)$ molecule and comment on the bond order of Be–F bond. 3+1

4. (a) Predict the shapes of the following species and describe the type of hybrid orbitals on the central atom: 4

$[ICl_4]^+$, $[SbF_6]^-$, $[BH_4]^-$ and SOF_4.

(c) Indicate the oxidation number of each P atom in the following compounds: 2

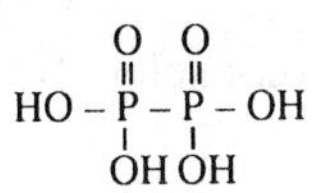

and

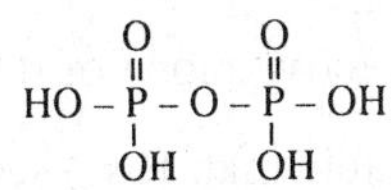

(e) Using M.O. theory, explain the observations that the bond length in N_2^+ is 0.02 Å greater than that in N_2, while the bond length in NO^+ is 0.09 Å less than that in NO. 2+2

UNIT—3

5. (b) Explain why O–O bond in O_2F_2 is much shorter than that in H_2O_2, while the O–F bonds in O_2F_2 are considerably longer than those in OF_2. 4

(c) Define "Ionisation Potential". Discuss the factors on which ionisation potential of the elements depends. 1+3

6. (a) Explain the basis of Pauling's electronegativity scale. Using the bond energy data, given in the parentheses in kJ mol^{-1}, determine the more ionic bond in H–O–Cl and suggest the mode of its ionisation in aqueous solution: H–H (437), Cl–Cl (239), O–O (143), H–O (462) and Cl–O (218). 2+3+1

(b) $MgCO_3$ is thermally less stable than $CaCO_3$—explain. 2

(c) State Slater's rules and apply these to find the effective nuclear charge acting on a valence electron in Cl^- ion. 2+1

(e) Why has CD_4 slightly lower boiling point than CH_4? 1

(f) $SnCl_2$ is reducing while $PbCl_2$ is neither reducing nor oxidising—explain. 2

UNIT—5

9. (d) The solubility of CaF_2 in water at 18°C is 2.04×10^{-4} moles $litre^{-1}$. Calculate. 2+2

(i) the solubility product and (ii) the solubility of CaF_2 in 0.01 (M) NaF solution.

(f) Comment on the relative acid or basic strength of the following pairs: 2×2

(i) NH_3, H_2O; (ii) H_3PO_2, H_3PO_4.

10. (a) Why does HNO_3 act as a base in conc. H_2SO_4 medium? 2

(b) $\frac{N}{10}Fe^{2+}$ (aq) is titrated with $\frac{N}{10}MnO_4^-(aq)$ at 1(N) acidity. Derive the expression for the equivalence point potential of Fe^{3+}/Fe^{2+} electrode at 25°C. 4

Given: $E^{\circ}_{Fe^{3+}/Fe^{2+}} = 0.76$ V and $E^{\circ}_{MnO_4^-/Mn^{2+}} = 1.51$ V.

(e) State the principle of quantitative determination of hardness/softness of a base. 3

(f) Draw a Frost diagram for mercury in acid solution from the given Latimer diagram:

$$Hg^{2+} \xrightarrow{0.911\text{ V}} Hg_2^{2+} \xrightarrow{0.796\text{ V}} Hg$$

Comment on the tendency of any of the species to undergo disproportionation. 2+1

COMPULSORY QUESTION

11. Answer the following questions: 2×10

(a) Aqueous $BeCl_2$ solution is acidic while that of $CaCl_2$ is neutral—explain.

(b) Explain the respective mode of decay of radionuclei $^{14}_{6}C$ and $^{13}_{7}N$.

(c) NaN_3 is more stable than HN_3—explain.

(d) Gold has high electron affinity—comment.

(e) "Dipole moment of CO is less than expected from electronegativity difference." Explain.

(f) A rosered precipitate is formed when $NiSO_4$ is treated with dimethyl glyoxime in presence of ammonia. What is its formula? State the nature of the compound.

(g) State what happens when bismuth hydroxide is treated with alkaline sodium stannite solution. Give the balanced ionic equation for reaction.

(h) State the theory by which the reaction

$$6CaO + P_4O_{10} \rightarrow 2Ca_3(PO_4)_2$$

may be regarded as an acid-base reaction.

(i) Establish Nernst equations for the following couples:

(i) $Cr_2O_7^{2-}/Cr^{3+}$ in acid medium; (ii) BrO_3^-/Br^- in alkaline medium.

(j) $\angle$ F–P–F bond angle in PF_3 molecule is 104° but $\angle$ H–P–H bond angle in PH_3 molecule is 93.3°. Why?

2006

(1ST SEMESTER)
PAPER—I H

Full Marks—50

Use a separate answer script for each group

GROUP—C

7. (a) Write electronic configuration beyond the inert gas core for Cr^{3+}, Fe^{2+}, Cu^{3+} and Zn^{2+} (example: Ti^{3+}: [Ar]$3d^1$). Atomic numbers are 24, 26, 29 and 30 respectively. 4

(b) For H-atom, $\psi_{1s} = 2(a_0)^{-\frac{3}{2}} e^{-\frac{r}{a_0}}$.

Imagine two this annular shells each of thickness, 0.001 Å, at (i) $r = a_0$ and (ii) $r = 0.1a_0$. Calculate the probability of finding an electron in each shell [a_0 = 0.529 Å]. 4

(c) In your answer book write the correct option(s) from each of the following sets: 8

(I) The electronic configuration of Ga :

(i) [Ar]$3d^{10}4s^24p^1$; (ii) [Ar], $3d^{10}$, $4s^2$, $4p^1$; (iii) [Ar] $4s^23d^{10}4p^1$; (iv) [Kr] $4s^23d^{10}4p^1$

(II) Number of unpaired electrons in the ground state of carbon:

(i) 0; (ii) 1; (iii) 2; (iv) 4.

(III) Number of nodes in a hydrogen 2*s*-orbital:

(i) 0; (ii) 1; (iii) 2; (iv) unpredictable.

(IV) The order of orbital penetration:

(i) $2s < 2p$; (ii) $3s > 2s$; (iii) $4s > 4f$; (iv) $6s > 6p$.

(1ST SEMESTER)
PAPER—II H

Full Marks—50

Use a separate answer script for each group

GROUP—C

9. Draw the graph of 2nd ionization energy (I, E_2) against the valence shell configurations of the elements of 2nd short period and explain the deviations in the graph, it any. 5

10. What is the principle of calculating electronegativity on Pauling's Scale?

Explain how the hybridizations of N in the following compounds affect their basicities:

$C_6H_5NH_2$, C_6H_5CN and C_5H_5N. 6

11. Calculate 1st ionization energy of He. 2

12. Compare: 3

(a) Electron affinities of O^-, $O^=$,

(b) Covalent van der Waals radii of hydrogen.

(2ND SEMESTER)

PAPER—III

Full Marks—50

GROUP—C

5. The bond angles in CF_4 as well as CH_4 are all 109°28′. But none in CH_2F_2 has a value 109°28′—explain. 3

6. Explain in which order, bond angle decreases for CH_4, NH_3 and H_2O. 3

7. Under what conditions the dipole moment of a molecule becomes zero? Illustrate with examples. 3

8. Derive an expression for determining the lattice energy of alkali halides. 5

9. Calculate the lattice energy of NaCl from the following data: 2

Madelung Constant = 1.748

$e = 1.6 \times 10^{-19}$ C

$n = 8$

Equilibrium interionic distance, $R_e = 281.4 \times 10^{-12}$ m

$E_0 = 8.85 \times 10^{-12}$ F m^{-1}.

(2ND SEMESTER)

PAPER—IV

Full Marks—50

Use a separate answer script for each group

GROUP—A

2. Answer any **three**:

(a) State the law of rational indices. What are the Miller indices of the face having intercepts on the three axes $^1/_2\, a$, $^1/_4 b$ and $\infty\, c$ respectively. Represent the plane in a diagram. 1+1+1

(b) Structure of KCl and NaCl are not same according to X-ray diffraction study—explain. 3

(c) The ionic radii of Na^+, Cs^+ and Cl^- are 95 pm, 169 pm and 181 pm respectively. Give the expected cation co-ordination numbers for NaCl and CsCl with reason. 3

(d) X-ray (wavelength 154.04 pm) diffraction study of crystalline copper gives the reflections at 21.65°, 25.21°, 37.06°, 44.96°, 47.58° and other angles. Determine the type of the cubic crystal formed by copper and radius of copper atom. 3

GROUP—C

7. Although H_2S is used both in precipitating cation of Group–II and Group–IIIB in detection of cations in wet way, Group II and Group IIIB cations are not precipitated simultaneously. On the contrary they are precipitated at the respective groups only—explain giving reasons. 3

8. Define solubility product and explain its difference with ionic product. 3

2007

(1ST SEMESTER)
PAPER—VII H (INORGANIC CHEMISTRY)

Full Marks—50

Answer Question No. **1** and any **three** from the rest

1. (a) Explain the paramagnetic properties of B_2 with the help of M.O. 4

   (b) For radioactive decay

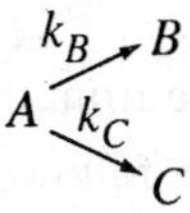

   prove that $t_{\frac{1}{2}}(A) = \dfrac{0.693}{(k_B + k_C)}$ 3

   (c) $BeCl_2$ is a polymer but $AlCl_3$ is a dimer. Explain. 3

   (d) $MnO_2 + 4H^+ + 2e^- \rightleftharpoons Mn^{2+} + 2H_2O$ ($E° = 1.23$ V)

   $Cl_2 + 2e \rightleftharpoons 2Cl^-$ ($E° = 1.36$ V)

   Based on the $E°$ values, it appears that MnO_2 should not oxidize Cl^- to Cl_2. But yet such oxidation takes place. Explain why. 4

2. (a) Write the M.O. electron configuration for the NO^- ion. 1

   (i) What is the bond order? $\frac{1}{2}$

   (ii) Will the bond length be shorter or longer than in NO? $\frac{1}{2}$

   (iii) How many unpaired electrons will be present? $\frac{1}{2}$

   (iv) Will the unpaired electrons be concentred more on the N or the O? Explain. 1

   (b) Which do you expect to be more stable linear H_3^+ or bent H_3^+? Explain. $1\frac{1}{2}$

   (c) A concentration cell is composed of

   $Zn\backslash ZnCl_2$ (1M)$\backslash\backslash$ZnS(Satd.)$\backslash$Zn

   Find out the cell potential (K_{sp} of ZnS = 10^{-29}) 2

   (d) Explain the following:

   (i) CO is a σ-base and a π-acid. $2\frac{1}{2}$

   (ii) SCN^- displays different mode of binding to Cu^+ and Cr^{6+}. $2\frac{1}{2}$

3. (a) Write notes on (i) Nuclear isomers, (ii) Magic number. 3+3

   (b) Define hybridisation and explain the structure and mode of hybridisation in XeF_4. 2

4. (a) State and explain the theoretical basis of Bent's rule. $2\frac{1}{2}$

   (b) Explain the colour properties of halogen molecule. 2

   (c) HF_2^- is more stable than HCl_2^-. Explain. $1\frac{1}{2}$

   (d) The $E°$ of Cu^{2+}/Cu half cell is 0.34 V and that of Cu^{2+}/Cu^+ half cell is 0.15 V. Find out the $E°$ of the Cu^+/Cu half cell. $2\frac{1}{2}$

   (e) State and explain the Lux-Flood theory of acids and bases. $3\frac{1}{2}$

5. (a) How would you establish the non-existence of electrons in the nucleus of the atom? 6

(b) Explain the following:

(i) Alkali metals cannot have bipositive oxidation states. 2

(ii) Tetrahedron is a regular structure. 2

(iii) Inert gases undergo liquefiaction. 2

(2ND SEMESTER)

PAPER—X H (INORGANIC CHEMISTRY)

Full Marks—50

Answer Question No. **1** and any **three** from the rest

1. (a) Applying Valence Bond predict the geometry of the following complexes (any ***one***): 2

$[Ni(CN)_4^{2-}$ diamagnetic or $[NiCl_4]^{2-}$ paramagnetic

(c) The two N–N lengths are same in N_3^- (azide ion). Explain with the aid of resonance. 2

(d) What do you mean by metallic bond? Briefly discuss with the help of band theory. 4

(e) Which among the following is most oxidizing? Justify your answer: 3

ClO_4^-, BrO_4^-, IO_4^-

5. (a) Identify fluorite, antifluorite, bcc and fcc structures in the following compounds: TlBr, Li_2O, ZrO_2, RbF, FeF_2 and K_2S. 2

(b) Calculate the r^+/r^- values of crystals possessing NaCl and CsCl structures with the help of hard-sphere approximation.

Osmania University Question Papers

2006

[FIRST YEAR]

OCTOBER—2006

4. (a) What is Planck's radiation law? Explain black-body radiation on the basis of Planck's theory.
   (b) What is Hamiltonian operator? Explain how Heisenberg's uncertainty principle is applicable for microsize particles and not macrosize particles.

   OR

   (c) What are Fajans' rule? How they are useful in explaining the covalent character of a compound?
   (d) Draw molecular orbital energy level diagram of CO and explain its bond order and magnetic properties.
5. (b) What is electronegativity and electron affinity? How is electronegativity determined by Pauling's method?
6. (a) Why BI_3 is a stronger Lewis acid than BF_3?
   (b) Why CO_2 is a gas whereas SiO_2 is solid?
10. (a) What are common defects in crystal? Write a brief note on stoichiometric defects.

    OR

    (b) What is a liquid crystal? What are its applications in LCD devices?
11. (b) What is bond order? Obtain the bond order of O_2 and NO, and explain their magnetic properties.

MARCH—2006

4. (a) Write Schrödinger's wave equation. Explain the significance of ψ and ψ^2.
   (b) State and explain Heisenberg's uncertainty principle.
   (c) Calculate the uncertainty in position of an electron if the uncertainty in its velocity is 5.7×10^{-5} m/sec ($h = 6.6 \times 10^{-34}$ Joule sec. and $m = 9.1 \times 10^{-31}$ kg).
   (d) Draw the molecular orbital energy level diagram for O_2 and explain the bond order.
5. Write structure and hybridisation of IF_4 and $XeOF_2$.
9. What are intrinsic and extrinsic semiconductor? How can silicon be made a p-type semiconductor?
12. State and explain Fajans' rules.

OCTOBER—2005

3. (c) What are Frenkel and Schottky defects? Explain.
   (d) Illustrate metal excess and metal deficiency defects in non-stoichiometric compounds. What are the consequences of these defects?
4. (a) Write short notes on:
   (i) Schrödinger's wave equation and significance of ψ and ψ^2

(ii) De Broglie's hypothesis

(iii) Heisenberg's uncertainty principle

OR

(b) Draw the molecular orbital energy level diagram of NO and explain its magnetic behaviour. What is bond order?

(c) Comment on the stability of the following species on the basis of MO theory: H_2, H_2^+, He_2^+.

5. (b) Give the structure and hybridisation of ClF_3 and SO_3.

6. (b) How are atomic radius ionisation potential and electropositivity related to each others?

MARCH—2005

1. (a) Discuss the structure of ICl_2, ICl_4 and I_3^-.

4. (a) Write notes on:

(i) Planck's radiation law, (ii) Photoelectric effect, (iii) de Broglies hypothesis.

OR

(b) Draw the molecular orbital energy level diagram of N_2 and explain its bond order and magnetic behaviour.

(c) Explain how He_2^+ can be formed but not He_2 on the basis of molecular orbital theory.

6. (a) Describe Mulliken's method for the measurement of electronegativity of an atom.

9. (a) Define: (i) Centre of symmetry, (ii) Space lattice, (iii) Unit cell.

12. (a) Discuss the hybridisation and shape of SF_6 and $BeCl_2$.

(b) What is lattice energy? Discuss the factors affecting it.

[SECOND YEAR]

OCTOBER—2004

10. (a) What are radioactive isotopes?

(b) Discuss the uses of radioactive isotopes in medicine and agriculture.

OR

12. (a) Define and draw the following elements of symmetry in a cube:

(i) Centre of symmetry

(ii) Four-fold axis of symmetry

(iii) Diagonal plane of symmetry

OR

(b) Write notes on Schottky and Frenkel defects.

MARCH—2004

9. (a) Mention two reactions each for transmutation by α-particle and proton.

OR

Write an account of bonding in metals.

12. Explain Schottky and Frenkel defects in crystal.

OCTOBER—2003

9. Describe band theory of metals.
11. Draw the sketches of (100), (110) and (111) crystal planes.

MARCH—2003

9. The amount of C-14 in a piece of wood is found to be $\frac{1}{10}$th of the amount precent in a fresh piece of wood. What is the age of the wood? $T_{\frac{1}{2}}$ of C-14 is 5770 years.

OR

What are lanthanides and actinides? What are the differences and similarities between them?

10. How are nuclear fission processes render self-sustaining? What is the significance of critical mass? What is the role of moderators in nuclear fission reactions?
11. Derive Bragg's equation for diffraction of X-rays by crystals.

OCTOBER—2002

10. Discuss the applications of radioisotopes in different fields of life.
12. Explain the symmetry elements found in NaCl crystal system.

MARCH—2002

3. What are transition elements? Give a comparative study of the first series with reference to (a) atomic and ionic radius, (b) oxidation states, (c) magnetic properties.
10. What is carbon dating? Write a brief note on the application of isotopes in the field of medicine and agriculture.

Ranchi University Question Papers

[PAPER-VII—2007]

2. Derive Schrödinger wave equation. What are importances of Schrödinger equation?
6. What is dipole moment? Describe one method for measuring the dipole moment of molecule.
8. Write notes on:
   (a) Defects of Bohr's model of hydrogen atom
   (b) Compton effect
   (c) Comparison of valence bond and molecular models
   (d) Quantum mechanical principles of hybridisation,
   (e) Photoelectric effect

[PAPER-VII—2006]

6. Explain—(a) Black-body radiation, (b) Planck's distribution of radiation.
7. (b) Discuss the concept on energy curves for bonding and antibonding M.O. Establish different relations between bond order and bond in a diatomic molecule.
8. Write notes on—(a) Hamiltonian operator, (b) Probability densities, (c) Uncertainty principle, (d) Valence bond model of H_2, (e) Eigenvalues and eigenfunctions.

[PAPER-VII—2005]

(NEW)

1. Derive the Schrödinger's wave equation. Derive the expression for the energy and wave function of particles in one-dimensional box with the help of Schrödinger's wave equation.
2. Outline the main postulates of quantum mechanics. Discuss de Broglie's equation. State and explain Heisenberg's uncertainty principle.
5. Explain dipole moment. How is it determined? Which of the following molecules have zero or finite dipole moment and why?

   H_2O, NH_3, CO_2, CCl_4 and Cl_2.

(OLD)

2. What is electrode potential? Discuss its origine. Discuss any two applications of E.M.F. measurement.
4. Derive Bragg's equation for the diffraction of X-rays by crystals. Discuss the crystal structure of diamond.
6. Discuss de Braglie's equation. State and explain Heisenberg's uncertainty principle. Discuss the physical significance of ψ.

Kanpur University Question Papers

[PART–I—2005]

1. Why isoelectronic Na^+, Mg^{+2} and Al^{+3} cations have different ionic radii?
2. Explain bond order, magnetic nature and stability of O_2 molecule with the help of M.O. energy diagram.
3. Derive Schrödinger's wave equation.
4. Explain Pauli's exclusion principle.
5. Explain electronic configuration of Cr–atom.
6. How does the ionization potential of elements change within a group and a period of periodic table?
7. Define electronegativity.
8. Write a short note on directional characteristics of covalent bond.
9. What do you understand by *sp*, sp^2 and sp^3 hybridisation?
10. What do you understand by molecular orbital theory?
11. What is the relation between lattice energy and solubility of ionic solids? Explain it.
12. What do you understand by polarisability and polarising power?
13. How do intramolecular and intermolecular hydrogen bond influence the physical properties of the compound?
14. Write a short note on diagonal relationship between Li and Mg.
18. Discuss the position of *p*-block elements in the periodic table.
22. Discuss the various theories of metallic bond.
23. Write the electronic configuration of Cu(29), As(47) and Au(79).

[PART–I—2006]

1. Calculate the effective nuclear charge of outermost electron in the atoms: (i) Li, (ii) Be, (iii) B, (iv) –C and (v) N.
3. Derive de Broglie's matter wave equation for dual nature of electron.
4. Explain Hund's rule of maximum multiplicity and give its important applications.
5. What do you understand by the term ionisation potential? Discuss the factors affecting the ionisation potential.
6. Define atomic radius and explain its variation in groups and in periods of the periodic table.
7. Explain hybridisation. Define the structures of $BeCl_2$ and IF.
8. Explain linear conbination of atomic orbitals (LCAO) in M.O. theory.
9. Write molecular orbital diagram of O_2 molecule. Explain its bond order and magnetic nature.
10. What are σ and π bonds, which one is stronger and why?
12. Explain ionic radius ratio.
13. What is an ideal crystal? Discuss various types of crystal defects.
14. Explain free electron theory or electron gas theory of metallic bond.
15. Define hydrogen bonding and explain its types with examples.
16. Define diagonal relationship and explain Li-Mg diagonal relation.
20. Write preparation, properties and structure of $[XeF_4]$.

21. Define quantum numbers. Describe all the four quantum numbers. Write the four values of quantum numbers for hydrogen atom electron.

[PART–I—2007]

1. Calculate the wavelength of an electron moving with a velocity of 2.5×10^5 m s^{-1}, mass of electron = 9.1×10^{-31} kg, $h = 6.625 \times 10^{-34}$ J s.
2. Explain paramagnetic nature of O_2 by drawing its M.O. diagram.
3. Write down the general form of Schrödinger wave equation and explain each term in it.
4. Draw the radial probability curves for 2*s*, 2*p*, 3*p* and 3*d* orbitals.
5. Explain Pauli's exclusion principle and its applications.
6. Define electronegativity. Explain the trends of increase or decrease of electronegativity in periods and groups.
7. Why is the electron affinity of Cl is greater than that of F? Explain.
10. Define hybridisation. Discuss the structure of $BeCl_2$ and IF_7.
11. How does the VSEPR theory explain T-shapes of ClF_3 molecule and linear shape of ICl_2^-?
12. Write down the limitation of V.B. theory.
13. Write down the molecular orbital electronic configuration of C_2 molecule and calculate its bond order.
14. What is screening effect? How does it affect the value of ionisation potential in an atom.
15. Why alkali metals are univalent, slrong reducing agent and have low ionisation potential?
16. Hydrides of alkali metals are basic but that of halogens are acidic. Why?
17. Why H_2O is liquid but SO_2 is gas?
18. NaI is ionic but AgI is covalent—explain.
20. Discuss hybridisation in XeO_4.
22. What is hydrogen bond? Explain its type. Discuss the effect of hydrogen bonding on the physical properties of the compounds containg hydrogen bond.

[PART–II—2005]

1. Which of the following reactions is possible and why?
   (a) $Cu + FeSO_4 \rightarrow CuSO_4 + Fe$; (b) $Fe + CuSO_4 = FeSO_4 + Cu$.
   Standard reduction of Cu is 0.34 V and of Fe = 0.44 V.
3. Discuss in brief about conjugate acids and conjugate bases.
4. Write down four characteristics of transition metals including electronic structure.
7. Write down the electronic structure of Europium(63) and corresponding element in actinide series.
13. The 4*f* series of elements are called Lanthanides although there is no 4*f* electron in case of Lanthanum(57). Why?
16. How redon potential data is useful in extraction of elements.
18. What do you know about trans-uranium elements.
19. Write down the general electronic structure and most common oxidation states of 5*f* elements.
20. Illustrate three physical characteristics of solvents.
23. Discuss Lux-Flood concept of acid and base. Compare this theory with Lewis concept.

[PART–II—2006]

1. Considering the following:

   $Ag^+ + e \rightleftharpoons Ag$, $E_0 = +0.799$ volt; $Cu^{+2} + 2e \rightleftharpoons Cu$, $E_0 = +0.337$ volt

   what will happen when

   (a) a silver wire is dipped into a cupric solution?

   (b) a copper wire is dipped into an aqueous solution of $AgNO_3$?

3. Why do lanthanides commonly exhibit +3 oxidation state?
4. Explain Lux-Flood acids and bases.
5. What is the difference between oxidation state and valency?
7. Explain—transition metals are less reactive than *s*-block elements?
9. On the basis of electronic configuration of Ti(22), give the magnetic properties of different oxidation states of the metal ion.
11. Label the Bronsted base of the following:

    (a) NH_4^+; (b) O^{-2}; (c) NaOH; (d) H_2SO_4.

13. What are trans-uranium elements—explain.
14. Explain amphoteric substances.
15. Explain the most common oxidation states of actinides.
16. What is lanthanide contraction—explain.
18. How redox potential data is useful in extraction of elements.
19. Give with reason the more common oxidation states shown by first transition series (3*d*) elements.
21. Explain Lewis acids and Lewis bases. How do they differ from Bronsted-Lowry concept of acids and bases.
23. Explain *f*-block elements. Why are they called inner transition series?

[B.SC. PART–II—2007]

2. Arrange the following in order of increasing acid strength:

   HBr, HCl, $HClO_4$, HNO_3, H_2SO_4.

4. The hydrogen ion (H^+) in aqueous solution is represented by H_3^+O. Why?
5. How is redox potential data useful in the extraction of elements?
7. Give the limitations of Valence Bond Theory.
8. Write a brief note on Lanthanide Contraction.
14. Actinides have greater tendency to form complexes than that of lanthanides—give reason.
15. Why $La(OH)_3$ is most basic while $Lu(OH)_3$ is least basic?
16. Name the elements which comprise second transition series.
19. All the actinides resemble one another very closely—discuss.
20. Distinguish between oxidation number and valency.
21. What are *d*-block elements? Explain the following properties of these elements:

    (a) Electronic configuration and oxidation states

    (b) Complex formation.

    (c) Magnetic prooperties

22. (a) All Arrhenius acids are Bronsted acids, but all Arrhenius bases are not Bronsted bases—discuss.

    (b) What do you understand by conjugate acid-base pair? Show that strong acid has a weak conjugate base and a weak acid has a strong conjugate base.

2005

PAPER—II

Full Marks—100

The figure in the margin indicate full marks.

Candidates are required to give their answers in their own words as far as practicable.

Question No. **1** is compulsory. Answer any *five* questions taking at least *two* questions each from groups A and B.

1. (a) The electronic configuration of an atom of an element is $1s^2\ 2s^2\ 2p^3$. Suggest its position in the Periodic Table and predict its probable oxidation state. 2

(b) Hydrolysis of $SiCl_4$ and CCl_4 takes different courses—explain. 2

(c) What basic information is obtained from Moseley's experiment? 2

(d) What would be the wavelength of emission spectrum when electron jumps back from $n = 2$ to ground state of He^+? (Given : $h = 6.625 \times 10^{-27}$ erg sec. $m_e = 9.1 \times 10^{-28}$ g, charge $= 4.8 \times 10^{-10}$ e.s.u., $c = 2.998 \times 10^{10}$ cm sec^{-1}.) 2

(e) Which equation is chemically correct? 2

(i) $CsI + LiF \longrightarrow LiI + CsF$

$LiI + CsF \longrightarrow LiF + CsI.$

(ii) $Zn^{2+} + Cu \longrightarrow Zn + Cu^{2+}$

$Zn + Cu^{2+} \longrightarrow Zn^{2+} + Cu$

(f) Diamond is hard and non-conductor of electricity, but graphite is soft and electrical conductor—explain. 2

(j) Explain the difference in physical state of H_2O and H_2S. 2

GROUP—A

2. (a) Complete the following reactions and point out the discoveries related with them:

(i) ${}^{14}_{7}N + {}^{4}_{2}He \longrightarrow$

(ii) ${}^{235}_{92}U + {}^{1}_{0}n \longrightarrow {}^{144}_{55}Cs + {}^{90}_{37}Rb + ...$

(b) It is given that 1g of Ra^{226} undergoes 3.7×10^{10} disintegrations/sec and its half-life period is 1590 years. Find the value of Avogadro number.

(c) Explain the significance of 'nuclear binding energy'. How does the binding energy curve give information about nuclear fission and fusion as energy resource?

(d) What is nuclear isomerism? 4 + 4 + 6 + 2

3. (a) Bohr energy (E) of each orbit is given by the equation:

$$E = \frac{2\pi^2 me^4 Z^2}{n^2 h^2}$$ (all terms have their usual meanings)

(i) Calculate the ionisation energy of hydrogen.

(ii) Calculate the energy of each of the first two orbits ($n = 1$ and $n = 2$) in He^+. Also compare the energies of the first and second orbits in He^+ with those in hydrogen. (Use parameters from Q. 1(d). 2+5

(b) Explain why Bohr-Sommerfeld's model of atom had to be replaced by the wave mechanical model. 4

(c) State the significance of ψ. What is $\int \psi^* \psi d\tau$? How does radial distribution function vary for 1*s*, 2*s* and 2*p* orbital electrons with distance from the nucleus? 5

4. (a) State the condition of formation of MOs from AOs. Deduce the energy value of stablest MO of H_2^+. 6

(b) Predict the bond length sequence of O_2, O_2^- and $O_2^=$ species using MO concept. 4

(c) Use the concept of electron pair repulsions to predict the geometry of the following molecules: AsF_6, $TeCl_4$. 3

(d) Explain: why metals are good conductors of electricity? 3

5. (a) Explain with reasons (any ***two***): 2×2

(i) The lower oxides of metal are more basic whereas the higher oxides are more acidic.

(ii) Acidity decreases in the sequence

$HClO < HClO_2 < HClO_3 < HClO_4$.

(iii) The bond angles show H_2O (104.5°) and H_2S (92.2°).

(b) Which of the following compounds is hardest and which is the softest? 2

LiF, LiCl, MgO, CaO.

(c) Calculate Pauling's electronegativity of bromine from the following data: 4

Bond energies: H_2 (104.2 kcal/mole), Br_2 (46.1 kcal/mole), HBr (87.4 kcal/mole). Pauling's electronegativity of hydrogen = 2.1.

(d) Distinguish the following (with examples): 6

d^2sp^3 and sp^3d^2 hybridisation, σ- and π-bonds, (2*c*–2*e*) and (3*c*–2*e*) bonds.

6. (a) The $E°$ values of Cu^{2+}/Cu^+ (0.15 V) and $\frac{1}{2}I_2/I^-$ (0.54 V) suggest the iodometric titration of copper (II) is not possible. Explain why it is found otherwise in practice. 4

(b) Is it possible to carry out permanganometric titration of Fe^{2+} in hydrochloric acid solution? Explain with reasons. 3

(c) Addition of phosphoric acid is essential in the titration of Fe(II) with dichromate solution in presence of BDS indicator. Discuss. 3

(d) Explain the following: 2×3

(i) The electron affinity of bromine is less than that of chlorine.

(ii) The melting points of NaX decreases in the order NaF > NaCl > NaBr.

(iii) Acetic acid is an acid in aqueous medium but as a base in nitric acid.

7. (a) Define chemical bond. Draw the stereograms of p_x, p_y and p_z orbitals of one atom and *dxz* orbital of another atom. Suggest which *p*-orbitals can overlap with *dxz* orbital. Draw all probable orbitals formed from them. 5

(b) Which of the following molecules should contain the least and which one the highest polar bonds? 2

$$OF_2, H_2O_2, SO_2, CO_2.$$

(c) What is radioactive equilibrium? Describe the conditions of transient and secular equilibrium. 4

(d) Write short note on magic number and mass defect. 5

10. Discuss the preparation and properties of XeF_2, XeF_4, XeF_6 and XeO_2F_2, and explain their structures using VSEPR model.

Write notes on oxidising property of perxenate. 3×4+4

2006

PAPER—II

Full Marks—100

The figure in the margin indicate full marks.

Candidates are required to give their answers in their own words as far as practicable.

Question No. **1** is compulsory. Answer *five* questions taking at least *two* questions each from groups A and B.

1. (a) Which has a higher energy—2*s*-orbital of H or 1*s*-orbital of He^+?

Which orbital screens better—4*f* or 4*d*? 2

(b) Which among CO, CO_2 and CO_3^{2-} has a dipole moment?

Which among O_2, O_2^- and O_2^{2-} is not paramagnetic? 2

(c) Which among (i)–(iv) is not a redox reaction? 2

(i) $K_2Cr_2O_7 + 2KOH \rightarrow 2K_2Cr_2O_4 + H_2O$

(ii) $Cl_2 + 2NaOH \rightarrow NaOCl + NaCl + H_2O$

(iii) $NH_4NO_3 \xrightarrow{\Delta} N_2O + 2H_2O$

(iv) 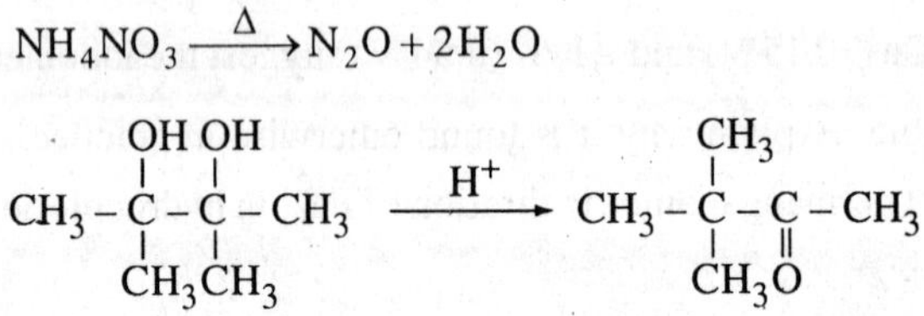

(d) Write the relation between half life and average life of a radio-element.

Starting from ^{101}Ra, is it possible to predict the exact number of radium atoms remain after one half-life period? 2

(e) Correct (if necessary) and justify—"Dynamic equilibrium is impossible with radioactive changes." 2

(h) Which is a stronger acid—(1) phosphoric acid or arsenic acid, (2) phosphorus acid or arsenious acid? 2

(j) Write the colour changes observed when chlorine water is added gradually to KI in water containing few mL CCl_4. 2

GROUP—A

2. (a) Write Pauli Exclusion Principle. Hence show that the *L*-shell can hold a maximum of 8-electrons. 6

(b) Write ground state electronic configurations for N^{3-}, Mg^{2+}, $Cr^{(o)}$, Mn^{2+} and Cu^{3+}. Hence predict the maximum number of unpaired electrons in each of the species. 4

(c) Write the mathematical form of uncertainty relation between energy and time. Hence show that there is no uncertainty in the energy of a stationary state. 6

3. (a) Complete the following nuclear reactions: 4

(i) ${}^{14}_{7}N + {}^{4}_{2}He \rightarrow {}^{17}_{8}O + ...$

(ii) ${}^{9}_{4}Be + {}^{4}_{2}He \rightarrow {}^{12}_{6}C + ...$

(iii) ${}^{23}_{11}Na + {}^{4}_{2}He \rightarrow {}^{26}_{12}Mg + ...$

(iv) ${}^{10}_{5}B + {}^{4}_{2}He \rightarrow {}^{13}_{7}N + ...$

(b) Explain why α-decay is common among heavy elements. 4

(c) Draw a qualitative diagram for the energy spectrum of β-particles emitted from a radioactive source. Explain the nature of the graph. 6

(d) Compare and contrast the 'radioactive equilibrium' and 'chemical equilibrium'. 2

4. (a) Why copper is a transition element? 2

(b) Atomic number of Zn is 30. What should be atomic number for Cd and Hg? 2

(c) Assign group, period and block to the element 56 in the periodic table. 3

(d) Why Ga is smaller in size than Al? 3

(e) Why electron affinity of fluorine is less than that of chlorine? 3

(f) Which has a larger radius Ne or F? Justify. 3

5. Write short notes on: 4×4

(iv) Diagonal relationship

(v) bent's rule

GROUP—B

7. (a) Arrange the following molecules, in the order of increasing acidity, with explanation:

(i) BCl_3, BI_3, BF_3 and BBr_3

(ii) HOCl, HOBr and HOI

(b) Unlike H_2SO_4, H_2SeO_4 can dissolve gold—explain. 2

(c) Complete the following reactions according to the SHAB principle— 2

(i) $Li^{+} + Cs^{+} + F^{-} + I^{-} \longrightarrow$

(ii) $Cu^{2+} + H^{+} + SO_4^{2-} + S^{2-} \longrightarrow$

(d) Calculate the heat of formation of $NaCl_2$ from the following thermochemical data with the aid of Born-Haber cycle: 4

Lattice energy of $NaCl_2 = -215$ kJ mole^{-1}

I_1 of Na = +494 kJ mole^{-1}

I_2 of Na = +4563 kJ mole^{-1}

Heat of sublimation of Na = +109 kJ mole^{-1}

Heat of dissociation of Cl_2 = +242 kJ mole^{-1}

Electron affinity of Cl = –347 kJ mole^{-1}

(e) K^+ and F^- have comparable size. Which one will have greater hydration energy and why? 2

9. (a) What do you mean by electronegativity? Discuss a method for evaluating electronegativity. 4

(b) "The geometry of ClF_6^- is octahedral while IF_6^- is not"—explain. 3

(c) "Thermal stability of $MgCO_3$ is less than $CaCO_3$"—explain. 3

(d) Explain the magnetic properties of B_2 and C_2 constructing their respective M.O. energy level diagram. 4

(e) A spectral line of the Balmer series of H-atom has a wavelength 4860 Å, indicate the electronic transition for this spectrum. 2

10. (a) Draw the qualitative M.O. energy level diagram of CN^-. Can CN^- act as ambidentate ligand? Discuss in the light of M.O. theory. 4+2

(b) Predict the structures of the following using VSEPR rule: 2+2

(i) $IO_2F_2^-$, (ii) XeF_5^-.

(c) Discuss the structure and bonding in B_2H_6 using the concept of M.O. theory. 4

(d) Explain why H_2O has higher boiling point than HF. 2

11. (a) Calculate potential for H^-/H_2 system at pH = 7 and at pH = 14. 4

(b) $FeCl_3$ oxidises iodide to iodine but oxidation does not occur in presence of excess fluoride. Explain. 4

(c) Construct the Frost diagram from the following Latimer diagram:

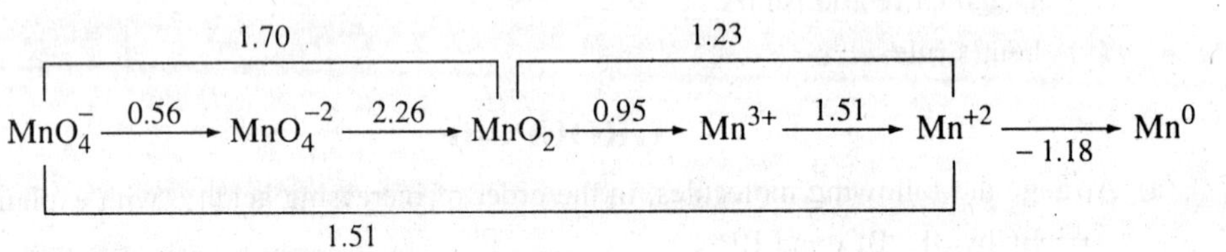

Show the calculations necessary for the evaluation of coordinate of each point of the plot. 2+3=5

(d) What is disproportionation reaction? What is the driving force behind a disproportionation reaction? Explain it with an example. 3

2007

PAPER—I
GROUP—B (INORGANIC)

Group—B (a)

Answer any **one** *of the following*

9. (a) Describe the Pauling's method to determine the univalent crystal radii. Hence calculate the univalent radii of Na^+ and F^- ions, if the internuclear distance between Na^+ and F^- ions in NaF crystal is 2.31 Å. 6
   (b) Explain why the ionisation potential of Li^+ is greater than He, although these are isoelectronic. 2
   (c) Calculate the electronegativity of lead following Allred-Rochow procedure.
   [Given : covalent radii of Pb = 1.53 Å and atomic no. of Pb = 82.] 4
   (d) Explain why the electron affinity of Cl is higher than that of F although the electronegativity of these elements is in the reverse order. 3

10. (a) Draw the MO energy level diagram for H_2O. Hence comment on the angular structure of H_2O molecule and compare with the structure of BeH_2. 8
   (b) Discuss the magnetic properties of B_2 and C_2 with reference to MO theory. 4
   (c) Write a brief account on 'defects in solids'. 3

13. (a) Discuss the structure of XeF_6 according to VSEPR theory. 4
   (b) Discuss the bonding in SO_3 according to VB theory. Indicate the hybridisation of S. 3
   (c) Explain why LiCl is soluble in organic solvents while the chlorides of other alkali metals are not. 3

14. (a) "All Arrhenius acids are also Bronsted acids, but all Arrhenius bases are not Bronsted bases"—comment. 4
   (b) "According to Bronsted theory CH_3COOH can either be an acid or a base." Explain with appropriate example. 2
   (c) Discuss the acid-base property of H_2SO_4 in the solvents, H_2O and HF. 2
   (d) Explain why $SOCl_2$ would behave as an acid in liquid SO_2? 2

15. (a) Write a brief account on Usanovich's concept of acid and base. 2
   (d) Complete the following reactions using the concept of HSAB principle: 3

   $BF_3H^- + BH_3F^- \rightarrow$

   $R_2SBF_3 + R_2O \rightarrow$

GROUP—B (c)

16. Answer any ***five*** questions: 2×5
   (a) Explain why Na-atom is larger than both Li and Mg.
   (b) First ionisation potential of C-atom is greater than B, whereas the reverse is true for 2nd ionisation potential. Explain.

(e) Comment on the C–O bond order in carbon monoxide according to qualitative MO theory.

(f) Assign the hybridisation of Xe in XeF_4.

(g) Arrange the following in the increasing order of acidity:

(i) HF, HCl, HBr and HI

(ii) $HClO_4$, $HClO_3$, $HClO_2$, HClO

(i) What will be the coordination number of each ion in NaCl, if the ionic radius of Na^+ = 0.95 Å and that of Cl^- = 1.81 Å. Predict the shape of NaCl crystal.

(j) What are the conjugate bases of HSO_4^- and $Si(OH)_4$?

Group—B

Answer **Question No. 8** *and any* **three** *from the rest*

7. Attempt any ***seven*** questions from the following: 2x7

(i) What would be the wavelength of emission spectrum when electron returns from $n = 2$ to ground state of He^+? (Given $h = 6.625 \times 10^{-27}$ erg sec. $m_e = 9.1 \times 10^{-28}$ g, charge = 4.8×10^{-10} e.s.u., $c = 2.998 \times 10^{10}$ cm sec^{-1}.)

(ii) Give two examples of mirror pair nuclei,

(iii) Explain why the acidity decreases in the following sequence: $HClO < HClO_2 < HClO_3 < HClO_4$.

(iv) What do you mean by isomeric transition?

(v) Explain why an aqueous solution of H_2O_2 decomposes spontaneously to give H_2O and O_2. (Given that $E^\circ_{H_2O_2/H_2O} = +1.76$ V and $E^\circ_{O_2/H_2O_2} = +0.68$ V.)

(vi) Explain why most stable oxyacids of S, Se and Te are H_2SO_4, H_2SeO_3; H_6TeO_6.

(vii) Explain—colour intensity of halogen increases from fluorine to iodine.

(viii) Molten ICl_3 has a high conductivity—explain.

(ix) $^{18}_{9}F$ nuclide is radioactive in spite of its *n*/*p* ratio is unity. Explain.

8. (a) Calculate the wavelength of the first transition in Lyman and Paschan series in the atomic spectra of hydrogen. ($R = 1.097273 \times 10^{-7}$ m^{-1}) 3

(b) Draw the radial probability distribution curves of 3*s*, 3*p* and 3*d* orbitals and compare their orbital electronegativity. 2+1

(c) Give two examples where Aufbau principle is violated in the stable electronic configuration of atoms. Mention the factors which cause such isolation. 3

(d) Calculate the de Broglie wavelength of 6*s* electron of Hg moving with a speed nearly $\frac{1}{6}$th that of light (velocity of light = 3.0×10^8 m s^{-1}). 3

9. (a) Estimate the thermodynamic feasibility of the following reaction:

$$Cu^+ + Cu^+ \rightleftharpoons Cu^{2+} + Cu^0$$

$$Ag^+ + Ag^+ \rightleftharpoons Ag^{2+} + Ag^0$$

Indicate the directions in which the reactions would proceed spontaneously. Given standard reduction potential E^0 values: 3+1

Couple	E^0 (V) at 25°C
Cu^{2+}/Cu^{+}	+0.16
Cu^{+}/Cu^{0}	+0.52
Ag^{2+}/Ag^{+}	+1.98
Ag^{+}/Ag^{0}	+0.8

(b) What is Zimmerman-Reinhardt solution? When and why is it used?

(c) Explain why Cl′ is oxidised by permanganate solution only at low pH below 1.5 but not in neutral medium.

Given: $E°\ (Ce^{4+}/Ce^{3+}) = 1.44\ V$

$E°\ (MnO_4^-/Mn^{2+}) = 1.51\ V$

$E°\ (Cl_2/Cl^-) = 1.36\ V$ 3

(d) From the following Latimer diagram, calculate the reduction potential of ClO_4^-–Cl^- couple: 2

$$ClO_4^- \xrightarrow{+1.2\ V} ClO_3^- \xrightarrow{+1.18\ V} ClO_2^- \xrightarrow{+1.7\ V} HClO \xrightarrow{+1.163\ V} Cl_2 \xrightarrow{+1.36\ V} Cl^-$$

10. (a) Nuclidic mass of $^{55}_{25}Mn$ is 54.988 mu. Determine its total nuclear binding energy and nuclear binding energy per nucleon. (Given: mass of electron = 0.0005486 mu, proton = 1.0072765 mu and neutron = 1.008665 mu.) 4

(b) State and explain Soddy's group-displacement law. Interpret the nuclear reaction given below: 4

$$^{10}_{5}B + ^{4}_{2}He \longrightarrow ^{13}_{7}N^{*} + ^{1}_{0}n$$
$$^{13}_{7}N^{*} \downarrow\ ^{13}_{6}C + ^{0}_{+1}I$$

(c) A sample of ^{18}F is found to undergo 90% radio active decay in 366 minutes. What is its computed half-life from this observation? 4

2003

PART—I

CHEMISTRY

Paper No. BCH-101 : Inorganic Chemistry

(New Course : 2001 Syllabus)

Full Marks—50

The figures in the right-hand margin indicate marks

Answer **five** *questions including Question No.* **1** *which is compulsory*

1. Explain the following: 2×5
   (a) F has highest electronegativity, while Cl has largest electron affinity.
   (b) Second ionisation potential is always greater than the first.
   (c) LiF has higher lattice energy than LiI.
   (d) Al forms $[AlF_6]^{3-}$ but B does not form $[BF_6]^{3-}$ though both B and Al belong to the same group.
   (e) Fe displaces Cu from $CuSO_4$ but Ag does not.
2. (a) Explain the significance of quantum numbers. Correlate the shapes of *s*, *p* and *d* orbitals with quantum numbers. 5
   (b) Write the electronic configurations of the following and indicate the number of unpaired electron in each case: 3
   (i) Cu^{2+}; (ii) Fe^{2+}; (iii) Cr^{3+}.
   (c) How many nodes are possible for a 3*s* orbital? 2
3. (a) What do you mean by electronegativity? How is it related with the ionic character of a chemical bond? Calculate the percentage ionic character in HCl if $X_H = 2.1$ and $X_{Cl} = 3.0$. 7
   (b) Predict whether the cell $Cu/Cu^{2+} \parallel Zn^{2+}/Zn$ will operate or not. Given that $E^o_{Cu/Cu^{2+}} = -0.34$ and $E^o_{Zn/Zn^{2+}} = +0.76$ V. 3
4. (a) Discuss the salient features of VSEPR theory and explain the shapes of ClF_3, ICl_2^- and NH_3 on the basis of this theory. 5
   (b) What is molecular orbital theory? How does it explain that the ionisation energy of O_2 molecule is less than that of O_2^+ ion? 5
5. Account for the following: 2×5
   (a) Viscosity of glycerol is greater than that of ethanol.
   (b) CO is diamagnetic while NO is paramagnetic.
   (c) He_2 and Be_2 molecules do not exist.
   (d) $SiCl_4$ is easily hydrolysed but CCl_4 is not.
   (e) HF is a weaker acid than HCl in aqueous solutions.
8. (a) What do you mean by lattice defects? Discuss different types of defects in ionic solids and their consequences. 5

CHEMISTRY

First Paper : Inorganic Chemistry

(Old Course : 2000 Syllabus)

Full Marks—50

The figures in the right-hand margin indicate marks

Attempt **five** *questions, including Question No.* **1** *which is compulsory*

1. Explain the following: 2×5
   (a) Complexation tendency of lithium is greater than that of other alkali metals.
   (b) The central atom of NH_3 and H_2O involves the same hybridization (sp^3), but the bond angle in the former is greater than that in the latter.
   (c) Alkali metals are soft, electrically conducting and potential reducing agents.
   (d) Fluorine shows the highest value of electronegativity but chlorine has the highest value of electron affinity in the halogen family.
   (e) Pauli's exclusion principle and Hund's rule.
2. (a) What is meant by the diagonal relationship? In what respects does lithium resemble magnesium? 5
3. (b) Explain the following: 6+4
   (i) Ge (II) compounds are reducing agents while Pb (IV) compounds are oxidizing agents.
   (ii) CCl_4 resists hydrolysis while $SiCl_4$ gets easily hydrolysed.
4. (a) What are non-stoichiometric compounds? Discuss the metal-excess and metal-deficiency defects quoting at least one example in each case.
   (b) State and discuss Fajans' rules in relation to polarization of ions.
   (c) Write the molecular orbitals for NO molecule in increasing order of energy and indicate the partially-filled and empty orbitals. 5+3+2
5. (a) Predict and discuss the geometrics of H_2O, NH_3 and SF_4 on the basis of VSEPR theory.
   (b) Write the band theory of metallic bond. State the type of energy bands obtained for (i) conductor, (ii) insulator and (iii) semi-conductor. 5+5
6. (b) Why is BBr_3 a better Lewis acid than BF_3? 2
7. (a) What is the difference between ionization potential and electrode potential? Mention the importance of electrochemical series.
   (b) Write and explain the difference between bond dissociation energy and average bond energy. 4+3
8. (b) Explain hydrogen bonding and discuss its origin and consequences. Differentiate between inter-molecular and intra-molecular hydrogen bonding with illustrations. 4

2004

PART—I

CHEMISTRY

Paper No. BCH-101 : Inorganic Chemistry

Full Marks—50

Answer any **two** *questions from Section—A, any* **four** *questions from Section—B and Section—C is compulsory*

The figures in the right-hand margin indicate marks

1. (a) Calculate the effective nuclear charge on the outermost electron of potassium ($Z = 19$) and copper ($Z = 29$). 4
   (b) What is bond energy? Explain the differences between average bond energy and bond dissociation energy. Mention two applications of bond energy. 6
2. (b) Explain radius ratio concept with suitable examples. What are the limitations of this concept? 4

SECTION—B

4. (a) What conditions a wave function should satisfy to be quantum mechanically acceptable? 3
   (b) Write any two limitations of valence bond theory. 2
5. What are crystal defects and how are they formed? Explain Schottky and Frankle defects. 5
6. What are sigma and pi molecular orbitals? Draw the molecular orbital energy level diagram of O_2. What is the bond order in O_2^{2-} ? 5
7. Discuss the shape of BrF_3, ICl_2^-, SF_4, XeF_4 and PCl_5 using VSEPR theory. 5
8. Giving examples, explain the concept of hybridisation with reference to sp^3d and sp^3d^2 hybridisation. 5

SECTION—C

10. Explain the following: 1×10
    (b) LiF has higher lattice energy than LiI.
    (c) Li^+ ion is most extensively hydrated among alkali metal cations.
    (d) Arrange C, N, O and F in the decreasing order of their ionization potential.
    (e) BCl_3 is monomeric while $AlCl_3$ is dimeric.
    (f) Cu and Ag do not liberate H_2 from acids.
    (g) BBr_3 is stronger Lewis acid than BF_3.
    (h) Liquid O_2 sticks to the poles of a magnet while liquid N_2 does not?
    (i) Pb(IV) is an oxidizing agent while Pb(II) is not?
    (j) Ortho-chlorophenol is more volatile than para-chlorophenol.

2005

PART—I

CHEMISTRY

Paper No. BCH-101 : Inorganic Chemistry

Full Marks—50

Answer any **two** *questions from Section—A, any* **four** *questions from Section—B and Section—C is compulsory.*

The figures in the right-hand margin indicate marks

SECTION—A

1. (a) What is Schrödinger wave equation? Explain the significance of the wave function. 4
   (b) How can the radius of a cation be determined by Lande's method? Calculate radii of Na^+ and F^- ions by Pauling's method. 6
2. (a) What is LCAO? Draw molecular orbital energy diagram for carbon monoxide molecule. 4

SECTION—B

4. (a) According to Bohr's theory, the velocity of an electron in the first orbit is 2.2×10^6 m s^{-1}. Calculate uncertainty in the position of the electron if uncertainty in its velocity is 0.2%. (Mass of an electron = 9.1×10^{-31} kg and $h = 6.6 \times 10^{-34}$ J s) 3
   (b) Out of the following, which configuration will be the most stable for chromium?
   (i) $4s^2 3d^4$, (ii) $4s^1 3d^5$, (iii) $4s^0 3d^6$.
   Justify your answer considering symmetry, exchange energy and pairing energy. 2
5. What do you mean by electronegativity? Describe any two methods of determining electronegativity. 5
6. (a) What are the limiting radius ratio for coordination number six? Describe the structure of TiO_2. 3
   (b) What are non-stoichiometric crystal defects? 2
7. (a) Describe different types of hybrid orbitals that can be formed by *s* and *p* orbitals in the valence shell of an atom. 3
   (b) Explain the structures of ICl_2^+ and ICl_2^- ions on the basis of VSEPR theory. 2
8. (a) Calculate N–N bond energy from the following thermochemical data: 2
   $NH_2–NH_2(g) \rightarrow 2N(g) + 4H(g)$, $\Delta H = 1724$ kJ mol^{-1}
   $NH_3(g) \rightarrow N(g) + 3H(g)$, $\Delta H = 1172$ kJ mol^{-1}
   (b) Write a note on bonding in metals. 3

SECTION–C

10. Answer the following: 1×10
    (b) Why is boiling point of water said to be anomalous among the hydrides of group 16 elements?
    (c) Why is the bond order in O_2^+ higher than that in O_2?

(d) Is it possible to oxidize As(III) to As(V) using I_3^- ion as an oxidizing agents?

$$(H_3AsO_4 + 2e^- \rightarrow HAsO_2 + 2H_2O, E^0 = +0.56 \text{ V}$$

$$I_3^- + 2e^- \rightarrow E^0 = +0.54 \text{ V})$$

(e) What kind of bond exists in $(LiR)_4$?

(f) Giving an explanation, arrange the following salts in order of increasing water solubility:

LiI, LiCl, LiBr, LiF

(g) Which is more stable between PbI_4 and PbI_2 and why?

(h) Explain Pauli's exclusion principle.

(i) PF_5 is known but NF_5 is not, why?

(j) Explain the difference in bond angles of H_2O and OF_2.

PART—II

CHEMISTRY–II

Organic and Inorganic Chemistry (New)

CHEMISTRY–IV

Organic and Inorganic Chemistry (Old)

Full Marks—50

2nd Year [MARCH–2005]

Organic and Inorganic Chemistry (New)

3. (a) Write MO configuration of NO, NO^+ and CO molecules. Calculate bond order in each case. Explain why NO^+ has bond stronger than NO. 9

(b) Answer ***any three*** of the following: 9

(ii) Give the brief account of Lux and Flood acids and bases.

6. (a) Answer ***any two*** of the following: 8

(i) Draw MO diagram of Li_2 molecule and calculate its bond order.

(ii) What are the ways for decreasing CO_2 concentration in air?

(iii) Explain Be_2 and He_2 molecules are unstable. 8

2nd Year [OCTOBER–2005]

Organic and Inorganic Chemistry

4. (a) Answer ***any two*** of the following: 12

(ii) Sketch the figures for the following combinations of atomic orbitals:

(i) s-s combination; (ii) p-p combination; (iii) s-p combination; (iv) p-d combination.

(iii) Give the classification of solvents.

(b) Or, Give a brief account of the oxidation states and ionization energies of zinc group elements. 4

5. (a) Draw molecular orbital diagram of O_2 molecule. Calculate bond order and stabilization energy. Explain why O_2 is paramagnetic. 9

   (b) Answer ***any three*** of the following: 9

   (i) Give a brief account of Lux and Flood acids and bases.

   (iii) Explain why NO^+ has stronger bond than NO itself.

6. (a) Answer ***any two*** of the following: 8

   (ii) Explain on the basis of MO theory why Be_2 and He_2 are unstable molecules?

   (b) Answer ***any four*** of the following: 8

   (i) What is the M.O. electronic configuration of F_2 molecule.

PART—III

CHEMISTRY–II

Inorganic Chemistry (New)

CHEMISTRY–VI

Inorganic Chemistry (Old)

Full Marks—50

3rd Year [MARCH–2005]

Chemistry-II [Inorganic Chemistry] (New)

1. (a) Attempt ***any one*** of the following: 8

   (i) What are '*d*' block elements? List the *d*-block elements of first transition series and give their electronic configuration. Comment on electronic configuration of Cr and Cu.

   (ii) What are assumptions of valence bond theory? How does VBT account for the following facts?

   (i) $[Ni(CN)_4]^{2-}$ is diamagnetic and square planar.

   (ii) $[NiCl_4]^{2-}$ is paramagnetic and tetrahedral.

   (iii) $[Ni(CO)_4]^0$ is diamagnetic and tetrahedral (Z for Ni = 28).

Chemistry-VI [Inorganic Chemistry] (Old)

3. (b) The inter-nuclear distance in NaF is 0.231 nm. Calculate the ionic radii of Na^+ and F^- by using Pauling's method.

   Slater Screening Constant = 4.5. 4

   (c) Give electronic configuration of $Cr_{(24)}$ and $Cu_{(29)}$. 4

2006

PART—I

CHEMISTRY

Paper No. BCH-101 : Inorganic Chemistry

Full Marks—50

Answer any **two** *questions from Section—A, any* **four** *questions from Section—B and Section—C is compulsory*

The figures in the right-hand margin indicate marks

SECTION—A

1. (a) What do you understand by dual character of matter? Obtain de Broglie relation. Using this relation show that the angular momentum is quantised. 5

   (b) What is screening constant? How does it affect the value of ionization potential of an atom? Calculate effective nuclear charge on the outermost electron of vanadium (Z = 23). 5

2. (a) Fourth shell may contain up to 32 electrons. Explain this fact with the help of quantum numbers. $2\frac{1}{2}$

   (b) What do you mean by LCAO? Draw the molecular energy level diagram of nitric oxide. $2\frac{1}{2}$

   (c) Arrange the following ions in the increasing order of their size, with reasons: C^{-4}, N^{-3}, O^{-2} and F^{-}. $2\frac{1}{2}$

   (d) What is Hund's rule of maximum multiplicity. Illustrate with two examples. $2\frac{1}{2}$

3. (a) Define lattice energy. How does the solubility of an ionic solid depend upon its lattice energy? 3

   (b) How is lattice energy of ionic solids evaluated using Born-Haber cycle? 2

   (c) What is radius ratio rule? How does it affect the arrangement of ions in the crystal lattice of an ionic solid? Explain with examples. Point out limitations of radius ratio rule. 5

SECTION—B

4. (a) Define electron affinity. How does it vary in a period and in a group of the periodic table? What is the difference between electron affinity and electronegativity? 3

   (b) What are the factors favouring the formation of a covalent bond in an ionic compound? 2

5. (a) Indicate the type of hybridization of the central atom and geometry of the following molecules: XeF_4 and SF_4. 2

   (b) Explain valence shell electron pair repulsion theory. On the basis of this theory, explain that ICl_2^- and H_2O have linear and bent geometries, respectively. 3

9. (b) What are semiconductors? Explain the behaviour of semiconductors on the basis of band theory. 3

SECTION—C

10. Answer the following in short:

(a) The electron affinity of chlorine is higher than that of fluorine but its electronegativity is less as compared to fluorine. Why. 2

(b) LiCl is more soluble in organic solvents than in water. Why. 1

(c) HCl is a gas while HF is liquid at ordinary temperature and also HCl is a stronger acid than HF. Why? 2

(d) Hydrides of alkali and alkaline earth metal ions act as reducing agents. Why? 1

PART—II

2nd Year [APRIL–2006]

Chemistry-II [Organic and Inorganic Chemistry]

4. (a) Attempt *any two* of the following: 12

(i) Discuss the bonding on the basis of molecular orbital theory for C_2 molecule. Calculate the bond order. Write electronic configuration and draw M.O. diagram for the same.

(ii) What is smelting? What are the methods of smelting? Discuss one method.

(iii) Discuss in brief about the following properties of the solvent:

1. M.P. and B.P. or liquid range
2. Dipole moment.
3. Dielectric constant.

5. (a) Sketch the figures of the following combinations of atomic orbitals to produce molecular orbitals:

1. s-s combination
2. s-p combination
3. p-p combination 9

(b) Answer *any three* of the following: 9

(iii) What are aprotic solvents? Describe three classes of aprotic solvent, mentioning example of each.

6. (a) Answer *any two* of the following: 8

(ii) Explain—Helium and Beryllium are monoatomic on the basis of M.O.T.

(b) Answer *any four* of the following: 8

(ii) O_2 molecule is paramagnetic in nature. Explain.

(iv) Draw the molecular orbital diagram for HCl molecule.

2nd Year [OCTOBER–2006]

Chemistry-II [Organic and Inorganic Chemistry]

5. (a) Draw molecular orbital diagram of O_2 molecule. Explain on the basis of MO. How bond order decreased from O_2 to O_2^{-2}. 9

6. (a) Answer *any two* of the following: 8

(i) Draw MO diagram of Li_2 molecule and calculate its bond order.

(ii) Explain Be_2 and He_2 molecules are unstable.

(iii) What are the major regions of atmosphere and give its characteristics?

PART—III

3rd Year [OCTOBER–2006]

Chemistry-I [Physical Chemistry]

6. (b) Answer *any two* of the following: 6

(ii) Define:

(i) Mean life

(ii) Curie

(iii) Isotone

(iii) Explain the use of radioisotope as tracer in esterification reaction and in decomposition of H_2O_2 by MnO_2.

(c) Half-life period of Na^{24} is 14.8 hours. Calculate the time required to loose 90% of its radioactivity. 4

Chemistry-II [Inorganic Chemistry]

1. (a) Attempt *any one* of the following: 8

(i) What are *d*-block elements? Write names and the electronic configuration of first series of *d*-block elements.

2007

PART—I

CHEMISTRY

Paper No. BCH-101 : Inorganic Chemistry

Full Marks—50

Answer any **two** *questions from Section—A, any* **four** *questions from Section—B and Section—C is compulsory*

The figures in the right-hand margin indicate marks

SECTION—A

1. (a) How would you establish that a diagonal relationship exists between lithium and magnesium? 3

2. (a) What are principal features of valence bond theory? With an example of a diatomic molecule comment on the success of VB theory in explaining its structure and bonding. 5

(b) Is it possible to explain the bonding and shape of NH_3 molecule by considering the involvement of pure $3p$ orbitals of nitrogen atom? Explain your answer. 3

(c) Mention important consequences of hydrogen bonding on the structure and properties of molecules with H-bonding. 2

3. (a) If the solution of the Schrödinger equation is to be acceptable, what should be the essential criteria? 3

(b) Obtain an expression to relate the mass of a particle with its wavelength. 3

(c) "Covalent radius of an atom can be directly obtained by measuring the interatomic distance in the molecule." Why is it not possible to follow the same procedure to determine the ionic radii? Describe a method to calculate the ionic radii. 4

SECTION—B

4. Explain the radial and angular functions for an atom. How shapes of atomic orbitals are related to these functions? 5

5. (a) Explain screening effect. How does it vary with type of orbital involved in screening? 3

(b) Calculate the effective nuclear charge felt by the outermost electron of a chromium atom (atomic no. 24). 2

6. (a) Describe the Mulliken scale of electronegativity. Calculate the electronegativity of fluorine if its ionization energy is 1681 kJ mol^{-1} and electron affinity is 333 kJ mol^{-1}. 2

(b) Indicate the thermodynamically feasible reduction step(s) from the following reduction potential diagram:

$$\overset{\text{VI}}{FeO_4^{2-}} \xrightarrow{2.20} \overset{\text{III}}{Fe^{3+}} \xrightarrow{0.77} \overset{\text{II}}{Fe^{2+}} \xrightarrow{-0.47} \overset{0}{Fe}$$

Calculate $E°$ for the reaction $Fe^{3+} \rightarrow Fe$. 3

7. (a) Give one example of each of three different AX type of ionic lattices and describe the NaCl structure with a neat diagram. 3

(b) Though CaF_2 and $CaCl_2$ are chemically comparable, their lattice structures are different. Explain why. 2

8. (a) Write a note on metallic bond. How is conduction behaviour of metals explained on the basis of band theory of solids? $2\frac{1}{2}$

SECTION—C

10. (b) In spite of its high ionization energy why is lithium a very good reducing agent? 2

(d) What is the shape of ClF_3 molecule? 2

(e) Draw the shape of a $3p$ orbital indicating the radial nodes. 2

(f) Sodium forms an unusual compound with crypfand-222 possessing two different kinds of sodium ions. What are the two different kinds? What is unusual in this compound? 2

(a) Why [illegible] good conductor of electricity? [illegible] on the basis of band theory of solids.

SECTION—C

(a) In some of its [illegible]

(c) Draw the shape [illegible] and indicate the [illegible]

(f) [illegible]